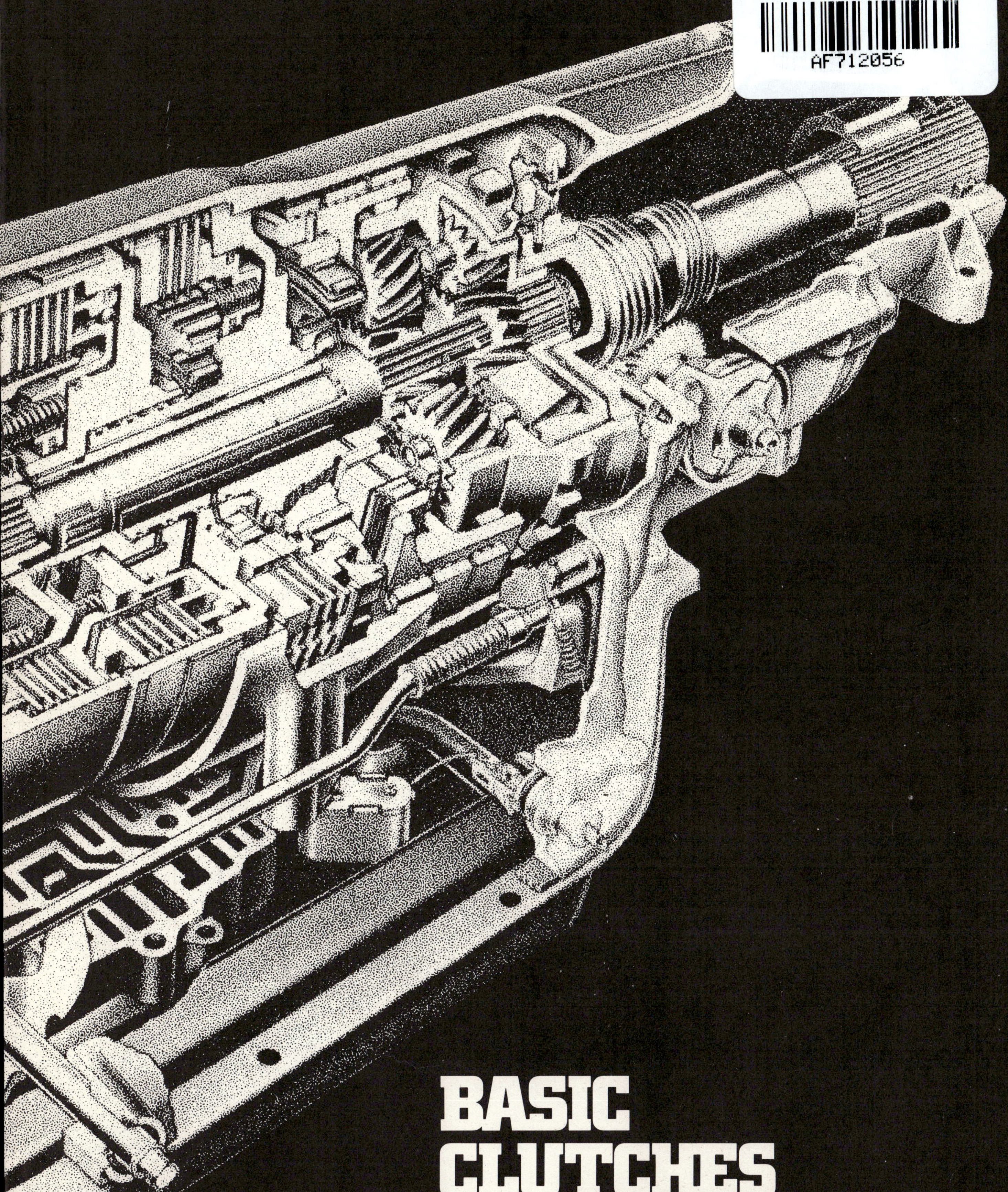

BASIC CLUTCHES AND TRANSMISSIONS

INTRODUCTION

It is somewhat unfortunate that a clutch and a manual transmission, or an automatic transmission, are not generally considered among the more glamorous parts of a car. You can't stand back and admire them as you can wheels and tires, or the sheetmetal body. You can't even toss up the hood to look at them as you can a chrome-bedecked engine. Nope, they're down and under, out of sight, where they're all but forgotten. Because they've been so highly perfected over the years, they usually work without a whimper, all the while collecting grease and grime and looking less appealing all the time.

This is a shame, for clutches and transmissions are actually pretty darned interesting. After all, they hook your engine up to the ground and allow you, your passengers, and the car itself to go wherever you steer it. It's all the more unfortunate when that awful day arrives when something goes haywire "down and under," for because clutches and transmissions are so seemingly foolproof, very few people are able to cope with their ills.

Oddly, clutches and transmissions are relatively simple to work on—once they've been removed from the car, that is, for they are h-e-a-v-y. But once out on the ground, or more properly up on the bench, most fix-its are merely a matter of tearing the unit apart and putting it back together again substituting new pieces where visual inspection shows they're needed. There are no super-miniscule tolerances to fool with, and no custom-fitting of parts. Say you get inside your 4-speed manual and find a gear with a broken tooth. Just yank the offending gear out, take it to the parts house for a matched part, then go home and put it in.

The table of contents opposite shows that in this 3rd edition we're taking a close look at all the more popular transmissions; not just the '74's alone, for we realize that not everyone has a '74 model car (especially one needing transmission work). The older but still-used Chevy Powerglides are in here, various versions of Chrysler's TorqueFlite and some others.

Because the gasoline shortage was at its peak (hopefully) while this book was being produced, we felt it timely to give you a few choice bits about transmission work that affects fuel economy. I won't get into the nuts and bolts of these here, but do suggest you flip back to the last several chapters if you're interested in wringing a few more miles per gallon out of your car.

Above all, now that you have this book in hand, feel free to tackle what might have seemed a mysterious job before. All you need are common automotive tools and the urge—or need—to dig in.

Spence Murray

BASIC CLUTCHES AND TRANSMISSIONS No. 3

Revision edited by Kalton C. Lahue and the Technical Editors of Petersen Publishing Co.

Library of Congress Catalog Card No. 74-78893

ISBN 0-8227-0066-2

COVER

A highly polished cutaway display model of a TurboHydro 400 by B&M Automotive Products, fitted with one of B&M's new Series 70 shifters with trans temperature gauge, makes a fitting cover subject. Photography by Pat Brollier of Petersen Photographic. Cover design by Pat Taketa.

CONTENTS

Clutch Design and Development

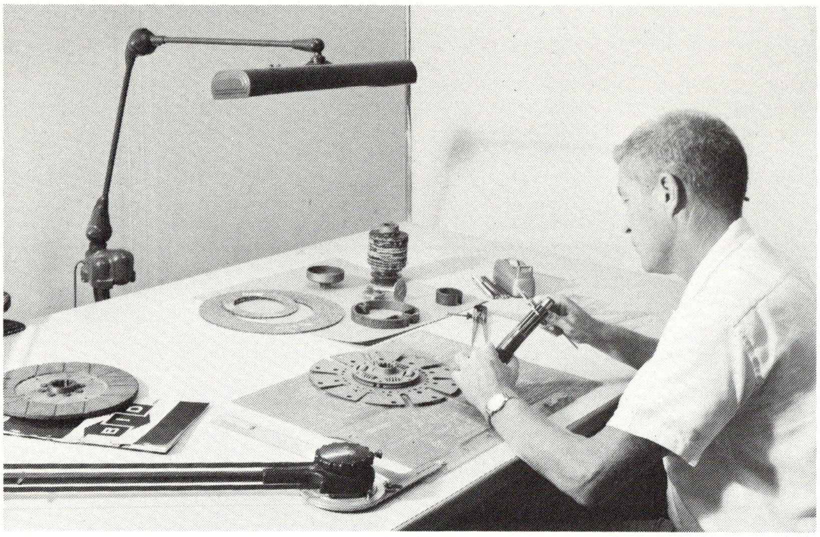

1

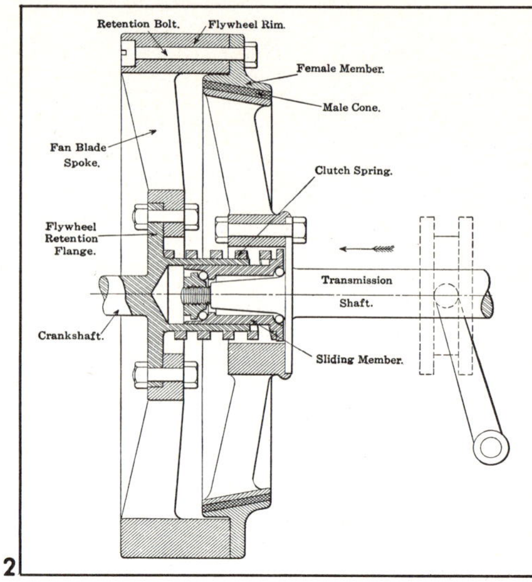

2

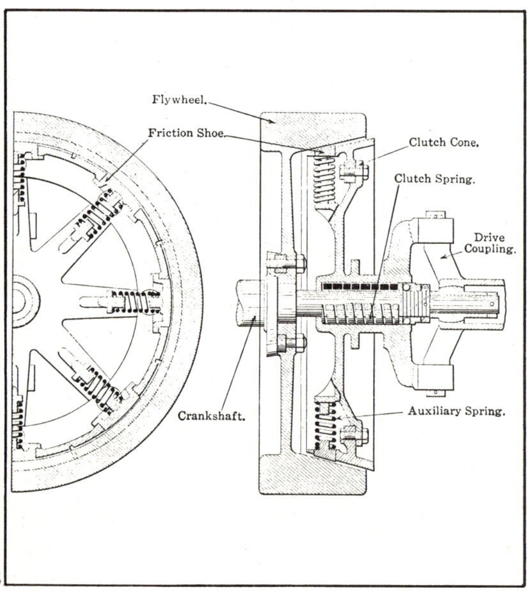

3

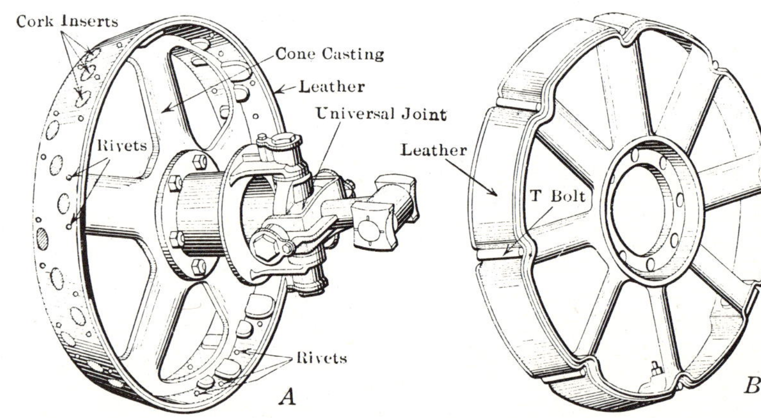

4

Several years ago someone called an automatic transmission "a brain beneath the floorboards." It's a great description for a precision piece of automotive machinery. And we're sure you'll agree that the "brain" is getting smarter all the time. But what about the manual transmission that needs such an unglamorous, unsophisticated and simple device like a "clutch" to make the drivetrain of an automobile workable? Well, maybe it sounds simple (we'll see), and it certainly looks unglamorous, but it's not without its own certain sophistication of operation.

The automobile clutch as we know it today has not basically changed in over 50 years. Certainly, construction materials have changed, with new metals and friction materials becoming available which have improved the breed, but in actual clutch design—not much change. This can better be described as clutch *evolution* rather than *revolution*.

In the early days of gasoline powered cars, if it weren't for the development of the clutch, the internal combustion engine wouldn't be practical for use in the automobile. To understand this, it's necessary for you to know that an internal combustion engine does not have much power at low rpm's. The clutch allows you to increase engine rpm without a load thereby gaining the necessary power to move the vehicle. It can be disengaged to allow the engine to continue to run while the car is stopped. You can slip it to start smoothly and shift gears smoothly once the car is moving. And it will lock up direct drive so that full power may be transmitted to the rear wheels. Today, automobile clutches can do all these things on vehicles that range from only a few horsepower to well over a thousand.

Remember, too, that while engine efficiency is important, the actual power of the car is not rated at the engine but at the rear wheels. In simple terms, a 100-hp engine with a 30% loss in power transmission is not as efficient as an 80-hp engine with a 10% loss in power transmission. By this comparison the lower-powered car will receive more useful driving effort at the rear wheels than the higher-powered vehicle, and the energy will be supplied with less stress on the various drivetrain components. So, when you think of the hundreds of horsepower available at the flywheel that must be transmitted to the rear wheels, the high temperatures generated when the clutch is slipped, the speed-shifts and jackrab-

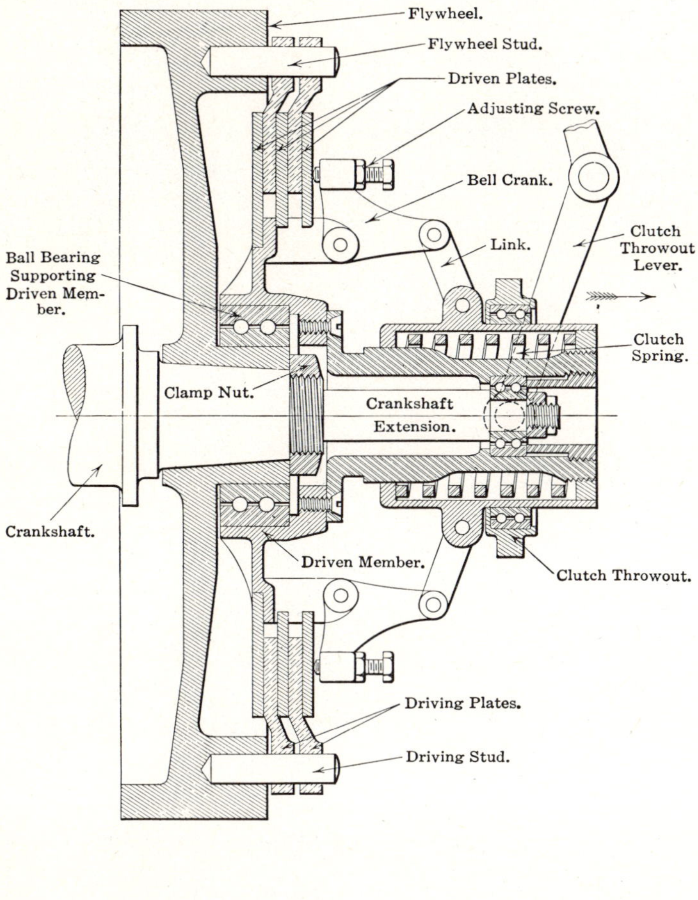

5

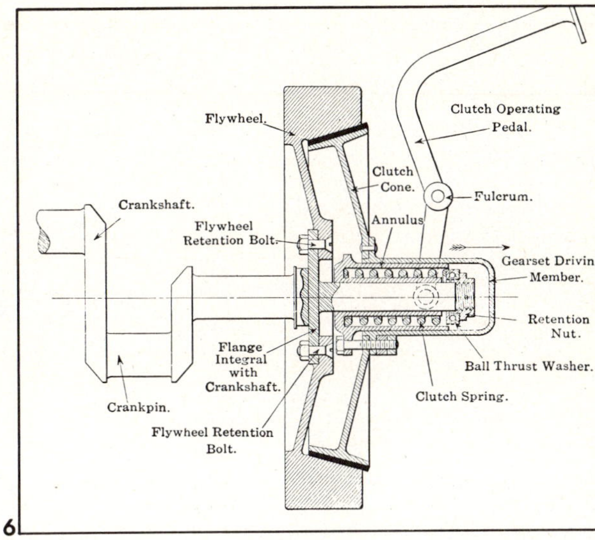

6

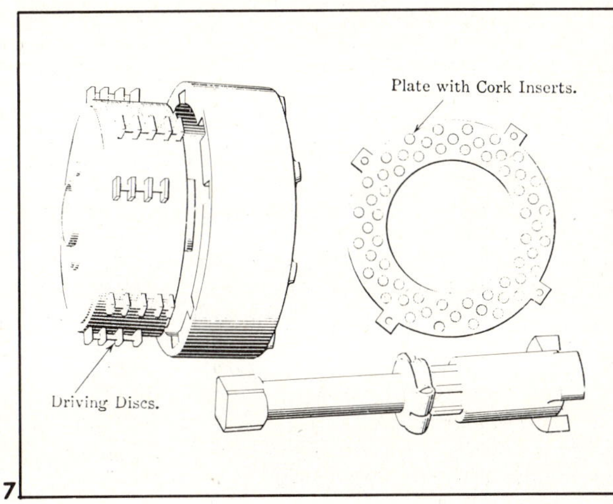

7

1. Clutch design hasn't changed much over the years, but sophisticated refinements are numerous.

2. One early design used a separate female member bolted to the flywheel to facilitate repair. Weight of this revolving mass made shifts difficult.

3. Further development of the cone clutch saw the use of friction shoes that were spring-mounted and contacted the female member first to ease action.

4. To provide for more slip and make a smooth engagement, cork inserts were added to the leather facing as shown. Still another type of cone clutch used "T" bolts to hold the leather in place while spacers smoothed engagement.

5. What could be called the forerunner of today's multiple-disc clutches is this early "5-plate" clutch which used spring and toggle links to engage the driving member. Notice the driving studs in the flywheel for the driving plates.

6. An alternate type of cone clutch used the flywheel as part of female member. Facing was leather and was difficult to slip smoothly into engagement, but when male member "wedged" into cone, it was a solid lockup.

7. Many cars used the multiple-disc clutch. This has cork-studded inserts on the driving members and plain plates as the driven members.

bit pop-the-clutch starts and how it keeps coming back for more—it's truly an amazing, wonderful and vital part of the drivetrain.

EARLY CLUTCHES

In the first years of the internal combustion powered car, many different clutch designs and ideas were tried for smoothly and efficiently transmitting power to the rear wheels. Most were adaptations of clutches used to power machinery in manufacturing plants. It was during these first years that four basic designs emerged at best fulfilling the needs of the motoring public: the cone, disc, band and plate types—all using some form of friction material for adhesion. As a result of the ensuing years of development, today all standard gasoline powered vehicles use some form of friction clutch. Other designs that were tried but did not meet with success were the hydraulic, pneumatic and magnetic types. However, even these may now be practical with space age developments in manufacturing methods and materials.

In its day, the cone clutch was one of the most widely used methods of drive, and the same basic idea is still in use today in various machinery applications. The cone clutch is simply a short section of a cone of large diameter fitting into a recess of precisely the same angle (about 12½ degrees) and pressed into place by spring pressure. The female member of the cone is usually machined into the flywheel or attached to it separately for easy removal and replacement. In operation, a heavy spring presses the male member into engagement with the female member and pressure on the foot pedal releases the clutch mechanism.

The cone clutch was simple to manufacture, its operation was easy for the public to understand, and it did the job of slipping and locked up extremely well for its time. This was due to two things: its friction material and its "wedging" action due to the conical shape. Leather, or a combination of leather with cork inserts, or an organic material such as that used on brake linings was employed. The full leather lining had a tendency to be grabby, but when used with cork inserts which stood up slightly higher than the leather, the cork

Clutch Design

made engaging very smooth. After processing, cork has a high frictional quality and its ability to "compress" made it ideal for clutch use. In this instance, the cork took the initial shock and made a smooth application of power. Another variation of the simple cork insert type was a series of spring-loaded friction inserts extending through the periphery of the cone and slightly above its surface. When the cone was first engaged, these friction shoes contacted the female member at many points and made a gradual application of power. When full spring pressure was exerted, the shoes depressed and the entire clutch surface was then engaged.

Other designers used a more simple form and inserted springs under the leather at several points around the periphery. The object of this was to have a limited area of leather surface engage the female surface just before full spring pressure is exerted. Again under full pressure the springs depressed and the entire clutch surface was engaged.

Besides a few small drawbacks, the chief disadvantage of cone clutches was that they were bulky and heavy. The size and weight of the male member made it "spin" after disengagement which caused difficulty in shifting. Remember, there were no synchromesh gearsets as we know them today, and stripping the gear teeth was a distinct possibility.

Still another variation of the cone clutch was a metal-to-metal friction surface that was housed in an oil-tight casing so that it worked in a bath of lubricant. The lower coefficient of friction existing between the lubricated surfaces was compensated for by increasing spring pressure. As you can see, this was but another phase of development in getting the proper ratio of initial slip and final lockup in drive.

The disc-type clutch, almost as popular as the cone type, transmitted power by using a large number of small discs to gain the required contact area rather than one big disc. The most widely used type of multiple disc clutch consisted of soft steel plates with alternate plates of phosphor bronze, housed in an oil-tight case. The oil served as a lubricant to prevent wear as well as providing a smooth delivery of power (slip) as oil was squeezed from between the discs. A large spring on the input shaft pressed one disc against the other until final lockup was made—metal-to-metal.

Other types of multiple disc clutches were designed to run dry and these used cork inserts or an

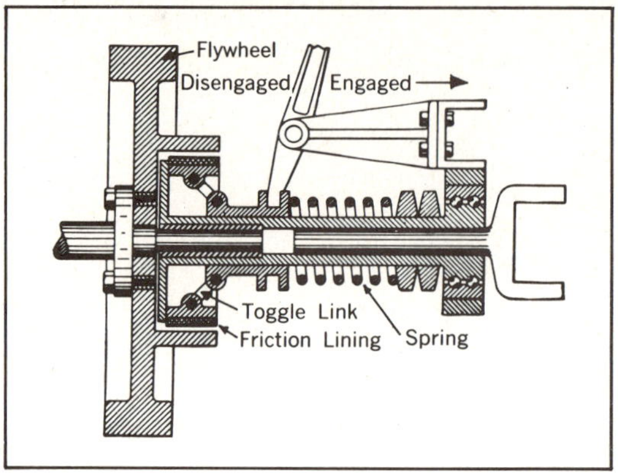

1

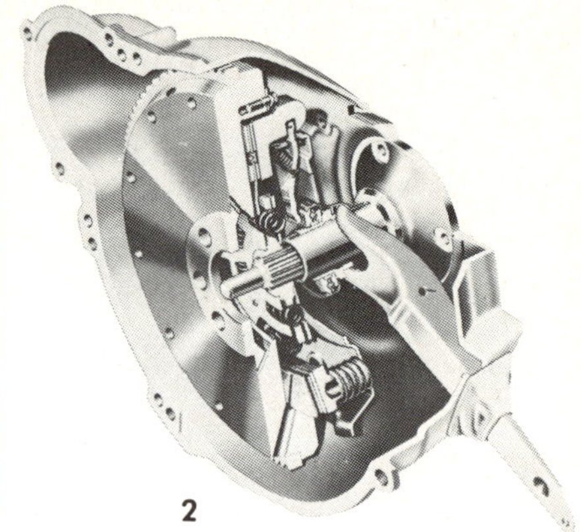
2

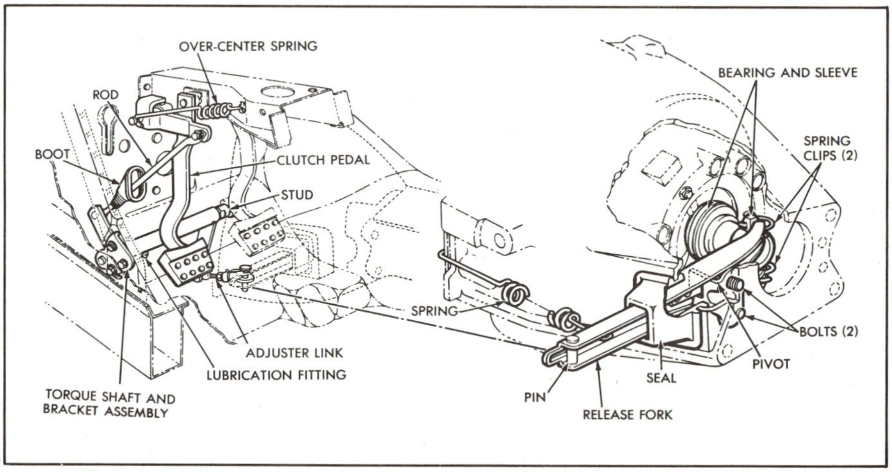

3

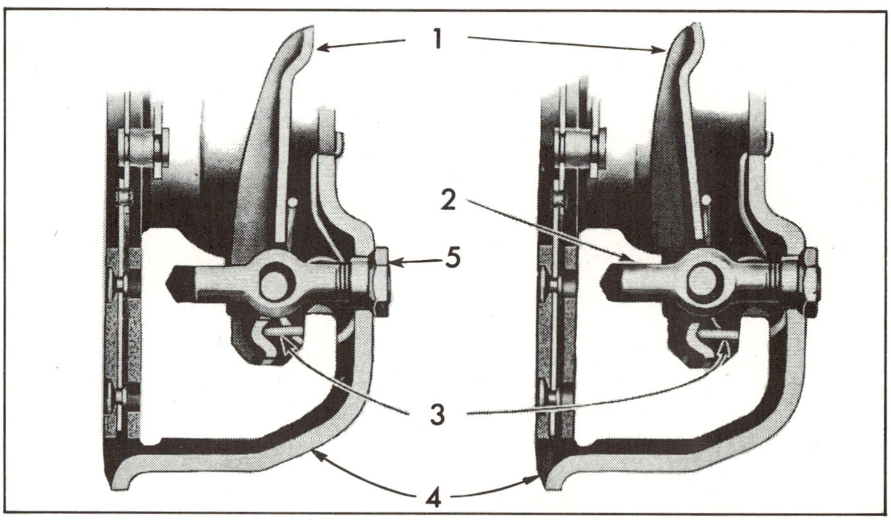

4

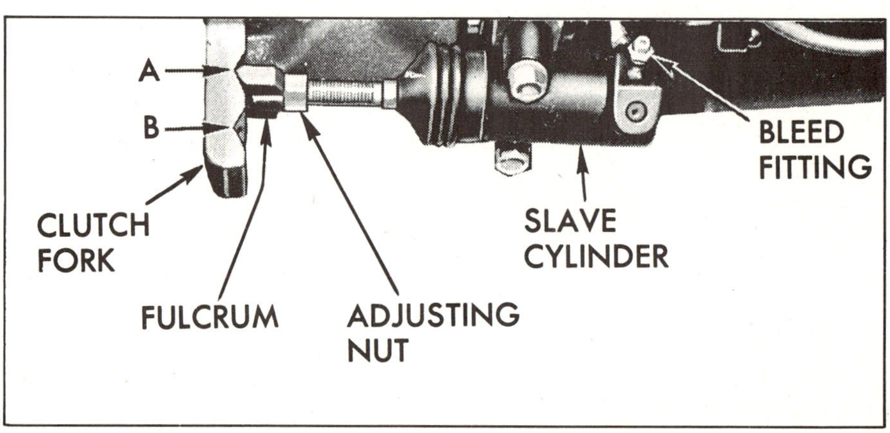

5

6

7

1. A second type of internal expanding clutch used a simple set of toggle links to engage the flywheel.

2. This view shows the relationship of the pressure plate, disc and flywheel assembly inside the bellhousing.

3. With clutch linkage set properly, you'd hardly know there were so many parts between your left foot and the flywheel of your engine.

4. This diagram shows the action of the clutch release levers. With lever in "out" position, clutch is engaged (left). When levers are pushed "in," pressure plate is pulled away from disc.

5. A hydraulic clutch will have two adjustments, one at the pedal and one at the clutch release fork. Note the adjusting nut and bleed fitting. The system is bled like a hydraulic brake.

6. Proper clearances and adjustments are important to the life of any type of clutch, diaphragm or Borg & Beck.

7. Pressure plates can be adjusted for proper release finger height, but this should be done by a clutch shop.

asbestos-wire fabric on alternate plates. On a typical multiple disc clutch, the clutch case is integrally cast with the flywheel which forms the hub. Slots in the flywheel hold the driving discs, while the driven discs are attached to the input shaft of the transmission.

While the multiple disc clutch does not have a tendency to spin because of inertia (due to its small diameter), the discs sometimes refused to disengage because of a partial vacuum produced when the oil film was forced out at engagement. Small spring tongues were later used to separate the plates when spring pressure was relieved. However, poor lubricating oil and weak spring pressure were the main drawbacks to this type of clutch.

Band clutches—external contracting and internal expanding—are nearly the same in general principle of operation as automotive brakes. The main difference is that a *centrifugal weight* was used to further tighten or expand the band according to its usage.

The external contracting design usually consisted of a flexible steel band with a leather liner which wrapped around the flywheel or a hub on the flywheel and was tightened by spring pressure against a wedge. A centrifugal weight further tightened the band around the flywheel as rpm increased.

Clutches of the internal expanding type were more popular and were used by several leading American automobile manufacturers. Expansion of the band was obtained by spring pressure acting against toggle links or a right- and left-hand quick-acting screw. The latter was also sometimes actuated by means of a rack and pinion, the rack being part of the throwout collar. This type of clutch had many advantages going for it—smooth, gradual engagement, positive centrifugal lockup, prompt disengagement, and easy service. But the one big drawback was that it was difficult to balance and the heavy flywheel kept the transmission input shaft rotating, as did the cone clutch, making shifting difficult.

This brings us up to the plate clutch, which is the forerunner of the design of the flywheel, disc and pressure plate assembly that you're familiar with today.

The first clutches of this type used a large flywheel with the disc and pressure plate set inside its diameter similar to the band and cone types. To help slow down its inertia for smooth shifting, a friction brake was employed that worked automatically when the clutch pedal was depressed. However, this metal was soon trimmed away in favor of a flat surface when this design proved unnecessary with the new plate clutches. Other early evolutionary changes saw a change in transmission input shafts where the shaft was splined and extended forward into a bearing in the flywheel hub or crankshaft flange. Then too, the big switch was on from leather and cork to wire-woven organic linings, and more-positive-acting toggle links were used for smooth engagement and final lockup.

The large single clutch spring was still favored, but one design utilized many small springs evenly spaced around the periphery of the flywheel which pressed against an inner driving plate. The disc was then sandwiched between two driving plates. Pressure of the springs against the pressure plate held the driven plate firmly against the outside driving plate. When you wanted to release the clutch, the pedal was depressed and it pushed the clutch throwout levers toward the flywheel so that the clutch spring throwout studs pushed the pressure plate away from the driven plate and allowed it to revolve independently of the clutch. The clutch throwout studs performed a double duty in that they also acted as driving members for the two driving plates.

The use of studs for driving members prompted other engineers to go one step further and use a dual disc clutch which was called a "five-plate design." The reasoning at that early time was to provide additional friction surface to handle the larger engines and bigger cars being produced, and to permit reduced spring pressure which made for easier clutch operation. Remember now, this was all back in the "mid-'Teens" and this same thinking and design is now the hot idea in "dual disc" or "floater" and "slider" clutch construction.

As the advantages of the plate-type clutch became more obvious, engineers turned their full attention to this design which has evolved into its present state of perfection—smooth power transmission, comfortable pedal pressure, instant disengagement

Clutch Design

for smooth shifting, positive lockup for full driving power, relative freedom from maintenance, and longer life under normal driving conditions.

HOW A CLUTCH WORKS

So that you will have a more complete understanding of how a modern clutch works, you must first understand why a clutch is necessary in a car powered by an internal combustion (I.C.) engine.

When we mention steam engines and electric motors, don't get all fired up and hot—we're not going to revive the fight at the turn of the century between automobile manufacturers as to which type of powerplant made the best prime mover for cars (even though they are both again being considered for possible near-future use). However, knowing their requirements will help you more clearly understand the I.C. engine's need for a clutch device.

If a steam engine or electric motor were used to power a car, neither a clutch nor a transmission would be necessary and the rear wheels could be driven by direct drive. Power from either source may be coupled or uncoupled by the simple movement of a throttle valve or switch.

When steam is the motive agent, it is generated and contained in a "boiler," and the amount of power delivered by the engine to which the boiler is connected will vary with the volume of steam admitted and its pressure. If the steam supply were shut off by the throttle valve, the engine would stop. Conversely, when the throttle is opened and steam is again allowed to flow from the boiler to the engine cylinders, the vehicle will begin to move. By admitting more steam, the engine will turn higher rpm, but maximum torque is delivered at the start of crankshaft rotation—just the opposite of the I.C. engine. To reverse the direction of the car, the steam flow is reversed by a simple mechanical movement and the engine will then run in the opposite direction.

If an electric motor were used to drive your car, power (electrical energy) would come from a group of storage batteries. Providing these batteries were fully charged, varying amounts of electrical current could be drawn from them and allowed to flow to the motor which then puts out the desired amount of power or speed. The vehicle is started or stopped by completing or breaking the supply of energy, and by reversing the flow of electricity the car can be driven backwards.

So you see, with an external power source or energy that is ready to use, a steam engine or electric motor can be started under load and develop full power immediately. It is these features that eliminate the need for a clutch, changing gears or a reverse.

A gasoline engine is radically different. Its power is dependent on the number of internal explosions per unit of time (rpm), and its energy or horsepower output increases as crankshaft speed increases. And, unlike the steam engine or electric motor, a gasoline engine cannot be started under load because its power is obtained by the combustion of a fuel/air mixture directly in the cylinder—there is no external power source to draw from.

Keeping what we have previously talked about in mind, it should now be easy to understand that a gasoline

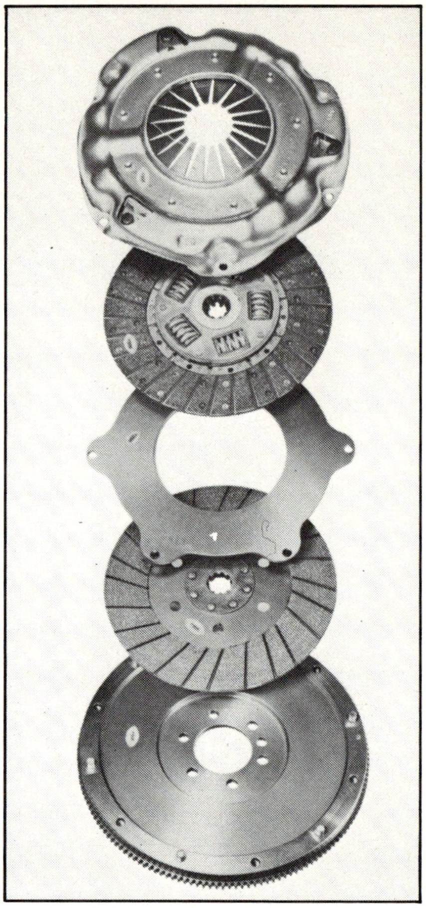

1

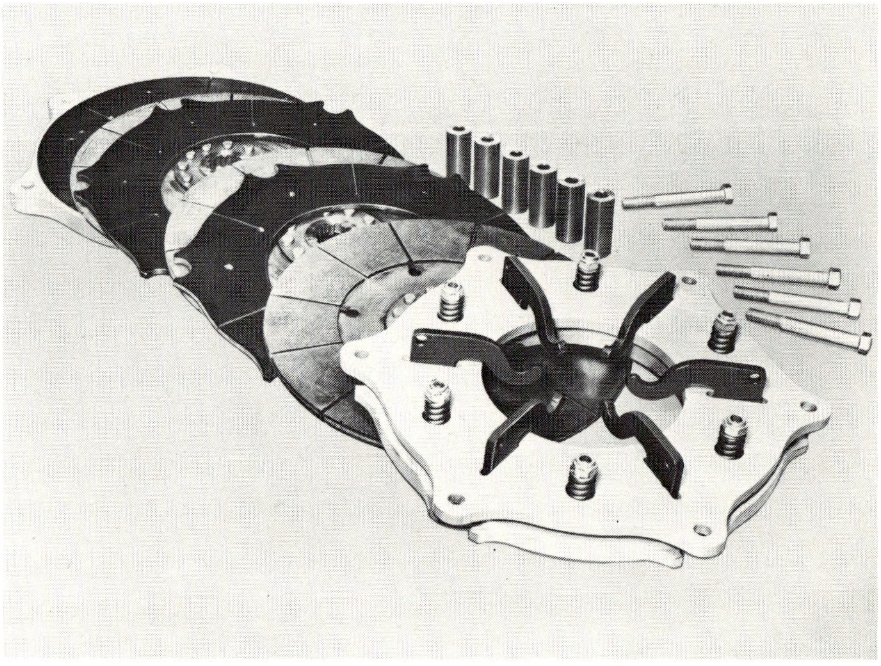

2

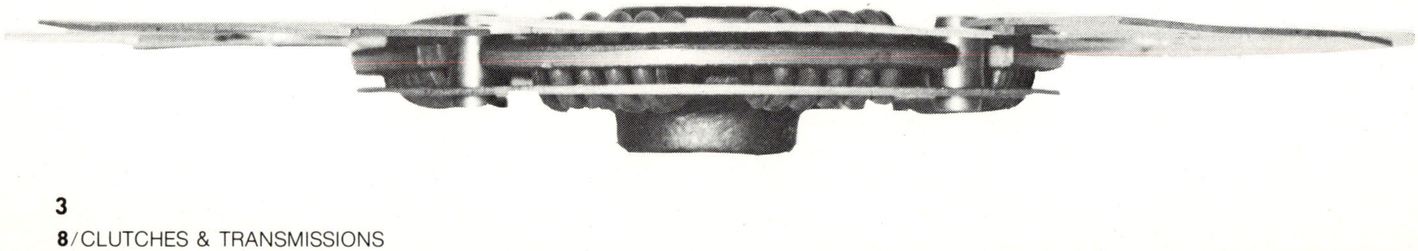

3

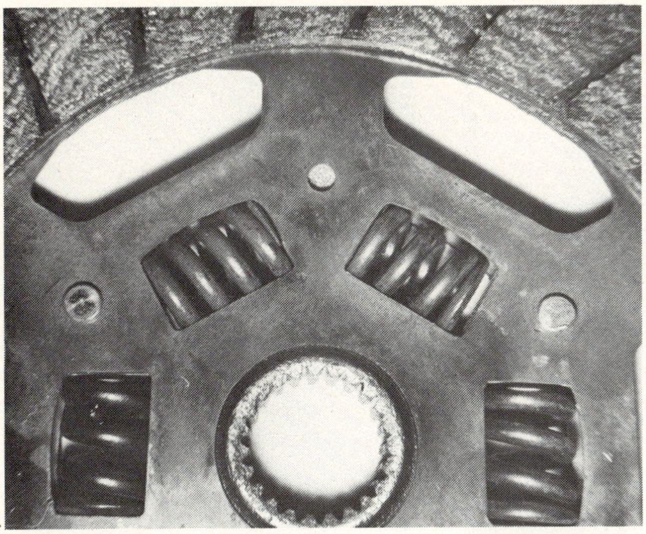

4

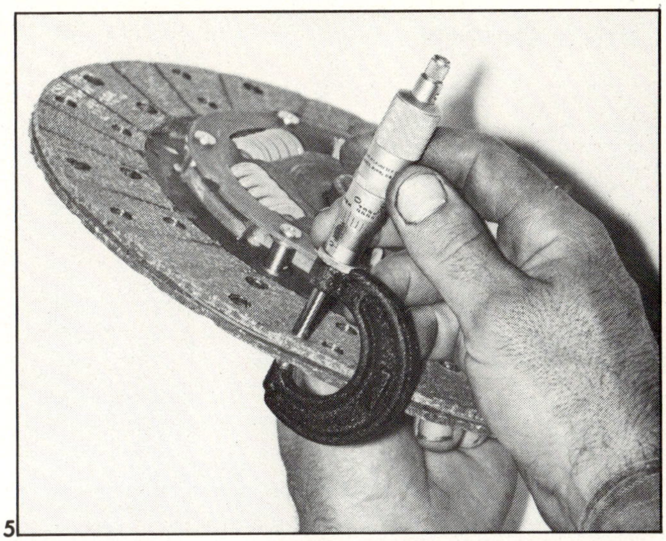

5

1. This racing setup actually uses two clutches for maximum friction, with a "floater" plate in between.

2. Here's another racing clutch that goes one better, it's got three discs. It uses two floater plates, and levers with adjustable centrifugal action.

3. Looking at a B & B disc from the side you can see the rivet "stops" and the pitch pressed into the wafers which cushion clutch engagement.

4. Springs like these around the hub mean this is a "damped" disc. These springs absorb the shock of engagement and dampen the engine's power pulses.

5. Proper disc thickness is vital to a smooth clutch operation. It must be thick enough to provide tight friction between the flywheel and pressure plate, but if it's too thick the clutch won't disengage fully.

6. Rebuilding car and truck clutches is a big business, but it must be done by a rebuilder with the equipment and know-how. This machine sandblasts cover plates in preparation for a preservative coating.

6

engine must run freely until it attains sufficient rpm to supply the necessary torque needed to overcome all resistance (vehicle weight and drivetrain) before it can be used to drive the vehicle. Also, the engine must be able to continue to run when the car is stopped, and because the I.C. engine develops its most useful power over a comparatively short rpm range, a transmission with three or four different gear ratios is necessary to enable the driver to maintain this effective rpm range. A clutch accomplishes all these things—let's see how it works!

As mentioned, the clutch must be able to transmit maximum engine torque without any loss of power due to slippage. And as the clutch takes hold, torque must be transmitted to the drivetrain in a gradual and uniform manner or the resulting "shock" on the drivetrain components can break the transmission gears, universal joints and/or rear axles. Then, when the clutch is released it is imperative that it positively disengages so that there is no continued rotation of the transmission input shaft.

A conventional automotive clutch has three main components: the flywheel, which forms one of the friction surfaces and is referred to as a "driving" member; the disc, which is referred to as the "driven" member and is splined to the input shaft; and the pressure plate, which can also be called a driving member because it is bolted to the flywheel. A throwout bearing operates the clutch but it's not a part of the clutch assembly.

ACTION IS SIMPLE

The action of these three components is quite simple. The flywheel is bolted to the engine crankshaft and revolves with it. The pressure plate is attached to a cover or housing which is bolted to the flywheel. Inside the clutch cover there are a series of small springs or a circular diaphragm spring which presses the pressure plate forward. The disc is squeezed between the flywheel and pressure plate and is kept centered as well as

Clutch Design

being splined to the input shaft and only revolves when the clutch is engaged. In addition, the splined input shaft is supported on the transmission end by a ball bearing and on the crankshaft end by either an Oilite bronze bushing or ball bearing.

The design of the driven member or disc is a little more complicated than it appears. It is actually made of many parts but the two main components are the inner splined hub and the outer ring to which the friction material is either riveted or bonded in some manner.

While the splined inner hub is riveted to the outer ring and can't pull away, it can rotate about ¼-in. An integral part of the hub is a ring of five or six stiff coil springs (some have inner springs as well) that absorb the shock of engagement. They also absorb engine vibrations (uneven pulsation regardless of flywheel weight) and very slight misalignment. The hub rivets act as "stops" for the rotational movement; however, coil bind (full compression of the springs) will result before the outer disc hits the stops. If this were to happen on a hard start, the disc would be ruined.

The outer ring is a series of stamped steel wafers that are bent slightly and to which the friction lining is riveted. These wafers also provide a cushioning effect when the clutch is engaged and will be squeezed flat at full engagement. Then too, the crimping helps prevent the disc from sticking to either the flywheel or pressure plate when the clutch is disengaged. Here again is another similarity between the early clutch designs.

When the disc is squeezed between the pressure plate and flywheel (engaged position), torque is transmitted to the rear wheels. To disengage the pressure plate (against the spring pressure), we have several release levers that pivot on the clutch cover and pull the pressure plate back. The release levers are pushed in by a throwout bearing that rides on a collar which extends about halfway up the transmission input shaft, and is attached to, and moved by what is called a "clutch release fork." The clutch fork or release fork pivots on a simple ball/socket arrangement at the edge of the bellhousing. In this case the bellhousing is referred to as the cover housing which runs from the rear of the engine block to the front of the transmission case and houses the flywheel and clutch mechanism. Since the clutch release fork is mounted to the bellhousing but pivots front to back, and the pressure plate with its release levers is rotated when the engine is running, the throwout

1

2

3

bearing must accommodate both of these movements at the same time. The throwout bearing looks like an ordinary ball bearing except that one side (the side that contacts the clutch levers) has a hardened steel surface which is backed by a ball race. It is this ball race which turns when it comes in contact with the rotating levers. The inner part of the bearing housing is pressed onto a collar which is in contact with the release fork.

Up to this point, we now know the action of the throwout bearing and the release fork as applied to operating the clutch. Actual control of the release fork may be accomplished by one of two means: by a fully mechanical linkage of levers and rods which extends to the clutch pedal in your car, or by what is called a "hydraulic clutch." A hydraulically operated clutch has a master cylinder (similar to the brake master cylinder) which is connected by linkage to the clutch pedal. When the clutch pedal is depressed or released, the clutch master cylinder operates a "slave cylinder" connected to the clutch release fork with a short adjustable rod. When pressure is applied to the slave cylinder, it in turn operates the release fork.

A spring is used to return the clutch pedal to its normal "out" position; however, this same spring actually has two functions. First, it returns the clutch pedal to normal position and secondly, it helps you push down the pedal against the pressure plate spring pressure. It's called an "overcenter spring." Here's how it works: One end is attached to the frame and one end to a point on the linkage so that its tension helps disengage the clutch when the pedal is depressed. Then, when the pedal returns to normal, the point of attachment on the linkage moves into a position where the spring is pulling in the opposite direction. So in effect it takes a certain amount of pedal pressure to push down the first few inches of travel, but the amount of required pressure does not increase proportionately as the pedal continues to the floorboards. True, additional pressure is required but it's not as great with the assistance of an overcenter spring arrangement.

TYPES OF CLUTCHES

Today, there are three basic clutch patterns or designs in use on American automobiles: Borg & Beck, Long, and diaphragm.

The Borg & Beck pressure plate design has been in use for about 20 years and is mainly used in Chrysler Corp. cars and some GM models. Conventionally, this is a 12-spring pressure plate assembly and is currently available in 10½- and 11-in. sizes (as measured across the friction face). It can be identified by its three "wide" release levers which do not have an end adjusting nut.

The Borg & Beck plate gets its total pressure from "spring pressure"—about 2800 to 3300 lbs. face pressure at lockup. The levers give some centrifugal assist, not much, but some—spring pressure is what this unit is based on. This makes the Borg & Beck tiring on your left leg because of the stiff spring pressure against the clutch pedal, but it affords a "clean release" at high rpm for smooth shifts.

To gain some additional centrifugal action, Borg & Beck installed needle bearing rollers on some plates which forced out against the pressure plate during rotation. They are housed in special machined grooves on the back side of the pressure plate, and centrifugal action wedges them between the pressure plate and the cover. In principle this works well, but these roller weights prevent shifting at high rpm because there is nothing to return the free-floating weights to a neutral position. While this is a distinct disadvantage in shifting, another drawback or weak point are the three eyebolts which must hold the high spring pressure. Most people don't realize that the spring pressure contained in the pressure plate housing isn't all exerted against the flywheel—an equal amount is also pushing back out against the cover. So here you can see the need for a strong cover and bolts, especially if you're going to use this on the Long-type unit on a hot street machine or for all-out drag racing.

The Long type of pressure plate has been around for over 35 years and is an offshoot of the early Ford style made popular by the V-8.

The Long is available in 10- and 11-in. sizes and is a nine-spring unit. This one gets its effective pressure from springs and a centrifugal assist. Looking at the photos you can see that the release levers protrude through the cover plate. It is this weight at the end of the lever which exerts centrifugal force against the pressure plate for positive lockup. Other advantages are that this type is impossible to overcenter—you just can't push the levers far enough forward. And for around-town driving, spring pressure on the clutch pedal is much easier than that of the Borg & Beck plate. However, at high rpm the weighted levers exert a high cen-

1. Clutch discs are rebuilt from scratch. The old lining rivets are drilled and punched out.

2. Proper pitch is restored by hand. Experience plays a large role here in knowing how much of this treatment each wafer needs.

3. New linings are riveted in place after torque springs are replaced.

4. Testing the disc for run-out is also a hand operation. At the same time, disc is checked for uniform thickness.

5. After installing new release springs, a quick filing removes any burrs from the cover.

4

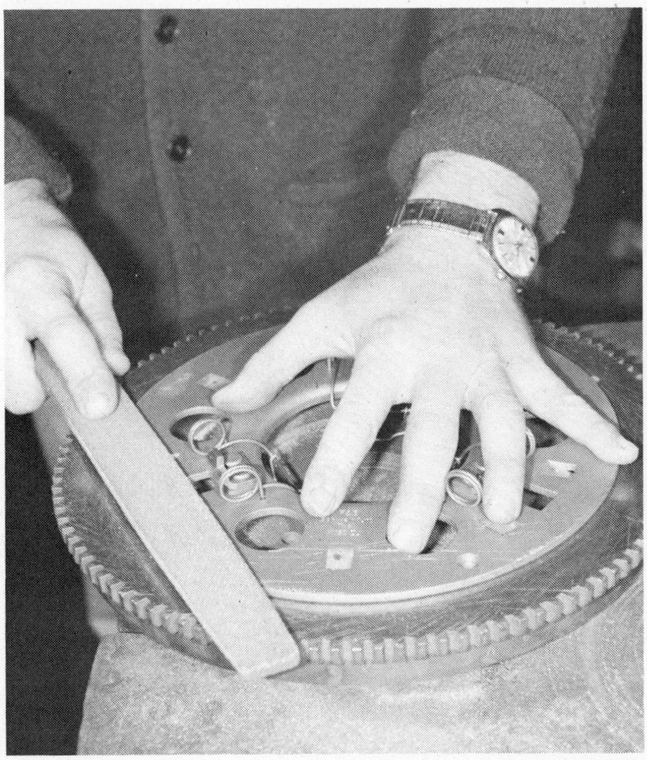

5

Clutch Design

trifugal loading and shifting is difficult. At 7000 rpm the weight exerts some 4500 to 5500 or more pounds of spring pressure.

The Rockford diaphragm type has been around for about 15 years or more. It's mainly used on GM cars coming in 10½- and 11-in. sizes.

There are two pressures associated with a diaphragm clutch in stock form—holding pressure (about 1800 lbs., pedal IN), and peak pressure (about 2500 lbs., pedal OUT). These of course can be changed by varying the length, width and thickness of the diaphragm plate in any one of multiples of changes in construction.

The diaphragm type is popular for stock and street machines because of the easy pedal pressure and smooth shifting qualities. In racing use, it has a quick lockup action that can break axles and U-joints—so be careful with this one. You can also catch some rubber going into each gear change.

This type of pressure plate is not recommended for use in cars where the engine will be turned over 7000 rpm because centrifugal force tends to hold the levers "in" after a "power on" shift. When the levers are pressed in, they are very near "center" position so centrifugal force holds them in this position until rpm's fall off enough to allow them to come back out. If this happens you can over-rev your engine very quickly and "lunch" it. Also, remember what we mentioned earlier about diaphragm plate deflection—it only moves about .075- to .100-in. at the edge so wear on the disc is critical. Keep checking it often and your assembly will last a long time. Remember, too, the correct way to adjust free-play pedal travel—this is very important to a diaphragm plate. If you push it past "center" it's ruined, like right now!

While the three styles of pressure plates—Borg & Beck, Long, and diaphragm—do not outwardly resemble each other, the Borg & Beck and diaphragm type have exactly the same mounting bolt pattern (providing it's of the same diameter). The Long type has a narrower pattern which is called "three sets of two's." If you wanted to change pressure plates it's possible to redrill your flywheel to accommodate both mounting bolt patterns. Most speed equipment manufacturers sell their special flywheels with this dual pattern already drilled and tapped.

Fundamentally, there are four types of discs available: damped (with springs around center hub), undamped (without springs, solid center section), and riveted or bonded linings. Add to this a variety of friction

1

2

materials available and you can just about design your own disc.

IF YOU ASK, "WHY?"

The reason for the damped disc is because an I.C. engine is a pulse-making machine regardless of flywheel and it needs a means of smoothing the pulse out before it gets to the drivetrain. At the start of engagement, the springs take up this pulsing and smooth out the shock. At lockup, the complete drivetrain—transmission, drive shaft, rear end and wheels—all become in effect a flywheel which further smooths things out. The springs in the disc are very important also—too little spring pressure will not dampen the shock sufficiently and tear out the hub. Additionally, too stiff a spring will make things too rigid and also tear out the hub. Spring pressure has to be just right according to the usage for which it was designed.

The solid or rigid disc, as it is sometimes called, is not recommended for street use. There is nothing to absorb the shock and it's dif-

1. All springs used in rebuilding are subjected to a compression test and rejected if they don't hold up to specs.

2. Final assembly is done on a pneumatic building jig. Center spindle of jig accepts dial indicator and other test devices to adjust clutch.

3. Assembled pressure plate is balanced using an electronic indicator. Holes in balancer allow a variety of clutch sizes to be checked on the same unit.

4. The sintered iron-type disc needs no special cover assembly but is very aggressive. It grabs quickly.

5. The Hays high-performance street and strip disc is organic with metal chips. A sintered iron facing for the pressure plate and flywheel is available for use with this particular disc.

3

4

5

ficult to slip for a smooth start. Nor is a lot of slipping desirable. In operation it seems like no matter how you try working the clutch pedal, the solid disc is either engaged or it's not—it's that positive. It's definitely best for the drags where the engine is really screaming and you can't feel a jerk anyway.

As for rivets or a bonded lining, it's up to intended usage as to which is best. A riveted lining is used with the wave design outer ring. The Marcel or pitch angle of the waves gives the disc compressability for further smooth engagement. However, at 7500 to 8000 rpm, a riveted disc becomes marginal and is quite apt to start chunking off pieces of material from between the rivet heads.

While the bonded disc has no compressability, it is superior to a riveted disc in the high rpm range. With increased contact area it is very doubtful that a bonded disc will ever chuck off its lining.

In conjunction with describing the various types of clutch components, it would be worthwhile to mention here that while throwout bearings are similar, they are not interchangeable; that is, from a lever-type pressure plate to a diaphragm type, and vice versa. Size is all-important and the face of the throwout bearing *must* contact the release levers at exactly the right spot. Obviously, if they don't meet just right you can understand that things will be pretty chewed up since the release levers are rotating and the throwout bearing won't be at contact point.

Up until now we have mainly talked about new, stock clutches as they come from the factory. But what about the rebuilt units that are available at almost half the price of a new part? Are they as good? And what is meant by "remanufactured"? For the answers to these questions we visited Adisco, a Gardena, California-based automotive parts rebuilder that services Southern California. Compared to some rebuilding operations Adisco might be rated a small company, but their quality rating in the rebuilding field is right at the top of the list.

A remanufactured part may be described as a used part that has been completely reconditioned to function like the original unit. And here's how Adisco does it.

First, all pressure plates which come in on an exchange basis and are reusable are washed in a "hot" degreasing bath. This removes all grease, oil and dirt. The plates are then disassembled. The pressure plate goes to the Blanchard grinding machine where its friction surface is ground to a perfect finish. This removes any burn marks, pits and scratches. You can tell if a part has been Blanchard-ground by what looks like a series of uniform cross-scratch marks all the way around the plate. This is the best method of finishing a friction surface.

Meanwhile the cover plate has gone through another operation that removes any rust spots and recoats the cover with a preservative coating. By this time, the cover and pressure plate have caught up to each other on the assembly line and are ready to be reassembled. But first each part is inspected again, the cover is checked for straightness and any burrs from the previous owner's rough handling are filed off.

To start with, new lever return springs are installed on the cover, and the pressure plate has new levers set in place with the pivot points lubricated with Lubriplate. From here the plate goes onto a building jig and new springs are positioned in their respective places. Next, the cover is carefully placed on top of the springs and is pulled down by a special air-operated clamp. Special nuts are now run down on the eyebolts and the cover clamp removed. Three other clamps are positioned over each set of mounting holes at the edge of the cover and pulled down tight via air pressure. A dial indicator is placed over a center spindle in the jig and each lever is individually set to an exact height.

From here the assembled pressure plate goes to the balancing machine for this final operation. It's securely clamped in place and run up in the machine. An electronic indicator tells the operator where the imbalance (heavy side) is and this side is drilled to lighten it. Rechecking and additional drilling is done as necessary until the unit is perfectly balanced.

Meanwhile, in another area of the plant the disc itself is being rebuilt from the ground up. First, the old lining is removed and the inner hub and

1. Installation of a clutch is made much simpler if you have an old, splined shaft with the gear cut off. The shaft will hold the disc in position while you wrestle the clutch plate into place and bolt it to the flywheel.

2. The Schiefer Super Rev-Lok uses aluminum pressure plate rings and flywheel faced with steel, applied by firing steel molecules into the aluminum surfaces at twice the speed of sound, achieving a permanent bond of steel into and on the aluminum for a stronger, longer lasting friction surface.

3. Oldsmobile introduced its 2-disc clutch with extra pressure plate forward in 1971. Called the W-37, the option came only on the 442.

4.Spinning the flywheel on this rig makes everything vibrate, and strobe light flashes to tell the operator which side is heavy.

1

2

3

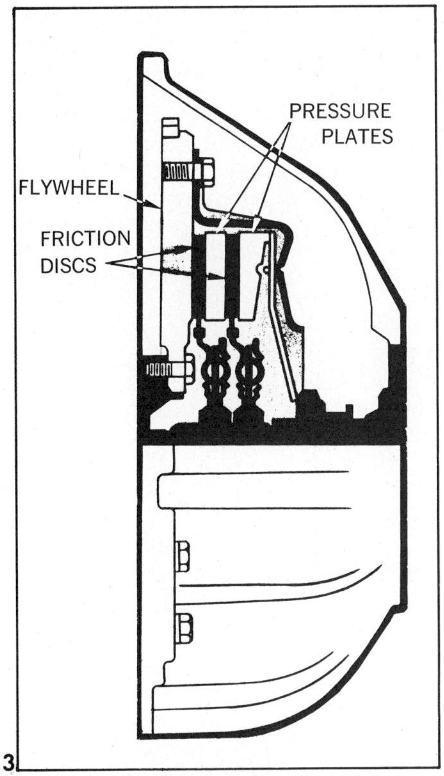

spline thoroughly cleaned. Next, the rivets which hold the inner and outer disc together are drilled and punched out. This leaves all the parts laid bare for inspection.

If the basic components are OK, new torque springs are installed and the discs re-riveted with new steel rivets. The proper pitch is then set in the waves and a new lining is riveted in place. A final check of assembly is made when the disc is tested for runout and uniform thickness around the circumference. Uniform thickness is an important factor in controlling proper engagement.

In a stock automobile that's used for everyday transportation a rebuilt clutch and disc are fine. All parts that wear or could wear have been replaced. Fatigued springs have been replaced, the friction surface has been ground like new and the unit rebalanced. True, there could be hidden metal fatigue, but in most cases a rebuilt clutch by a reputable manufacturer should be more than satisfactory for a stock automobile. If, on the other hand, you have a high-performance stock factory car or plan on making some horsepower modifications, then the best and safest way is to use only new factory parts or a special racing clutch and flywheel assembly.

There hasn't been as much stock clutch swapping lately—installing an 11-in. pressure plate in place of a 10-incher or installing heavy-duty springs—as was the case five or 10 years ago. At that time you had to build up your own if you needed a beefy clutch, but now you can buy just what you need from one of the many specialty equipment manufacturers. It doesn't pay to purchase a new 11-in. clutch and flywheel to replace a 10 and then install heavy-duty springs and have the whole thing balanced. You would be safer and money ahead to purchase the racing type that's ready to go and engineered for extremely hard use.

DISC, PRESSURE PLATE AND FLYWHEEL MATERIALS

To illustrate the transmission of torque by frictional adhesion we can

4

Clutch Design

take a simple case of two metal plates in contact with each other, the pressure existing between the surfaces being due to the weight of one plate bearing upon the other. If the top plate is not too heavy, it will be comparatively easy to turn one upon the other, but if weight is added to the plate more resistance will be felt which will increase directly as the weight increases. Consequently, it may be possible to add enough weight so that it will be practically impossible to move one plate without turning the other. This is called friction adhesion.

The main objective of engineers in designing a clutch is to increase the amount of frictional adhesion between the parts as much as possible. The power-transmitting efficiency of a clutch will vary with the coefficient of friction between the surfaces. On a modern automotive clutch, a metal forms two of these surfaces: the flywheel and pressure plate facings. *Metal* (iron or copper) or *organic* (asbestos) materials are common friction surfaces used on clutch discs today. Additionally, special iron- or copper-based friction materials are being attached to the faces of aluminum and steel flywheels and pressure plates to improve their friction.

Organic friction materials are commonly used on most cars and light-duty trucks. The term organic refers to the main manufacturing ingredient—asbestos. However, most organic discs contain a high percentage of woven copper wire or brass chips to increase the frictional quality. As we all know, asbestos is highly heat resistant, which has made its use widespread in clutch and brake linings.

Powdered metals are the newest friction materials to be used in clutch and brake applications, in the passenger car field. As far back as 1924 the S. K. Wellman Co., now a division of Abex Corp., has been producing metallic friction materials. It wasn't until World War II that their use became widespread in military vehicles. It was found that the metallic friction discs lasted longer, needed less adjustment, dissipated heat faster and generally took a beating that other materials couldn't.

While metallic friction material is commonly used in the military, aircraft, trucking and heavy equipment fields, it has been slow in being applied to the passenger car field. Metallic brake linings were the first breakthrough, whereas metal clutch facings are still generally used in racing only. It must be remembered that it is more costly to produce a metallic friction surface than to manufacture an organic type.

Due to the ability of metal powder friction material to absorb energy at higher rates than organic friction materials, it can be used at higher temperatures and at greater pressures and speeds. This quality is due to the fact that the metals can stand higher surface temperatures than organic materials before melting or disintegrating. Also the ability to absorb a greater portion of the heat of the friction application means more wear resistance and longer life for the disc and the pressure plate and flywheel. By dissipating heat in two directions there is less chance of heat distortion, heat checking and scoring. These factors, combined with the greater stability of ingredients making up powdered metal friction surfaces, offer a material with a more constant coefficient of friction. This is particularly important in the design of the "slip" or "slider" clutch.

Copper-based powdered metal friction material can take the most abuse—a lot of slip—and it is very smooth. However, you need a pressure plate that is about 20% heavier in the spring pressure department. This means more pedal pressure.

The iron-based lining is more aggressive or "grabs" quicker than the copper, but its major advantage is that it needs no special cover plate. This sintered iron disc is the type you will find in most racing applications.

Up until now, manufacturers have not been recommending metallic clutch lining for street use because of the reasons stated. Basically, you need your engine screaming coming off the line to take advantage of this high coefficient of friction. In fact, metallic linings really don't work at their best efficiency unless they're worked hard. And while you still wouldn't use a sintered iron or copper disc on the street, the Wellman Co. has developed a solid hub disc with a bond-on lining made up mainly of iron and graphite. Its advantages are that it's street usable because it's lightweight (in many cases lighter than organic), it has a friction level similar to organic material, and no special cover assembly is needed. It's also very smooth in operation as the

1

2

center hub is undamped. This would make it an ideal street and strip combination disc setup.

Metal clutch linings are made up of many different materials. According to usage, chemical composition can be related to wear, coefficient of friction and operating temperatures, etc. A typical "mix" for a copper-based metal lining might include carbon 6.0%, silicon dioxide 4.4%, tin 5.4%, lead 8.7%, iron 7.4%, and copper 68.1%. These materials start out as a dry powder and are ground to a specific size after blending. From here the powders are fed into a mold, and a press capable of several hundred to several thousand tons pressure forms the powders into a solid briquette for ease of handling.

Next comes the blanking operation which punches out the exact size of disc. The discs are then copper-plated in preparation for the brazing procedure which bonds the lining to the steel backing. After brazing and/or punching the rivet holes and assembly, the disc goes to a precision grinder which finishes the disc to an exact parallel and thickness. This assembly is driven on a splined shaft such as used in a transmission, thus giving a plate with friction surfaces perpendicular to the driving spline.

The final operation is balancing. Balancing can be done easily with a metal lining as small amounts can be ground away at the edges where it counts. You'll notice in the photos that the metal linings are also grooved. These grooves vary in width and depth, and in application serve to aid release and to keep the friction face clean of wear particles.

Stock factory flywheels and pressure plates are generally made of cast iron with cast steel or billet steel used on the high-performance models. Under no circumstances should a cast iron flywheel be turned over 5000 rpm. Nor should they be lightened as this materially reduces their strength. Then too, consider that the engineer who designed the flywheel knew what he was doing and that a particular weight is needed to operate your car more smoothly on the street. A steel wheel is the best for a combination of street and strip operation.

Clutch covers are die-stamped steel pressings. For the most part these production parts are made by several suppliers. Clutch manufacturers buy these stampings and assemble their own brand units. However, while the specialty equipment manufacturers buy stampings from these same companies, they specify special metals and their covers are usually about .025-in. thicker than stock.

Since heat is the biggest problem with a friction-type clutch, the springs and levers must be of high quality to resist this destructive heat. Again, specialty equipment manufacturers have suppliers furnish them with parts made from superior metals.

With the rate at which engines, tires, gears and other components are improving, in the not-too-far-distant future we will see a completely specialized clutch evolve to meet the demands of the quarter-mile.

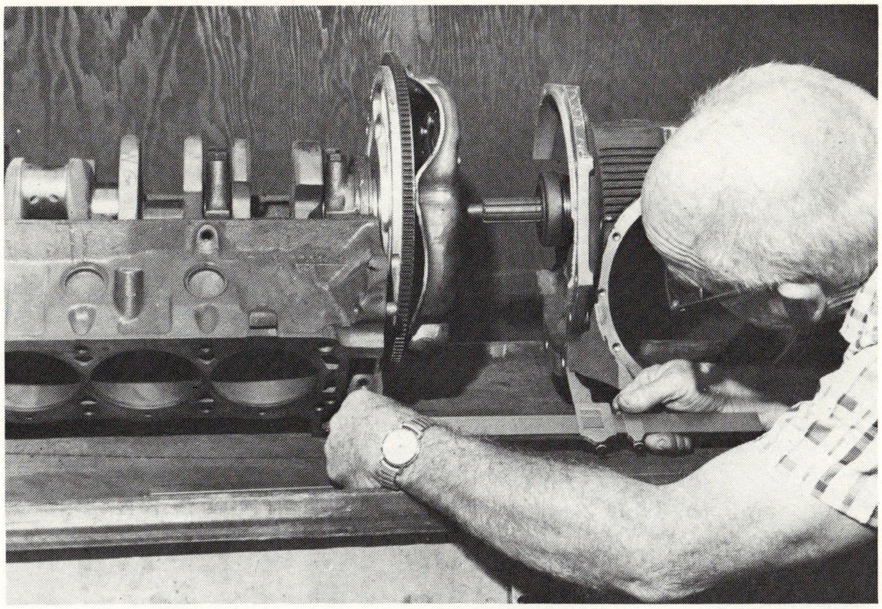

3

4

5

1. This flywheel casting was quite obviously a long way from being in balance, because of all the metal that had to be removed to get it right.

2. Balancing the flywheel is easy, you just drill shallow holes where the flywheel is heavy. Knowing where and how much to drill is the tricky part.

3. The item responsible for keeping the transmission aligned with the clutch and flywheel is the bellhousing. Designing one requires great accuracy.

4. Flywheels and automatic converters should be checked for runout with a dial indicator.

5. Clutch and/or flywheel explosion is always a danger. At 6000 rpm they can cut a car in two or cause serious injury. Just ask Don Garlits. If you're racing, don't skimp on safety equipment.

Diagnosing Clutch Problems

1

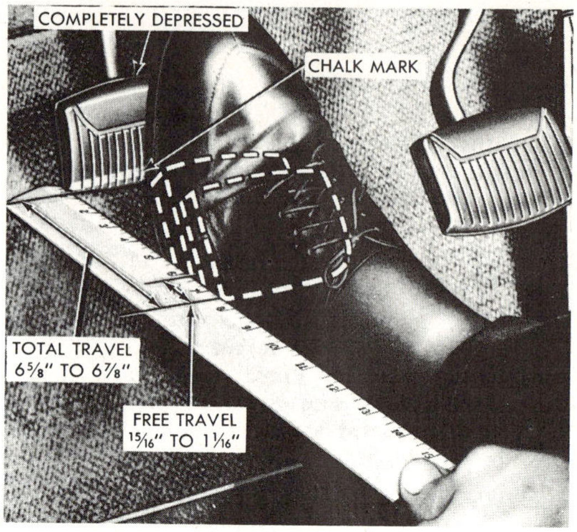

2

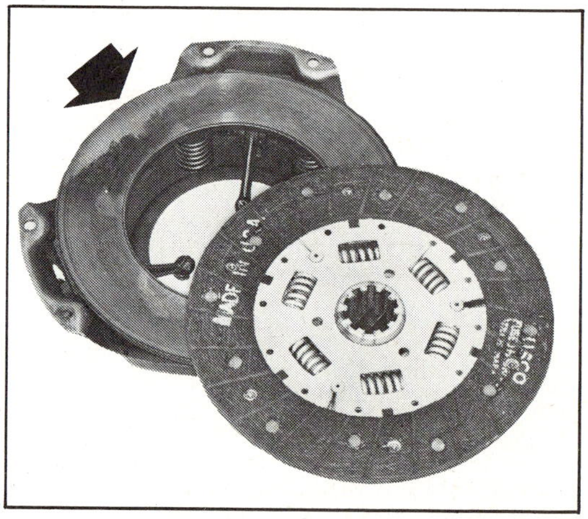

3

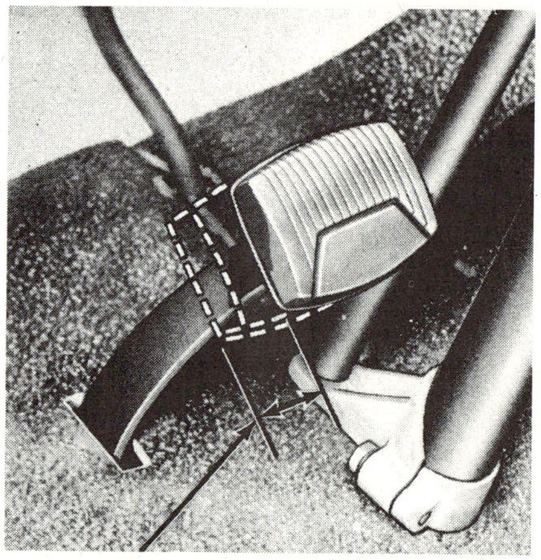

4

The actual operation of a clutch-equipped car is not difficult and most people reading this book surely know how to throw a fast shift. But there are a few good tips to remember about operation, care, service and troubleshooting in maintaining your unit.

Some of the first things we learn about driving a standard shift car is to let the clutch pedal out smoothly without over-slipping the clutch while simultaneously increasing engine rpm with the acceleration pedal. Then, as the car picks up sufficient speed to relieve the load, you shift gears by momentarily depressing the clutch pedal and backing off on the accelerator pedal. The clutch can be let out quicker after a shift because the car is already moving so there is much less chance for any roughness or shock to the drivetrain. The most important thing to remember is not to ride the clutch, causing undue slippage. Slippage of friction surfaces causes heat, and heat is destructive to automotive parts. Heat will burn the friction lining and/or score the flywheel and pressure plate facing. It can even affect spring pressure resulting in a complete overhaul or replacement. Allowing your foot to rest on the clutch pedal is also bad practice. This slight depression of the pedal brings the throwout bearing in contact with the clutch release levers, causing the bearing to spin unnecessarily and wear out prematurely.

With a good knowledge of clutch design and operation behind us, now let's look into care, service and troubleshooting the many common clutch complaints which have plagued the automobile since its inception.

CLUTCH SLIPS WHEN ENGAGED

You can easily detect a moderate amount of slip, as the acceleration and pep of your car will go "soft." You'll hear the engine rev up when you gun it, but the car doesn't speed up. It sounds a little like a fluid drive at low speeds. Never let clutch slippage go, or not only will your clutch be ruined but the surface of the flywheel and pressure plate as well.

The first place to look for the cause of clutch slip is in the linkage between the pedal and release fork. It may only be out of adjustment. (The release levers inside the pressure plate could be out of adjustment and cause this, but this is very unlikely unless the pressure plate is completely worn out). A misadjustment in any part of the linkage between pedal and release fork will try to hold the pressure plate from squeezing in fully against the clutch disc, thus causing the slip. Normal wear of the friction

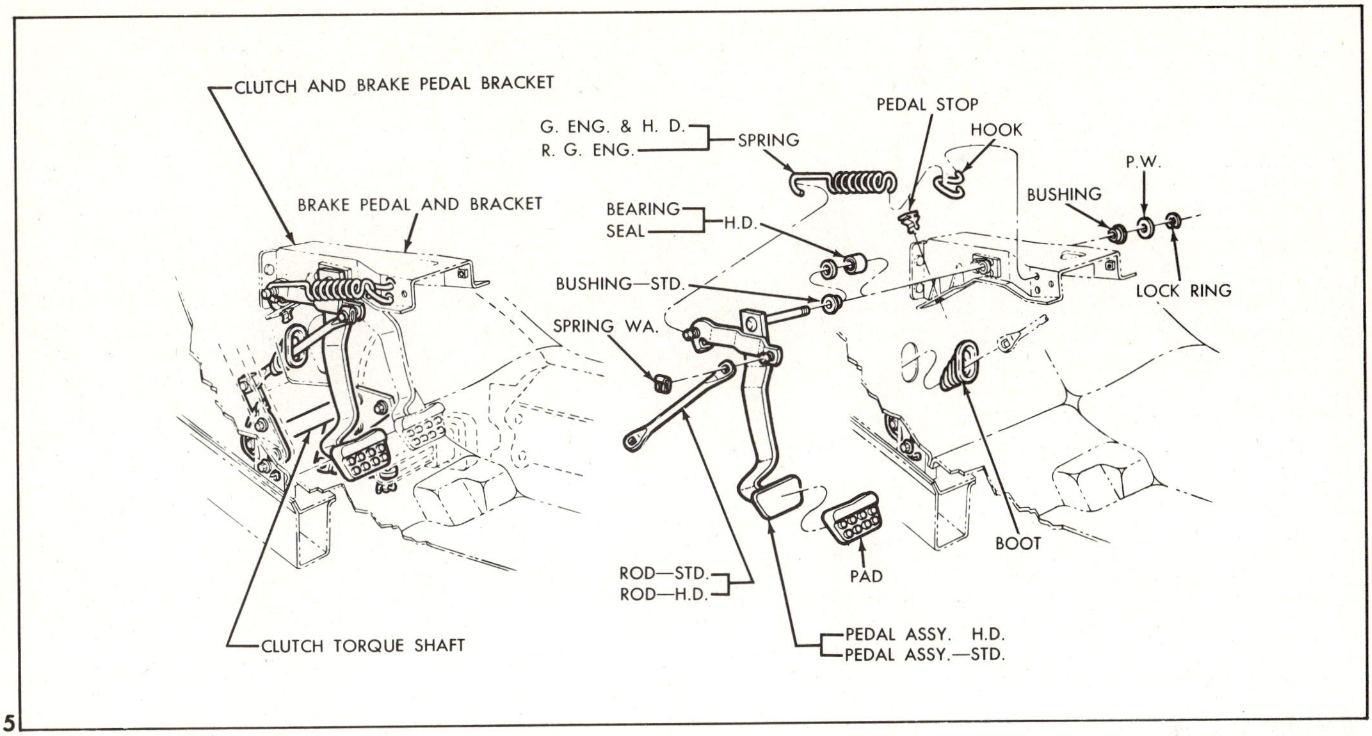

5

1. Keeping on top of the condition of your clutch and flywheel could save you a lot of grief someday.

2. Different manufacturers have different specifications for free-play in the clutch linkage. Consult the manufacturers' manuals for specific dimensions.

3. You'll seldom have any trouble with your pressure plate, but if it looks rough or chipped like this, you'll have to replace it.

4. Free-play is essential if long life is expected from the clutch and the throwout bearing. It can be checked from the inside of the car if you depress the clutch pedal until an increase in the pressure indicates that the throwout bearing is contacting the clutch.

5. Pivot bearings and bushings can become worn with use and must be replaced to restore proper adjustment.

surfaces can cause slip because this is compensated for in the linkage adjustment. Always keep a close watch on pedal adjustment because it can save you many dollars in the long run. It means the difference between replacing the entire clutch assembly or just the disc, which should be considered expendable because it's the cheapest part.

You can determine if the linkage is at fault by checking the amount of "play" (free pedal travel) at the clutch pedal. However, you *must* first determine which type of pressure plate your car is equipped with—the lever type or diaphragm type. This is extremely important because each type is adjusted differently and you can ruin a brand new diaphragm pressure plate by adjusting it according to lever-type operation.

Lever-type adjustment: Using your fingers and holding a ruler next to the clutch pedal, depress the pedal. It should move very easily inward for about one inch. This is designated as "free-play." After that it will seem to strike something and you will have to push much harder. The hard point was where the throwout bearing and the clutch release levers contacted. If you have about one inch of play, the slippage cause is probably in the clutch itself. If there is no play in the pedal, the trouble is most likely in the linkage. This is a quick adjustment. A mechanically operated linkage will have one or possibly two points at which the rods can be adjusted. The thread adjustment rod(s) usually have locking nuts (or jam nuts) which must be loosened, and the pivot point may have to be disconnected before the rod length may be adjusted. Make two turns on the free travel adjustment, reconnect the clutch rod and check pedal travel again. Repeat this process until you have only about one inch of pedal travel.

Diaphragm-type adjustment: There are two ways to make a correct adjustment, and both require the help of an assistant as well as the removal of the inspection plate or lower half of the bellhousing. One person must get under the car with a .030 to .040 feeler gauge and the necessary tools to adjust the linkage. The feeler gauge is to measure the "air gap" between the disc and flywheel when the disc is against the pressure plate; or it can be measured between the disc and pressure plate when the disc is against the flywheel. The person in the car slowly depresses the pedal until the feeler gauge can be inserted as described above. At this point the linkage is adjusted so that when the pedal is on the floor it has the required air gap. An alternative is to install a block under the clutch pedal to limit any additional travel. Or a bolt with two adjusting nuts can be welded under the pedal for future easy adjustment. The second means of adjustment is to back off the linkage until it is fully *out*. Then, with the clutch pedal held on the floor, you can start adjusting the linkage until the feeler gauge can be inserted as described above.

The reason for adjusting a diaphragm pressure plate this way is because it can only pull back the pressure plate about .075 at the edge. If you were to adjust it using the one-inch free-play method, the pedal will have enough travel to push the fingers *past center.* When this happens the "set" or "spring" of the diaphragm plate will be ruined, and it can no longer exert the pressure it was intended to against the disc. You can tell by a very much softer pedal pressure and slip.

The adjustment of a hydraulic clutch is the same process as above, with the adjusting link between the slave cylinder and the clutch fork. However, before adjusting a hydraulic clutch, check to see if the master cylinder is full and "bleed" the line to remove any air. Air in the system will prevent the clutch from operating correctly and smoothly. A hydraulic clutch is bled the same way as a brake system—have someone pump

Clutch Problems

up pressure and "hold" until you allow fluid to escape from the bleeder fitting and retighten the fitting. You should do this three or four times or until all air has been let out of the system. Keep the master cylinder full during bleeding so that new air does not enter the system, and fill to the recommended level when bleeding is complete. Also check fittings for fluid or air leaks.

If, after closely inspecting the complete linkage, you're sure it's not at fault, then your trouble may be inside the clutch. The most likely trouble is oil or grease on the friction disc. It can come from the transmission by way of a leaky input shaft seal, the pilot bearing or throwout bearing. Oil can also come from a leaking rear-main-bearing seal, but this oil is usually deflected by the engine side of the flywheel.

Slip can also be caused by broken springs behind the pressure plate, or by springs that have lost their "spring" due to overheating. Some bellhousings have an inspection plate which can be removed for a visual check to determine clutch slippage. Another possibility is that the friction material that is bonded or riveted to the disc has broken and chunked off. However, in either case, the disc must be replaced.

Try this simple test to determine clutch slippage: Start the engine, set the brakes hard, shift into high gear—and then release the clutch pedal slowly as you speed up the engine slowly. If your clutch is in good condition, it will stall the engine as you disengage it.

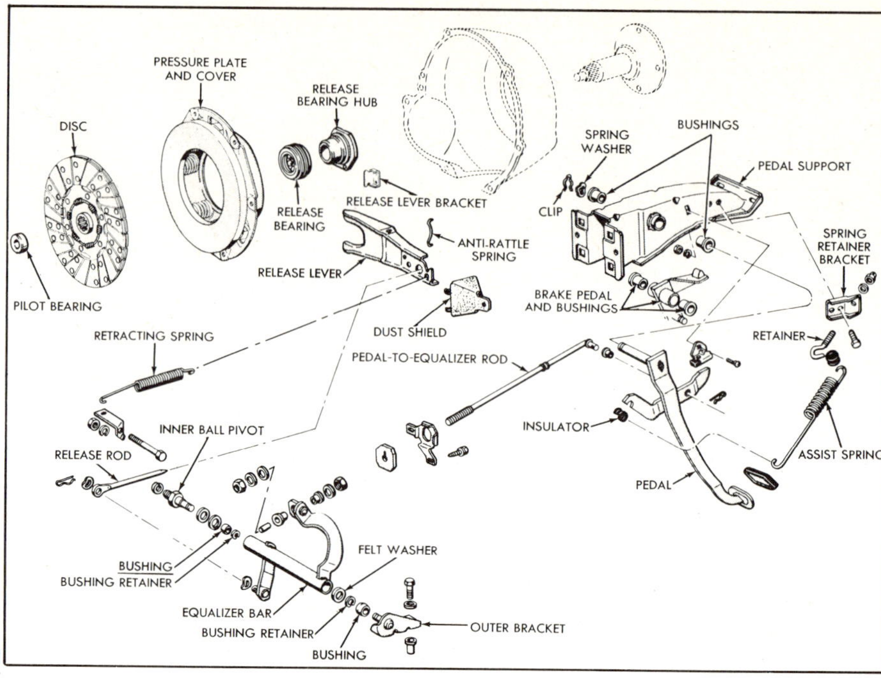

1

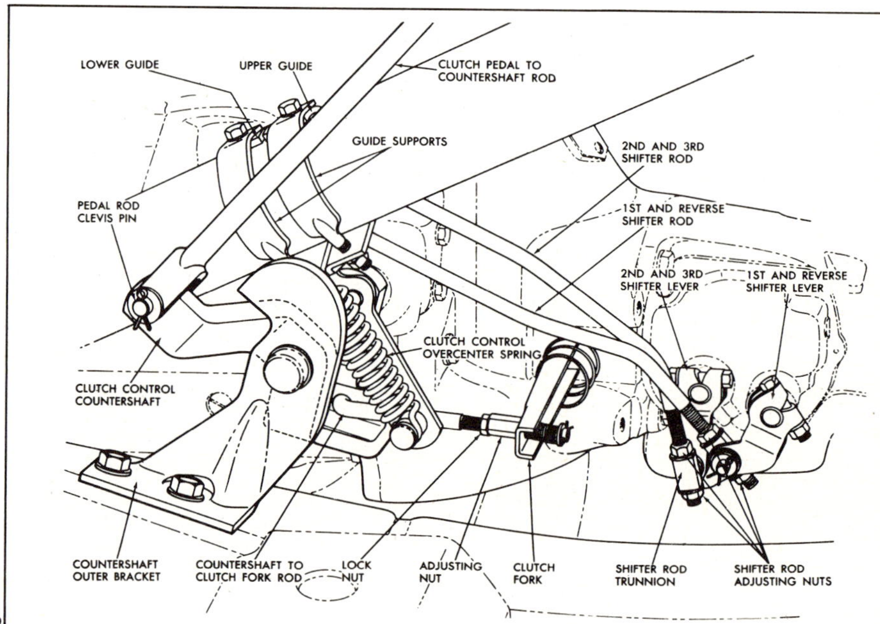

2

CLUTCH CHATTERS OR GRABS WHEN ENGAGING

This trouble is very easily recognized. No matter how slowly and carefully you release the clutch pedal, the clutch will engage with a hard jerk or else a chattering—or maybe a series of smaller jerks.

The first thing to check here would be to see that the pedal linkage is not hanging up and then suddenly releasing, so that the clutch is engaged with a jerk when you let the pedal out slowly.

Clutch chatter can also be caused by the disc binding on the input shaft, so that it cannot slide back and forth freely (even though it only has to move a few thousandths of an inch). This can be caused by dirt, rust, scratches or corrosion on the splines which intefere with proper engagement and release. Visualize the input shaft splines. These splines are made very accurately, as are the splines inside the hub of the disc. They must fit snugly to maintain perfect alignment and at the same time permit rapid engagement and disengagement. Another important area of alignment for the input shaft and thus the clutch disc is the pilot bearing. The pilot bearing is located on the centerline of the engine driveline and is mounted in the flywheel or in the end of the crankshaft. Since the tip of the transmission input shaft rides in the pilot bearing, clutch chatter can be caused when this bearing becomes worn and causes misalignment. Alignment is very important.

Another very important area to look for clutch chatter is in the engine and transmission mounts. The bolts must be tight and the rubber shock pads firm and in good condition. If you were to install a new clutch in an older car and it still had clutch chatter, suspect the mounts. Most people blame the clutch when in effect it is the rotted mounts that are at fault. You will have to inspect the front engine-mount rubber very carefully on all late-model cars as the rubber is bonded to the bracket and it's sometimes difficult to tell if it's torn loose. A spongy feeling on the accelerator pedal when accelerating hard can be your tip-off to a broken engine mount.

Sometimes the clutch friction disc will not be fully released by the pressure plate when you push the pedal to the floor. This will cause a clutch to drag and make the car want to "creep" when standing still—and you

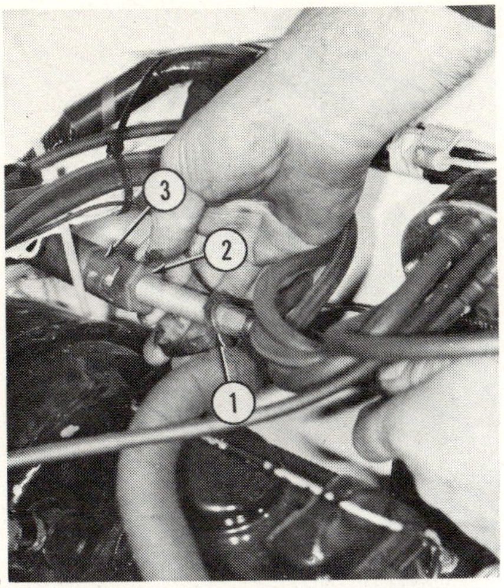

3

4

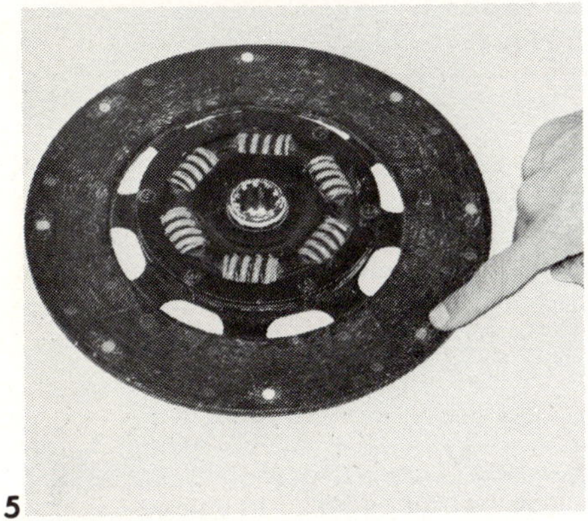

5

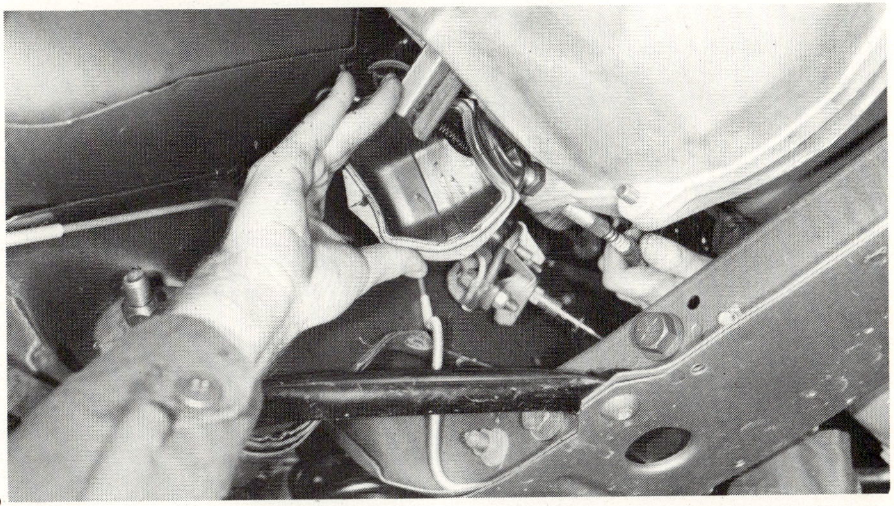

6

1. Typical of the clutch release mechanism on modern cars, this system provides an adjustment point at the threaded rod leading from the pedal to the equalizer bar. An adjusting nut and a lock nut are used to adjust the play at the throwout bearing by lengthening or shortening the pedal-to-equalizer bar.

2. Clutch and shifting linkage adjustments can sometimes be confused. A clutch that is not completely disengaging can cause a clashing of gears that could be wrongly diagnosed as a fault of the transmission.

3. Mustang II clutch adjustment can be done from under the hood. Remove cable retaining clip (1), pull the cable toward you and rotate nut (2) away from adjusting sleeve (3) ¼-in. Pull cable again to eliminate free movement, rotate adjusting nut until it touches sleeve, then index into next notch.

4. Like on this Chevy, the clutch and flywheel are accessible for repairs and inspection through the bellhousing. To replace them, though, you have to take out the transmission.

5. This disc is worn right down to the rivet heads and must be replaced. With bonded linings, it must be replaced if it measures less than 1/32-in. on a side.

6. Clutch adjustment on '74 Pinto is simple. Pull rubber boot off, loosen cable lock nut and adjusting nut at flywheel housing boss and pull cable toward front until release lever free movement is gone. Run adjusting nut to ¼-in. of boss and torque lock nut from 40 to 60 ft.-lbs.

will grind the gears no matter how carefully you try to shift.

The most likely trouble here is nothing more than an adjustment in the linkage between the pedal and release fork. This adjustment is actually the reverse of the clutch-slip trouble we talked about earlier. In other words the full travel of the pedal does not push the pressure plate far enough to fully release the disc. A quick linkage adjustment will usually take care of it. If this doesn't do it, the trouble is inside the clutch. Most likely it's a warped disc that's dragging, requiring replacement. Binding of the disc on the input shaft splines is a possibility, but misadjusted external linkage or a warped disc are the two chief causes of clutch drag. Obviously as the facings wear down and get thinner, the pressure plate will have to move farther and farther to fully engage the disc. Eventually you reach a point where the linkage is against the outer stop before the clutch is engaged and you have a slip condition.

With a lever-style pressure plate you should check free-play periodically to be sure that it is adjusted properly. Checking a diaphragm-type pressure plate is even more critical because deflection (movement) at the edge of the diaphragm plate is only .075 to .100 and therefore cannot make up much wear. A new disc is generally .300 to .330 in thickness, so the allowable wear should not go to less than .260 to .275 if the pressure plate is to perform properly—the correct amount of slip and then positive lock-up. When installing a new disc, with a diaphragm plate, it's advisable to measure its thickness and keep track of wear. This way you can replace the disc before you ruin the pressure plate—it saves a lot of bucks in the long run, especially if you're a racer. Remember that plates, flywheels and balancing are far from being cheap.

Even if you've never adjusted a clutch before, it's so simple that you can figure it out yourself. Get under the car and trace down the linkage from the clutch pedal to the release fork. Your adjustments are usually threaded rod(s) going directly to the release fork—usually there are jam nuts that must be loosened before turning the rod one way or the other to make an adjustment.

Clutch Removal and Repair

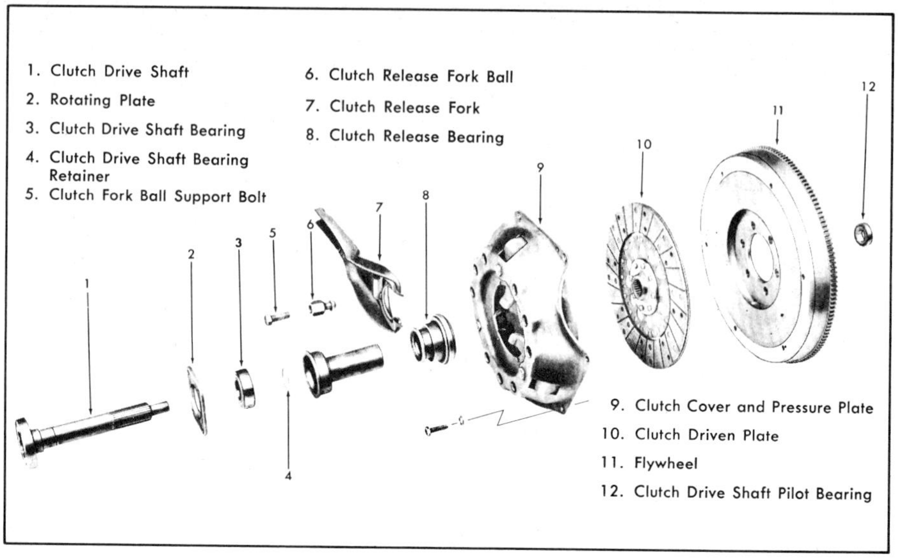

1

1. Clutch work is not as complicated as you might think—here's the entire assembly laid out for you. Getting to the assembly to work on it is more than half the battle.

2. The spring in this system is an over-center type. When the clutch pedal is depressed a short distance, the pivot point goes "over-center" and the spring assists clutch disengagement by pulling the pedal toward the direction of the floor.

3. This type of linkage uses a return spring on the clutch release fork.

If you have more serious clutch problems than those covered in the previous chapter, that clutch assembly is going to have to come out of the car for repair. While this can be a big job, it's really not as difficult as you might think, nor is it impossible for a beginner to tackle, providing you understand exactly what you're up against.

Clutch removal can be done in two different ways: by pulling the engine out of the car or by removing the transmission. Which one you choose to remove may depend upon whether additional work is necessary on the engine or transmission, but on some cars like Volkswagens and early model Fords (through 1948), pulling the engine is preferable, as it's far easier than pulling the transmission.

If you elect to pull the engine, start by disconnecting all wires and cables leading from the chassis/body to the engine. Disconnect throttle linkage, exhaust manifolds and radiator hoses, then remove the radiator and shroud. Disconnect and plug the fuel line and unbolt the front motor mounts. Hook up an engine hoist and start removing the bolts around the bellhousing.

Block the transmission from underneath to support it and lift the hoist enough to take the engine's weight off the car, then pull the engine straight forward until it clears the input shaft. Hoist the engine the rest of the way up and out and the complete clutch assembly comes too, making it easy to work on out in the open.

Begin pulling the transmission on cars with an open driveline by disconnecting the universal joint(s) and remove the driveshaft. Cover the end of the transmission housing with a plastic sandwich bag and secure it tightly with a large rubber band. This precaution will prevent the transmission oil from leaking during removal and prevents dirt from getting in.

Disconnect the speedometer cable, transmission shift levers and parking brake cable, then remove the rear transmission mounting bolts and support the rear of the engine with a jack or blocks. Now remove the bolts holding the transmission to the flywheel housing and you're ready to pull the transmission, but be careful—some transmissions weigh over 100 lbs., so if you don't have access to a transmission jack, it's best to have a friend along to help you lift it out. Draw the transmission straight back until the input shaft clears the clutch splines, then lower it. This is where you should exercise care, as the high weight leverage makes it easy to have the splines bend in the disc, making removal difficult and possibly bending the clutch disc.

After the transmission is out of the way, it's possible to remove the bellhousing to completely expose the flywheel/clutch unit on some cars (or the bellhousing may be part of the transmission case). On other models, only the lower half or an inspection cover will come off. This particular design slows down your work because you have to rotate the flywheel to expose each set of pressure plate mounting bolts.

Check to see if there are any brake or clutch linkage brackets that should be disconnected before removing the bellhousing. A factory service manual comes in handy here as it tells you exactly what needs to be disconnected. Also disconnect the clutch release fork at this time; it's only held in place by spring or snap-spring clip.

Once the clutch and flywheel are exposed, make one mark at the edge of the pressure plate cover with a center punch, and a second mark directly opposite on the edge of the flywheel. This gives you a pair of alignment marks that will let you reassemble the pressure plate and flywheel in exactly the same position as they are before you remove them, and is necessary as the two are balanced as a unit. If this isn't done and the pressure plate is simply remounted, the engine will be out of balance.

REPLACING PRESSURE PLATE

If it turns out that the pressure plate must be replaced entirely, the new plate and flywheel should be rebalanced as a complete assembly. Once the machine shop has balanced the clutch and flywheel, they'll mark the pressure plate cover so that it can be installed in the proper position on the flywheel. Although many exchange or rebuilt clutches are installed without balancing as a unit, and work fine on a stock installation, the pressure plate assembly itself is always balanced by the rebuilder before sale. But it's always better and safer to have the complete flywheel/clutch assembly balanced as a unit.

While the same procedure for removing the pressure plate is employed for all types, there are again two ways to do it correctly. The first and probably best method is to push down each clutch release finger one at a time using a steel bar. Insert a large nut (about ¼-in. high) between the release arm and cover as you do this to relieve all the spring pressure from the pressure plate. This lets you remove all the cover bolts quickly and in any order.

The other method is to back off each cover bolt in order around the cover about one full turn at a time. You must loosen the bolts evenly to prevent warping the pressure plate. Because of the spring pressure, they'll turn hard at first, but once

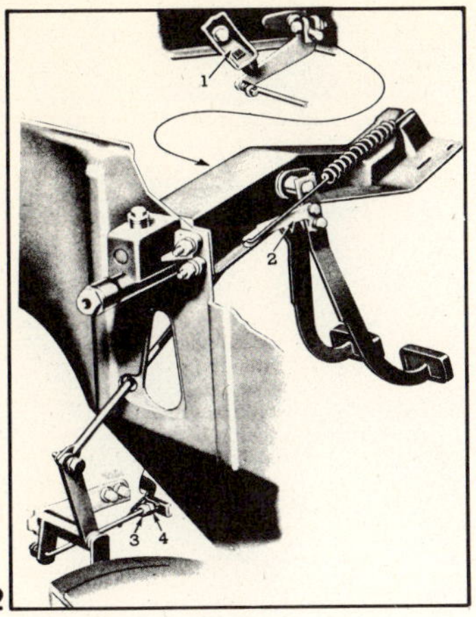

2

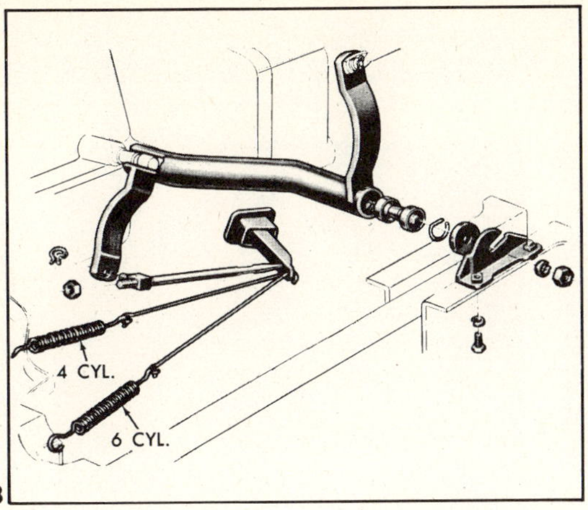

3

they're about half-way out, the release levers will contact the cover and relieve the pressure, a point at which you can just run the bolts right out at random.

If it's necessary to work through an opening at the bottom of the bellhousing, rotate the flywheel a littie at a time to loosen the bolts. But be careful when that last bolt comes free or you'll end up with a mashed hand—that pressure plate is heavy and it'll come down through that opening with a force than can damage anything in its way. Once the last bolt is free, ease the pressure plate and disc out together through the access hole.

The shop manual for your car gives the proper procedure for clutch removal, but you can usually tell by looking at the bellhousing. If it has a cover plate on the bottom, you can be pretty certain that the clutch comes out that way after the transmission is dropped. Never attempt to remove the bellhousing from the back of the engine unless absolutely sure that it's necessary for clutch removal—in most cases, it isn't.

Once the clutch is out, inspect all the components. If the disc is oil soaked, has a cracked or broken facing, is warped or has loose rivets at the hub, or is worn down to the rivet heads (bonded lining less than 1/32-in. thick on either side), then the disc must be replaced. While loose springs will not cause noise when the car is in operation, neither will they transmit torque smoothly. Even if the pressure plate is the source of your clutch trouble, a half-worn disc should be replaced if you expect to get the full-life span from the complete clutch unit.

Check the pressure plate friction surface for scoring, burn marks (blue color spots) or ridges. If necessary, the pressure plate can be resurfaced by Blanchard grinding, but minor burn marks and scratches can often be removed by polishing with crocus cloth. If the problem is not readily apparent to you (broken spring(s), warped or cracked facing, etc.), take the pressure plate to a good clutch repair shop. They'll tell you what's wrong and also check the lever release height, operation and runout—never try to adjust lever height or to dismantle the pressure plate yourself. If repair involves a lot of work or a cost close to the price of a new unit, you might as well spend the few extra bucks and replace it.

If the flywheel is scored, ridged, has burn marks or is highly polished, it should be resurfaced. A light cut can be taken off the face and you'll get like-new performance, but do this before rebalancing. Wipe the clutch release bearing clean; it's prelubricated and sealed, and should not be cleaned in solvent. Hold the inner race and rotate the outer one while applying pressure to it. If bearing rotation is rough or noisy, replace it. As mentioned in the previous chapter, most release bearing failures are caused by improper clutch pedal adjustment, or by riding the clutch. The release bearing is lightly pressed onto the bearing hub and can be knocked off easily. Carefully tap a new bearing on with a hammer using a block of wood as a driver to avoid any possibility of damage to it. Coat the release bearing hub with a thin application of Lubriplate before installation on the transmission input shaft cover.

The best rule of thumb concerning the pilot shaft bearing (a ball bearing or bushing) is to replace it—it's far easier and less expensive to do it now and then forget it. If the flywheel is off the engine, there's no problem in knocking out the pilot bearing but if it's still in place on the crankshaft (or mounted in the crankshaft end), the pilot bearing must be removed with a puller. There are several ways to do this but a slide hammer puller with a hook on its end is usually used. Insert the hook in the hole of the pilot bearing and knock it out with the slide hammer.

Check a new disc on the input shaft splines—it must slide smoothly. If it doesn't, rust or a very small nick may be hanging it up. A fine file will clean up small marks and crocus cloth should be used to give the splines a final polish. Clean the splines well and apply a very thin coat of Lubriplate before assembling.

REASSEMBLY WARNING

If an oil leak has ruined your clutch disc, you'll have to find the cause—engine or transmission—and have the leak repaired before assembly, or the new parts will also be ruined in a very short time.

Everything is now ready to go back together. Before attempting the initial assembly, make sure that your hands are clean and free from grease. A greasy fingerprint or two on the flywheel or pressure plate facing will cause chatter and ruin an otherwise good clutch job. Clean the pressure plate and flywheel faces with a non-oil-based solvent just before beginning reassembly.

Start by installing the flywheel. Torque the bolts to factory specifications and put a small amount of Lubriplate inside the pilot bearing or ball bearing inner race. You'll need some kind of pilot shaft to align the clutch disc. While there are special tools for this purpose, the most inexpensive and practical way is to cut an input shaft from a junked transmission—it'll always be handy for future jobs.

Hold the pressure plate and disc in close contact while pushing the alignment shaft through the disc spline and into the pilot bearing. Line up the two punch marks you made before disassembly and start bolting up the pressure plate cover using the same procedure you followed during removal. Torque the bolts to factory specifications, remove any spacers used and pull out the alignment shaft. Lubricate the clutch release fork at its pivot points and complete the assembly. Take care not to bend the disc when replacing the transmission, as this will also ruin your clutch job.

After installation of your clutch as we've described, there should be no further need of lubrication or service of its life. The only maintenance required is to periodically check and adjust (if necessary) free-play and to lube the outside clutch linkage pivot points at regular intervals with lightweight motor oil. While it may sound difficult to the beginner, it really isn't; and to help you visualize the high points of the procedure, we've included a how-to sequence. It should help to clear up any confusion that may develop along the way. Keep on Clutching!

1. If wear is not excessive and the clutch is a straight replacement job, mark the flywheel and the cover for easy and proper reassembly.

2. Back off the retaining bolts one turn each in rotation around the cover. This avoids warping an otherwise good pressure plate—and your wallet.

3. Once all tension is off the bolts, you can just run them right out in rotation and then lift off the pressure plate and clutch disc.

4. If the disc is worn near the rivets, or a bonded lining is worn to 1/32-inch thick on either side, it must be replaced, or you'll ruin the flywheel.

5. Examine the pressure plate for scoring, burn marks or ridges. This one is highly polished and has very light marks on the surface.

6. Crocus cloth or 320 or 400 grit sandpaper may be used to remove polish, small marks and blue spots. Note new finish in front of the hand.

7. Test new disc on input shaft splines—it should slide back and forth freely. File or sand away any small nicks or scratches that cause hesitation.

8. With the pressure plate and disc serviced, remove the pilot bearing from the flywheel. A slide hammer works best for this job.

9. The flywheel also has a highly polished finish and contains a few small grooves. The condition is not beyond salvage by sanding.

10. If you replace the disc before it damages the friction faces, the flywheel can be refinished by hand and will look like this when it's finished.

11. The flywheel must be torqued to factory specs if it has been removed, but even if it hasn't, checking the torque is always a good idea.

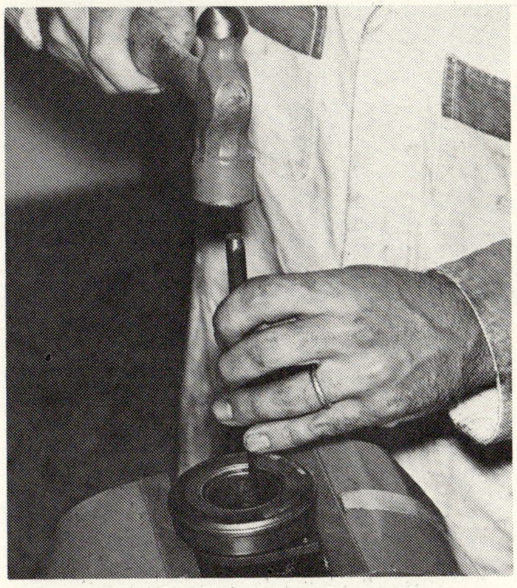

12. Even if the old throwout bearing looks good, replace it to be safe. Knock the bearing off the hub by supporting the bearing edges in a vise.

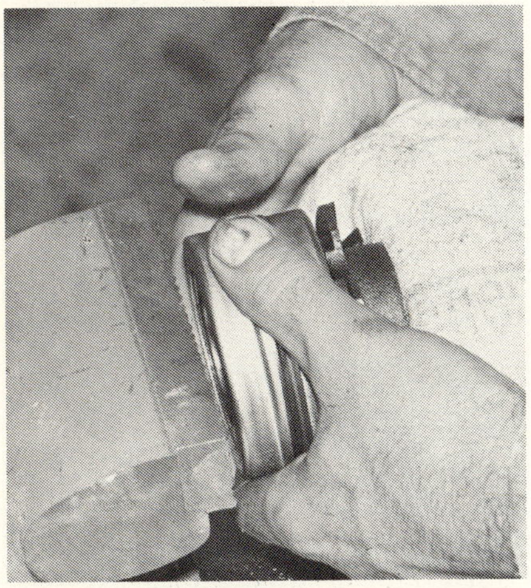

13. After thoroughly cleaning the collar, press the new bearing on by using the old one (or a block of wood) as a shield and tightening the vise.

14. Check the throwout bearing collar on the trans support shaft to be sure it slides freely. Polish out any marks, then lubricate shaft and splines.

15. Put a small touch of Lubriplate on the release fork pivot points and then slip the throwout bearing into place on the fork.

16. The disc must be properly located relative to the flywheel—the damping springs face to the rear. Hold disc and pressure plate with marks lined up.

17. Align the unit with the pilot bearing using an old input shaft. Install the bolts, pull down evenly and then torque to specifications.

18. Remove input shaft, replace remaining parts in reverse order of disassembly, and be sure to put a little grease on the pilot bearing. Replace trans.

High Performance Clutches

The hot rod application of clutches and flywheels has come a long way since its beginning. In the early days the theory was to beef-up stock components—heavy springs in the pressure plate, as heavy as you could find, and lighten the flywheel as much as possible to bring up rpm quickly. But with the *big* cubic inches available today this has all changed—it's no problem for big cubes to bring up rpm quickly with a heavy flywheel, and it's now necessary to have the weight to maintain momentum whenever a shift is made.

Today, you can't get away with just beefing-up stock parts. Engines are too powerful and high rpm makes centrifugal force a potential time bomb. Special metals, sophisticated machinery and careful assembly are now all part of the scene. Names like Crower, BID, Hays, Schiefer and Weber are manufacturers who build complete matched and engineered assemblies. These special assemblies are the *only* parts that should be considered for use on a high-performance car—your competitiveness and safety depends on it. In fact getting through the quarter-mile has become so sophisticated that equipment manufacturers need to know what type car you have, the weight, tire make and size, engine horsepower and rpm, gear ratio, etc., to recommend the best type of pressure plate, disc and flywheel to get the job done for you. It's as important as matching any other piece of speed equipment to your engine for total performance.

The Borg & Beck type pressure plate has been adapted to racing by specialty manufacturers by increasing the thickness of the cover. This enables the cover to withstand high rpm and the additional heavy spring pressure required to transmit torque from the "stump pullers."

Since this pressure plate doesn't have a centrifugal assist, shifting is smooth at high rpm. The only disadvantage is that you wouldn't want to use it for street driving due to the high spring pressures. Your left leg would soon be numb if you had to drive through traffic.

BID, Hays, Schiefer and Weber offer this type of unit. BID and Weber both have forged steel facings, while Hays offers it in forged steel or with a rivet-on heat shield facing of sintered bronze. Schiefer uses a forged aluminum with a special patented metal-sprayed surface of copper and steel finished by Blanchard grinding.

While the Borg & Beck type units aren't "totally" suited for racing, they do have a place and are particularly adaptable to the smaller bellhousings. The Borg & Beck has smaller overall dimensions than the Long, so it's particularly suited where space limitations are a problem.

DIAPHRAGM PRESSURE PLATES

The diaphragm-type pressure plate has received similar beefing with a thicker cover plate and heavier diaphragm spring. The diaphragm plate utilizes the Belleville washer principle. Besides giving equal pressure around the entire 360°, it permits light clutch pedal holding pressure. It works this way: Pressure applied to the diaphragm fingers increases as they are depressed until an "over center" position is reached. After this point is passed, very little foot pressure is required to hold the pedal down. This feature alone makes the diaphragm popular for street use. The only disadvantage is that at very high rpm, centrifugal force tends to hold the fingers in. Schiefer uses helper springs to help eliminate this undesirable problem.

Hays, Schiefer and Weber make this unit available. Hays with a forged steel or sintered copper heat shield, Schiefer with a forged aluminum plate and thin layer of metal-sprayed facing. The Weber unit is steel faced and is also made into a dual disc assembly (sold as a unit) which consists of steel billet or cast steel flywheel, two organic clutch discs and an alloy steel floater plate. Special mounting bolts are provided and the complete unit will fit into a stock drive setup.

The Long is probably the most widely used pressure plate as it can be used with a single disc or dual or triple disc assembly.

SLIPPER CLUTCHES

Spring pressure is not objectionable for street use and where power shifts must be made, the plate is provided with a standard type of lever *without* the big barrel weight at the end. By increasing or decreasing the weight at the end of the levers you

1

can control the amount of centrifugal assist. When ordering the Long unit, be sure to specify the kind of intended use, and the manufacturer will supply you with the proper levers on your pressure plate. This is particularly important on competition cars where no shift is made and lock-up is required about three-quarters of the way down the strip. So, by varying the amount of centrifugal assist (lock-up) you can control the initial slip coming off the starting line.

With these big slicks running at rock bottom pressures and staying glued to the strip, something has got to give. This slipping clutch idea has been coming on for several months now and it has caused considerable change in dragster thinking. The problem is simple: If the tires are getting maximum traction, and the engine is putting out maximum torque and horsepower, something, somewhere has to give. In this case it's the clutch.

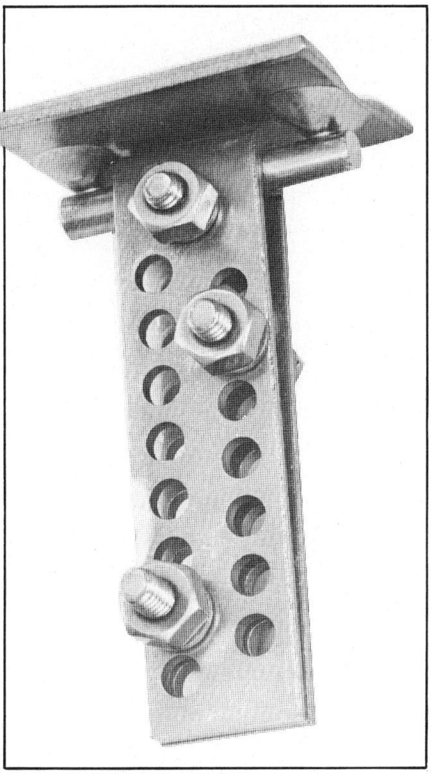

2

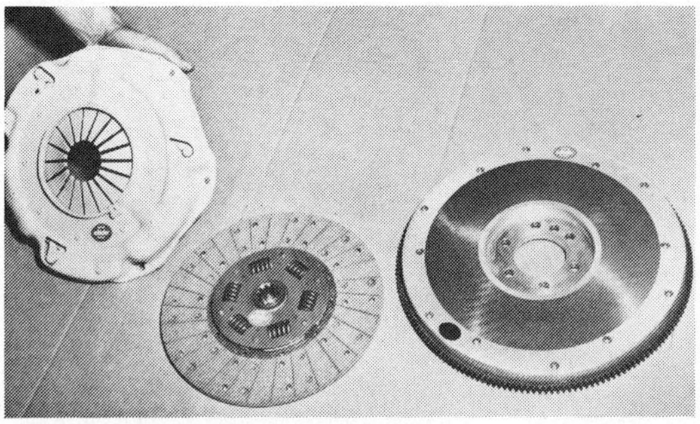

3

4

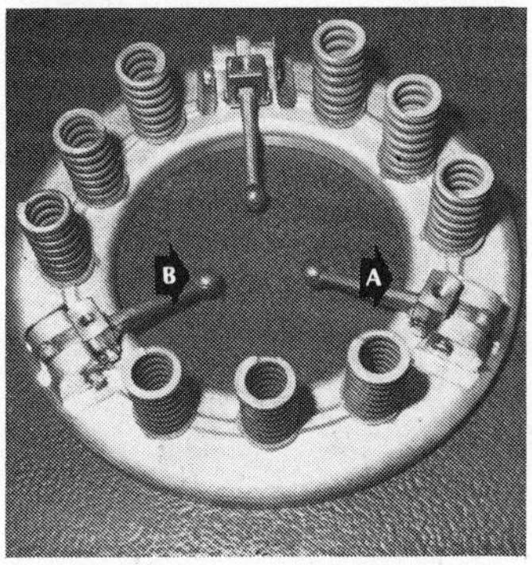

5

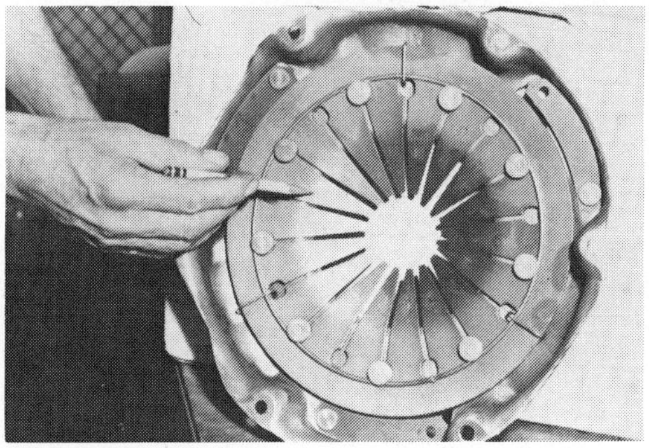

6

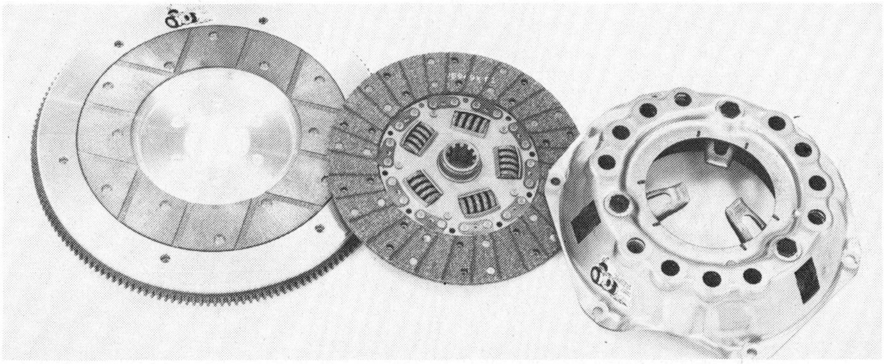

7

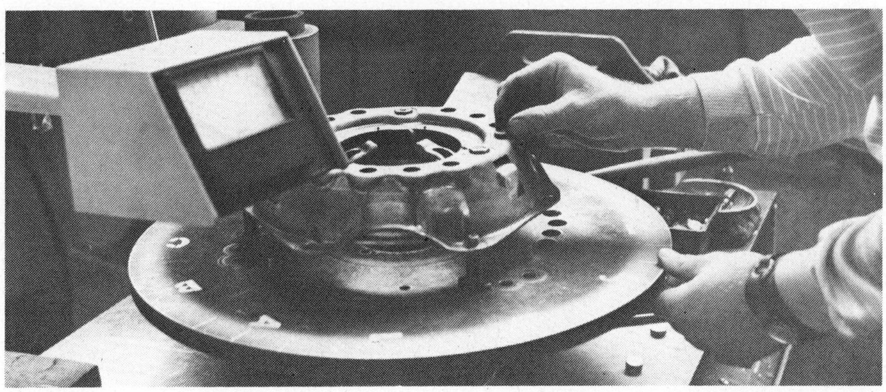

8

1. A powerful engine won't do you any good if you can't transfer all of it to the ground with an efficient driveline, which includes the clutch package.

2. Pushing the clutch pedal all the way to the floor for shifting can be a waste of energy and time on the racetrack. With this Dzus device on the arm of your clutch pedal, you adjust it so it hits the floor just after you reach full disengagement. This saves wasted motion in clutching.

3. Typical street and strip combination by Schiefer includes aluminum flywheel with friction facing, non-cushioned bonded lining with damped hub and diaphragm pressure plate.

4. The Rev-Lok "hat" holds operating diaphragm, diaphragm anchor point strips (A), and special Schiefer wire ring (B). These clutches must be set to proper .040-in. release clearance or unit can incur high-rpm pedal sticking and even pressure plate damage.

5. For competition use on dragsters, heavy-duty, Long-type pressure plate ring and hardware are used. Release levers carry the stands (A) and old Ford-style adjusting buttons (B) to aid in setting clearance from the arm to the throwout bearing.

6. Pressure of a diaphragm spring can be varied by the width of the fingers and by actual thickness of the metal.

7. The Hays street and strip assembly might include an aluminum flywheel with a heat shield facing, a damped and cushioned disc, and the beefy Borg & Beck pressure plate.

8. Balancing is critical on all racing components that move. Here a pressure plate is ready to be spun at high speed. Machine will indicate heavy side, which will then be lightened by drilling.

Clutches

The new idea is to never let the tires break traction, which means keeping wheelspin completely out of the picture, which in turn means that under current dragster construction practices, the clutch must be slipped. Clutch slippage is the only way to get a direct-drive machine off the starting line without bogging the engine. With a 3.23 rear-end ratio and the engine idling up around 1000 rpm, the car would be traveling around 40 mph with the clutch engaged. To get out of the hole at over 4000 rpm, therefore, the clutch is allowed to slip and then lock-up at a point that won't break tire traction. And if the tires are smoked, they won't recover. Setting up a clutch to slip to the precise point and then to lock-up is no amateur process, and must be carefully undertaken.

All this leads to a situation that's uncomfortable at best. A clutch is designed to utilize controlled friction in moving a car, but heat in the clutch is its own worst enemy. Tom McKewen experienced clutch assembly failure under such a condition, and reported the exposed flywheel was cherry red immediately following clutch disintegration. Which all proves that too much slip can generate enough heat to melt metal and cause serious difficulties. Already, the problem has lead to a series of changes in safety rules.

Until recently, the average dragster would be running a double disc assembly with a Long pressure plate and a total of 2000 lbs. static face pressure. Since the clutch was designed to make the connection between engine and driveline as rapidly as possible, there was minimal slippage and at the end of a run, clutch temperatures seldom would exceed 300°. All this from over 1000 bhp in a 1200-lb. car.

In AA/Fueler practice, the clutch mechanism has become a kind of

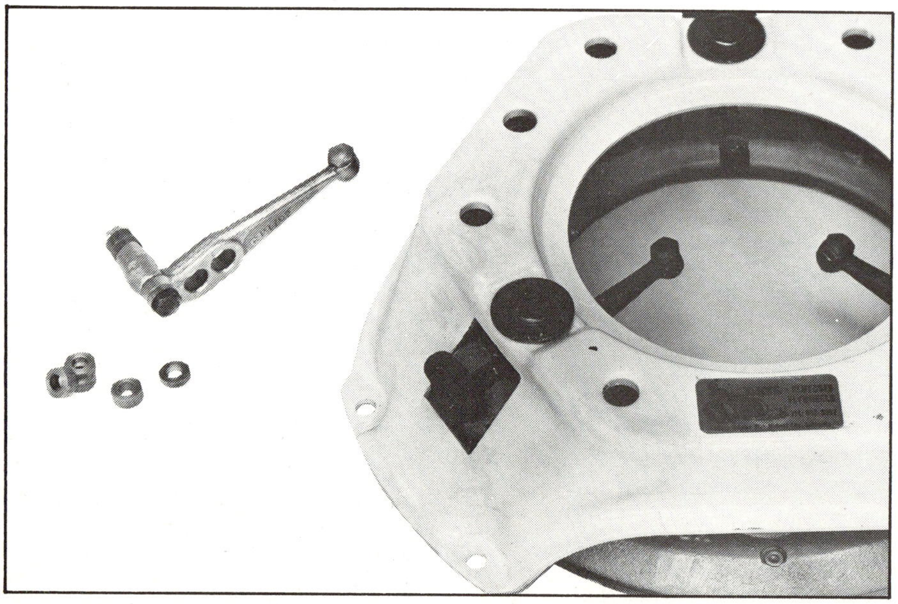

1

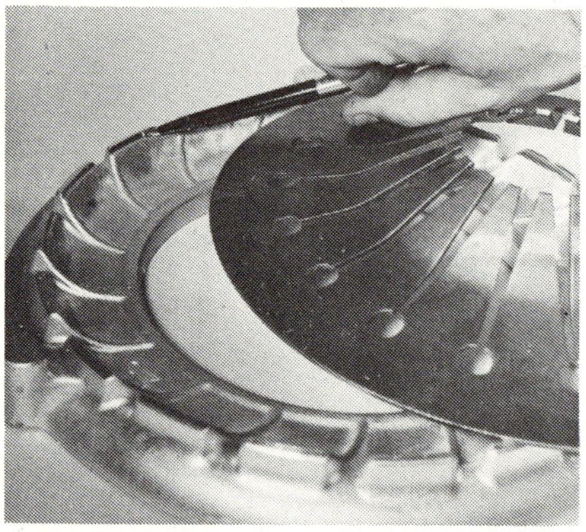

2

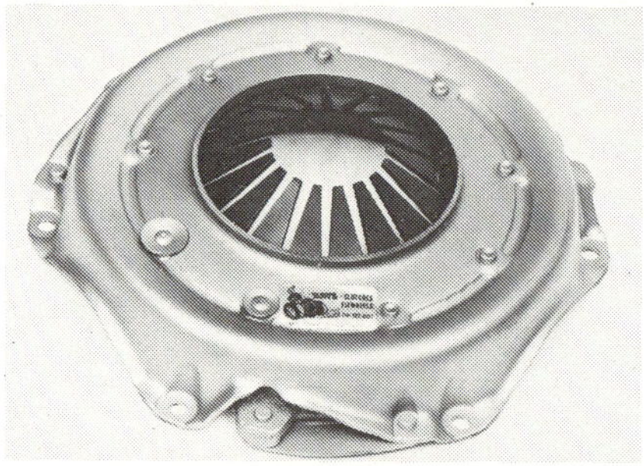

3

4

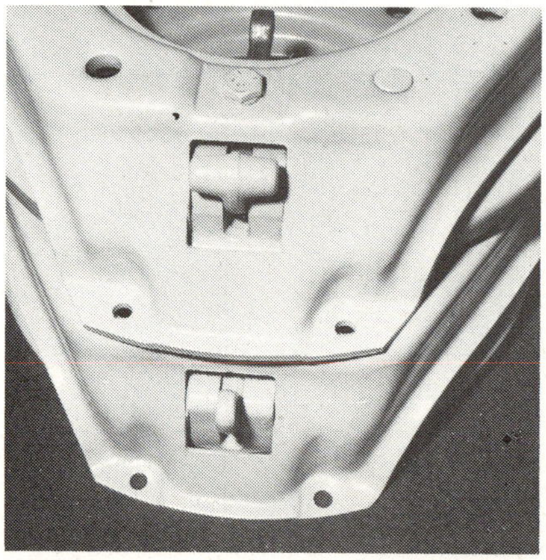

5

1. Shown is the Hays Long with adjustable weighted levers. Note various sizes of steel washers that can be added to bolt. You can vary the centrifugal assist according to the strip conditions.

2. A special retaining ring is machined into the Schiefer pressure plate on the diaphragm unit to keep spring in place and to distribute pressure evenly around the periphery.

3. Hays diaphragm pressure plate is new design with ultra-high-pitch levers; notice retention strap located outside the pressure plate. This adds strength to unit under torque transmission.

4. Both Keith Black and Ed Pink build up special clutch units for the cars they work on. This one by Pink is a 3-disc unit with parts built by Schiefer.

5. The bob-weights at the end of the release levers are the secret to slipper clutches. Stock weight is shown at top, bottom modified for bolts and weights.

6. These two lines compare slip vs. lock-up. The new "slipper" clutch units come close to lock-up but maintain that fine margin to keep the tires on the strip. Note the greater slip at start-up until centrifugal assist comes in.

7. This chart shows the power or torque transfer of the three popular types of pressure plates. The diaphragm comes on strong or "grabs" at start-off. The Borg & Beck has slightly more holding power at top end. At low end the Long is about middle-of-the-road until centrifugal assist starts to come in.

8. Here you can see the smooth, even power flow with centrifugal assist levers.

9. This line shows the pedal pressure associated with a diaphragm-type pressure plate. Pressure increases evenly as you push down on the pedal until you hit a peak; then holding clutch on the floor is easier as you reach over-center position.

10. The Pink pressure plate features adjustable spring pressure in conjunction with variable weighted release levers. Spring pressure is adjustable with unit in the car. Note bolt and nut at lever end.

11. The difference between torque transfer of damped and undamped hub design is great at initial start-up; it's due to the damped design having to take up cushioned wafers in addition to the shock-absorbing springs.

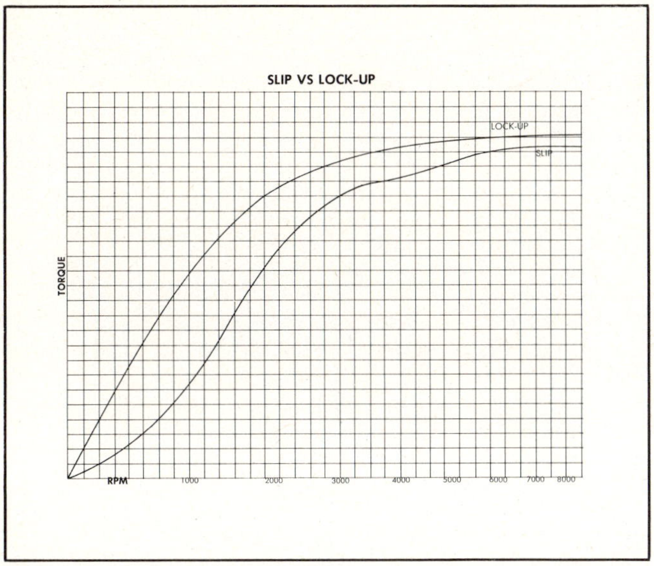
SLIP VS LOCK-UP
LOCK-UP
SLIP
TORQUE
RPM
1000
2000
3000
4000
5000
6000
7000
8000

6

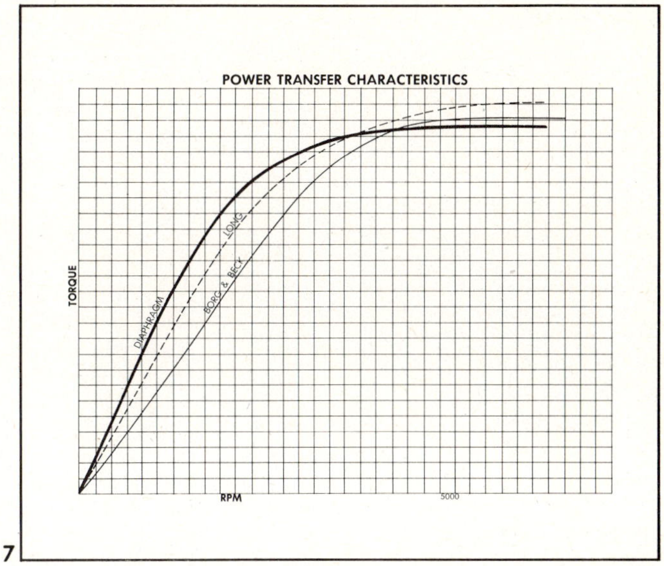
POWER TRANSFER CHARACTERISTICS
DIAPHRAGM
LONG
BORG & BECK
TORQUE
RPM
5000

7

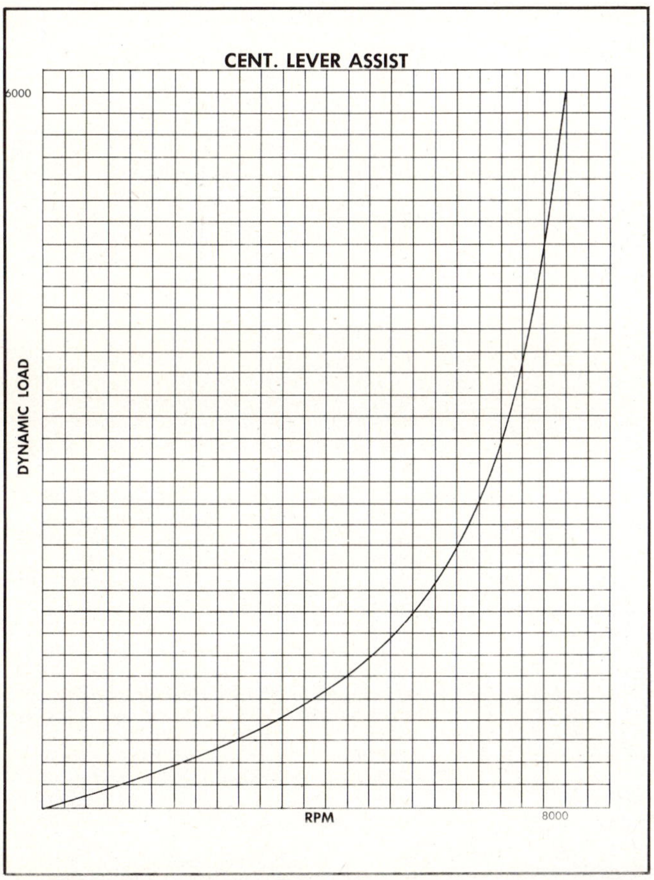
CENT. LEVER ASSIST
6000
DYNAMIC LOAD
RPM
8000

8

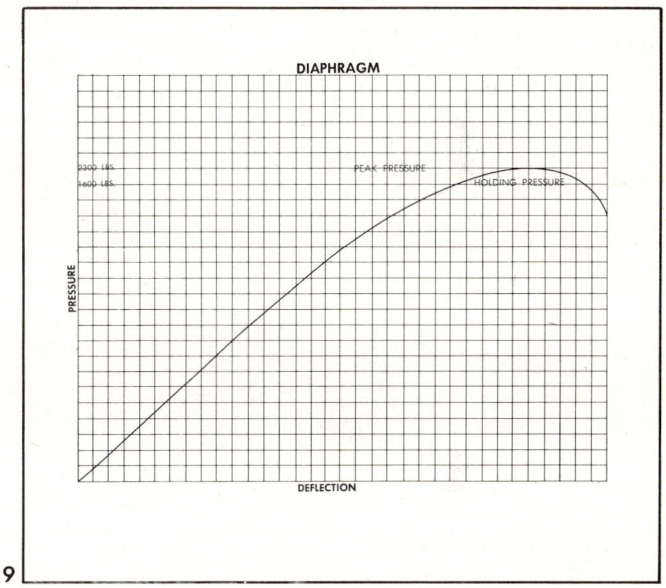
DIAPHRAGM
2300 LBS.
1600 LBS.
PEAK PRESSURE
HOLDING PRESSURE
PRESSURE
DEFLECTION

9

ED PINK RACING ENGINES
CALIF.

10

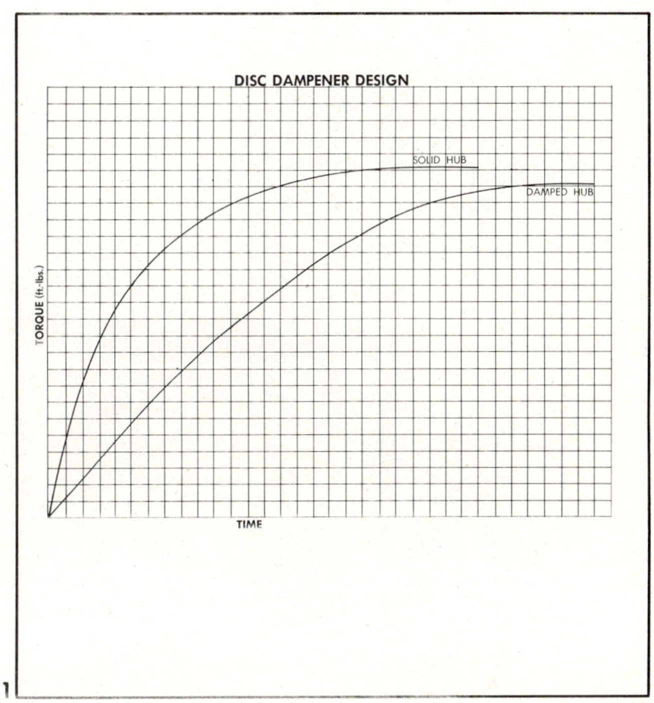
DISC DAMPENER DESIGN
SOLID HUB
DAMPED HUB
TORQUE (ft.-lbs.)
TIME

11

heat-sink converter. You must have parts that will withstand heat as well as dissipate it quickly. You see, under no condition could a current rear end take all of the punishment dished out by the high horsepower and sticky tires if the clutch did not slip. There just isn't any differential strong enough, and if there were, something else would have to break. Realizing this, clutch manufacturers have poured countless hours into making a clutch that will live under such abnormal conditions.

The way it works: decreased spring pressure increases slip, and increased lever weight decreases slip. At low rpm coming off the line the counterweights will not exert too much pressure, so slippage is possible. With the driver feeding just the right amount of power, and the counterweights increasing pressure, the tires will get maximum traction. This means no more quarter-mile smoke runs. Now the tires will just barely wisp trails of smoke on the sub-seven-second blasts.

Initially, dragster drivers and mechanics experimented with the slipper clutch by making up components supplied by leading manufacturers. Two- and three-plate units were commonly run. Name engine builders like Keith Black and Ed Pink worked out units for the engines and cars they built. The Pink unit has adjustable spring pressure and of course adjustable lever weights. His theory is to make a total adjustment according to strip conditions. But you have to know your car and, more importantly, know what you're doing.

It wasn't long before Hays and Schiefer were supplying the slider clutch for general sale. The Hays Long Twin Invader uses a billet aluminum flywheel with a sintered bronze facing riveted in place. The dual discs are sintered iron on an undamped center plate, and the floater is of high-alloy steel, precision ground and drilled for ventilation and to pick up wear particles. The Long-type plate has adjustable weighted levers and features the riveted heat shield facing over a steel plate.

The Schiefer unit is very similar to the Hays with solid-hub sintered iron discs, Blanchard-ground floater plate, and the Long-type pressure plate. Both the forged aluminum flywheel and pressure plate surfaces are coated with their exclusive metal sprayed process. Neither the Hays nor the Schiefer assemblies have spring pressure adjustments—both use varying centrifugal weights on the levers to control slippage.

SOMETHING DIFFERENT

Crower Cams' new super clutch called the "Crowerglide" is completely custom-made from special parts. The Crower unit consists of an aluminum flywheel and pressure plate faced with steel inserts secured by Allen capscrews. In between are the three sintered iron discs and two steel floater plates.

So far this all sounds normal enough; however, the pressure plate is not of the type we are familiar with. The six small springs located around its periphery do not exert pressure against the discs but just the opposite—they pull the pressure plate away from the discs. The whole theory behind the Crowerglide is *centrifugal force.* The easiest way to understand its operation is to drive a dragster equipped with a Crower unit. Let's take an imaginary ride and see what happens. . . .

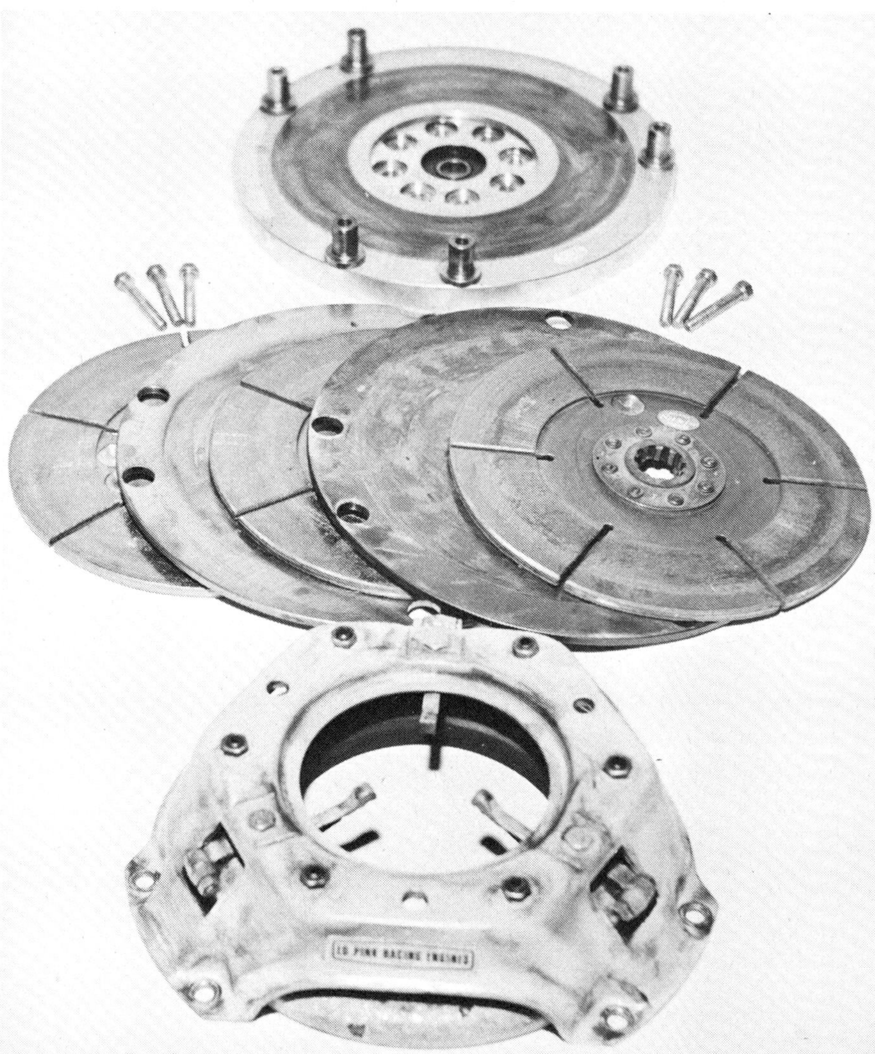

1

1. This is the assembly from Art Linkletter's "House Party" dragster and is made of sintered iron discs, steel floater plates and a Long-type pressure plate.

2,3. Pressure plate cover development continues under dragstrip testing, with the result being a change to reinforced driving lug openings (cover at left) and an increase in gauge thickness of the cover material.

4. The Hays Long Twin invader uses the sintered bronze facings on flywheel and pressure plate. This heat shield facing is riveted in place and can be changed when needed. Discs are sintered iron on undamped hubs and the hi-alloy steel floater is drilled for cooling and to remove any wear particles. Flywheel is made of billet aluminum.

5. Hays floater plates are replaceable or you can convert your present setup. Stands are chrome-moly and the bolts are of special high strength material.

6. The Weber dual-disc clutch is for competition only, with metallic clutch discs and single floater. Discs are recessed into the flywheel and held by a Long-type pressure plate. Weber also offers setups for the street.

7. The Schiefer dual-drive unit uses the Long pressure plate and forged aluminum flywheel. Both facing surfaces are coated with a hot metal sprayed bronze and steel process, Blanchard-ground and shot-peened. Steel floater plate with square stands and sintered iron discs are a part of the total package.

2

3

4

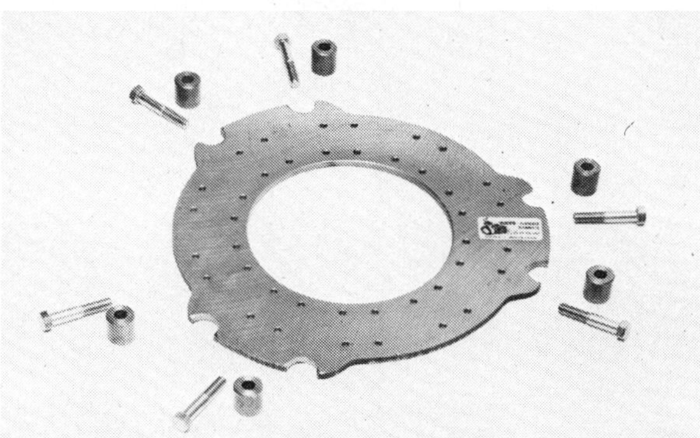

5

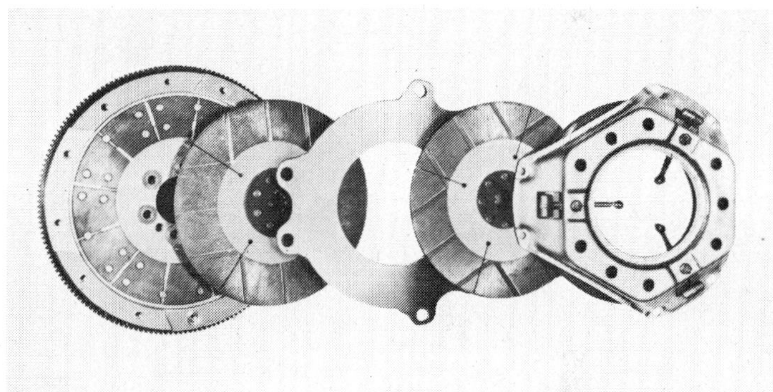

6

The push car moves your race car up to *fire speed.* Now you *push the clutch in* to engage the engine with the rear wheels and all of a sudden you have fire. The car lurches forward under power and you let up on the clutch pedal (neutral) to disengage the engine. Nearing the starting line, you ease in the clutch pedal to bring the car into the stage lights. Once staged, you remove your foot from the clutch pedal and press hard on the hand brake to prevent the car from creeping. As the tree begins to blink down toward the green light, you slowly apply the throttle until the clutch is felt attempting to grip. As the last amber light goes off, you nail' the throttle and release the brake—centrifugal force does the rest. As engine rpm increases, so does the grip of the clutch. By altering the amount of weight attached to the outboard end of each arm and varying the spring pressure, the clutch may be altered for stall speed, ultimate clutch pressure and lock-up.

The clutch arms or levers are designed with two heels equidistant

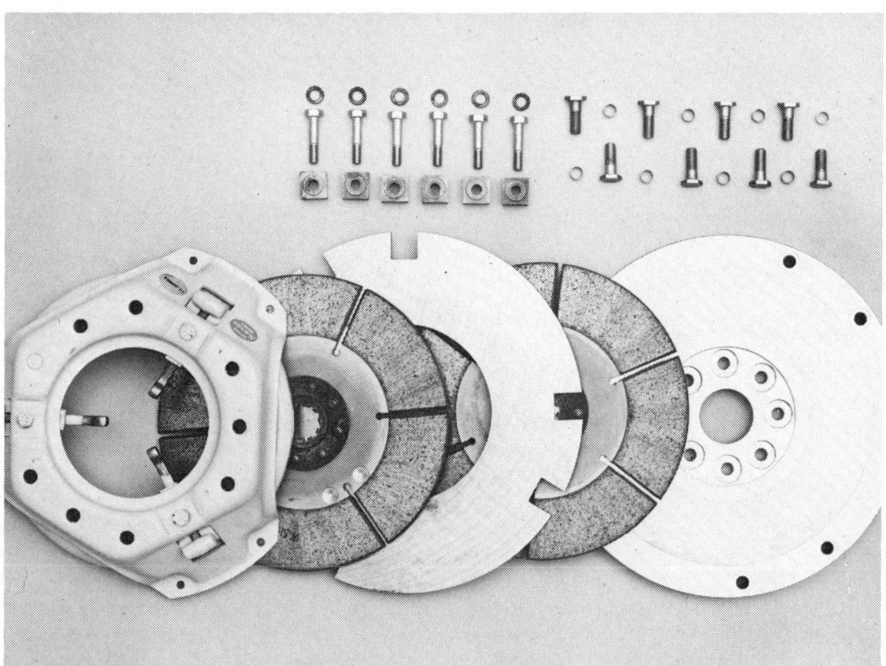

7

from the arm's pivotal point. Thus the clutch may be engaged when the levers are pushed in (engine stopped), disengaged when in a neutral position, and engaged with levers out (engine running) to take advantage of centrifugal force.

At the end of the run, you back off the throttle and the clutch disengages as engine rpm drops. Also you may affect clutch disengagement at any time by pushing in on the clutch pedal to a neutral position.

HIGH GEAR ONLY

It should be mentioned here, that except for the Weber Twin Power Drive utilizing the diaphragm clutch, the Crower, Hays and Schiefer units are for cars without transmissions—high gear only. The high centrifugal assist needed for operation makes them unsuitable for shifting. Another point to remember, when using heavy spring pressure in a cover plate assembly you also apply whatever pressure it takes to depress the clutch pedal against spring pressure to the thrust bearing of the engine. Most people don't think of this as it seems as though all pressure is applied to the flywheel. But the flywheel being attached to the crankshaft means that the crankshaft thrust bearing eventually takes the load. This is another reason not to use too great a spring pressure for street driving as this constant pressure will soon cause damage to vital engine parts.

Generally, the riveted linings are not as strong or blowout-proof as are bonded linings. And the bonded facings and solid discs allow a crisper engagement needed to break loose a set of big slicks. Bonded facings are also better supported along this *full* surface to improve burst resistance.

The type of disc (damped or undamped) and lining you use depends entirely on how you use your car (street, strip or combination)—and car weight, engine, gears, etc., all play an important part in selection. As a rule of thumb, organic or the new Velvetouch bond-on is best for street and strip use. The full metallics with undamped centers are best for drags-only use. In this area most specialty equipment manufacturers have about the same high quality hubs and friction materials available as they are supplied from the same O.E.M. suppliers. Besides the things we've already mentioned, always check the thickness of a new disc if you're changing from one type to another. If the replacement disc is of a different thickness, the pressure plate will have to be readjusted to compensate for the extra meat. This can be done by

1

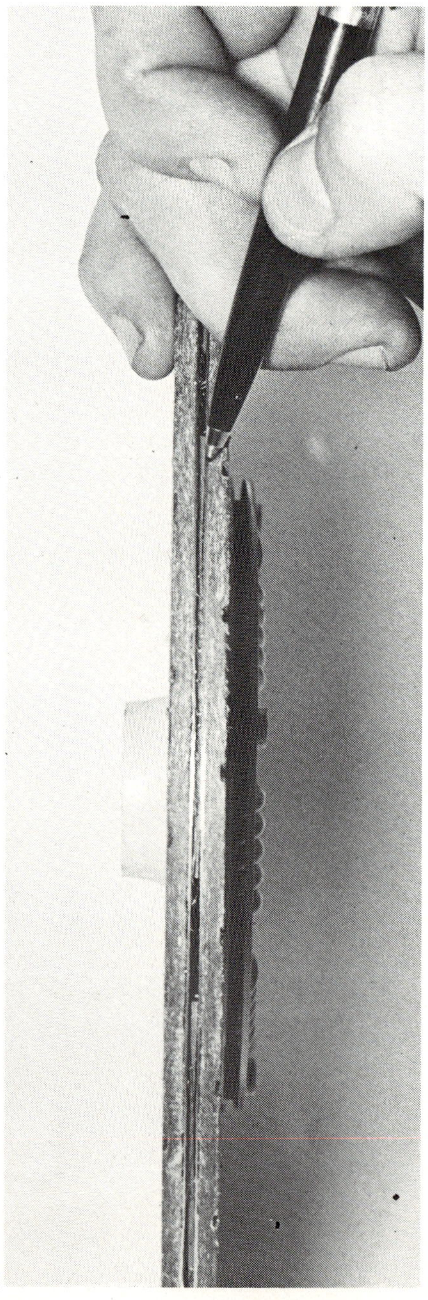

2

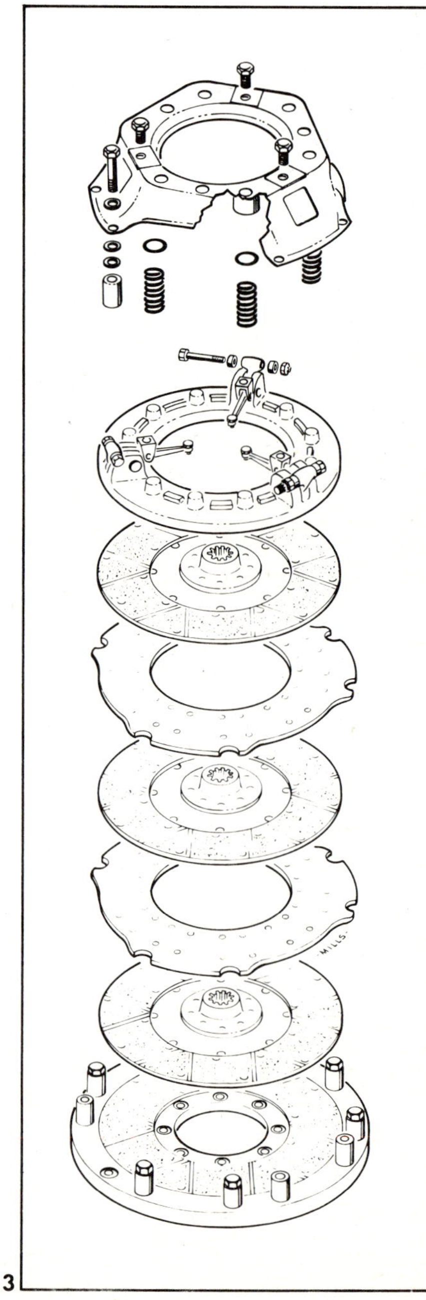

3

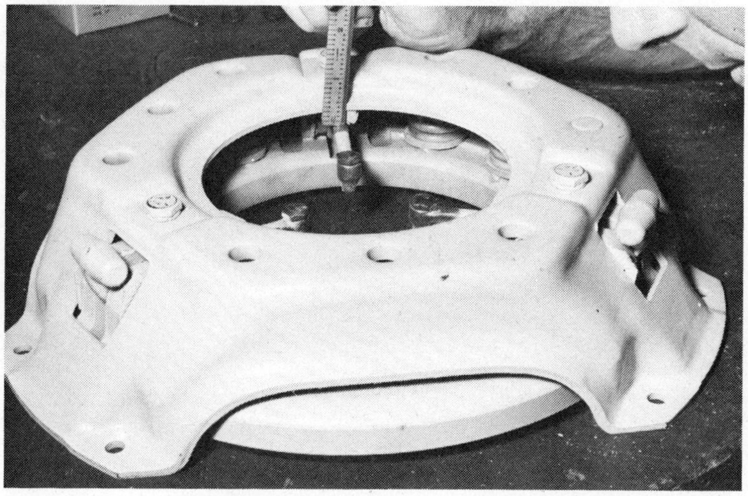
4

5

6

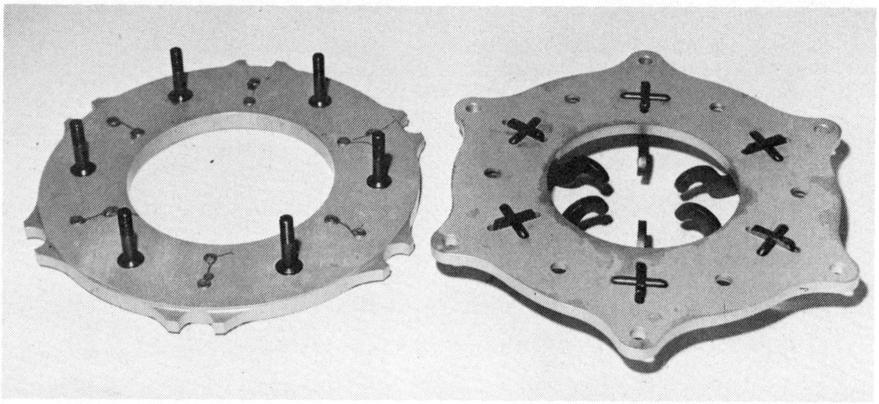
7

1. The complete Crowerglide unit weighs 43½ lbs., which includes the flywheel that is part of the assembly and is the thick plate at the bottom.

2. One of the newest Schiefer discs uses linings bonded to separate thin steel backing plates and then riveted to the wafer plates for smooth engagement.

3. This exploded view shows the new Hays 3-disc setup. Basically the same as the 2-disc, the extra floater and disc offers increased friction area to handle the biggest of engines.

4. Adjustment of finger height is very important to clutch life and efficiency, especially with a racing clutch.

5. Floater plates fit between dual and triple disc clutch designs, with drive transferred to the flywheel by lugs (also called stands) on perimeter.

6. Steel facings are set into aluminum flywheel and pressure plate facings and secured by flathead Allen screws. Facings may be changed in field if necessary.

7. With the pressure plate separated facing at left and release arm plate at right, you can see how arms pivot on recessed pins and the pivot "buttons" which the heels of the arms push against.

8. A side view of the top plate shows clutch arm at "in" position. This engages the clutch for starting engine.

9. Shown in the "out" position, full centrifugal weight would now make lock-up with discs. Note that hole at end of lever is to increase centrifugal weight.

8

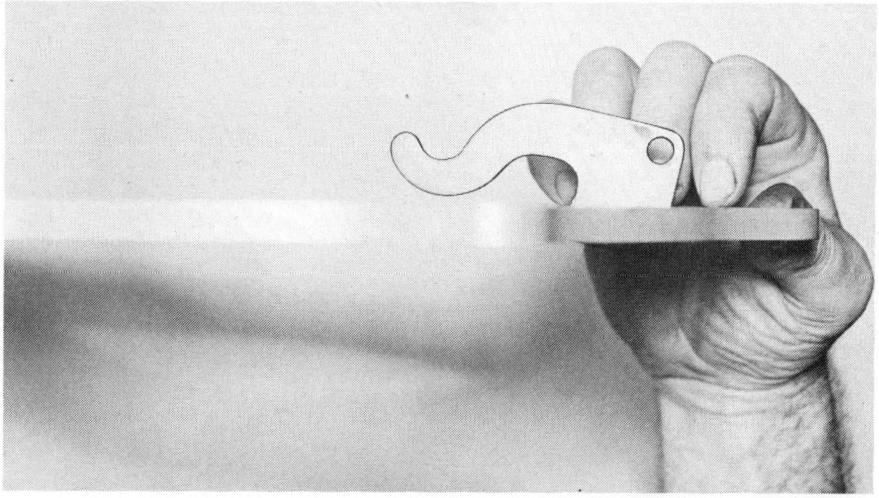
9

a competent local clutch rebuilder. If you don't adjust the plate, it may either fail to release fully, or worse yet, fail to engage. When the latter happens tremendous heat is generated in milliseconds which can warp or crack pressure plate and flywheels—or even cause them to explode.

SPECIAL FASTENERS

High-tensile steel bolts for cover plates and flywheels are a must and they should always be torqued to manufacturer's specifications. Use the factory specs on a stock unit and the specialty equipment manufacturer's specs on special flywheels. Special bolts are either furnished with full assemblies or can be purchased separately. The use of Loctite on flywheel and cover bolts is also recommended. This sealant is drawn into the threads and is self-hardening to form a lasting bond. It's been proven superior to lock washers and will not loosen from vibration, but can be easily removed with standard tools. Additionally, these bolts should be replaced with new ones after they have been removed more than about four or five times.

FLYWHEELS

In the flywheel department, Hays offers steel and aluminum billet wheels with the aluminum wheel faced with a sintered bronze facing. This facing is first bonded to a thin steel disc which acts as a heat shield to protect the flywheel and to dissipate heat evenly at a controlled rate. The special steel rivets which hold the heat shield to the flywheel also aid in dissipating heat.

Schiefer produces flywheels in billet steel and in forged aluminum with the metal-sprayed copper and steel facing. This surface is then Blanchard-ground and shot-peened for stress relieving.

Weber has flywheels in forged aluminum, cast and billet steel. The aluminum wheel uses a steel facing riveted in place, then Blanchard-ground.

All the above-mentioned flywheels are balanced individually and may be used as is on stock engines. But with engine modifications, your pressure plate, disc and flywheel should be balanced as a unit.

THE "SORT OF" AUTOMATICS

The B & M Clutch-Flite and TurboFlite transmissions should be mentioned here because these two automatics are "clutch" driven. Basically, the torque converter has been removed, and a special front plate, clutch release fork and throwout bearing installed in its place. The automatic transmission is then coupled to the engine crankshaft with a standard or high-performance (preferred)

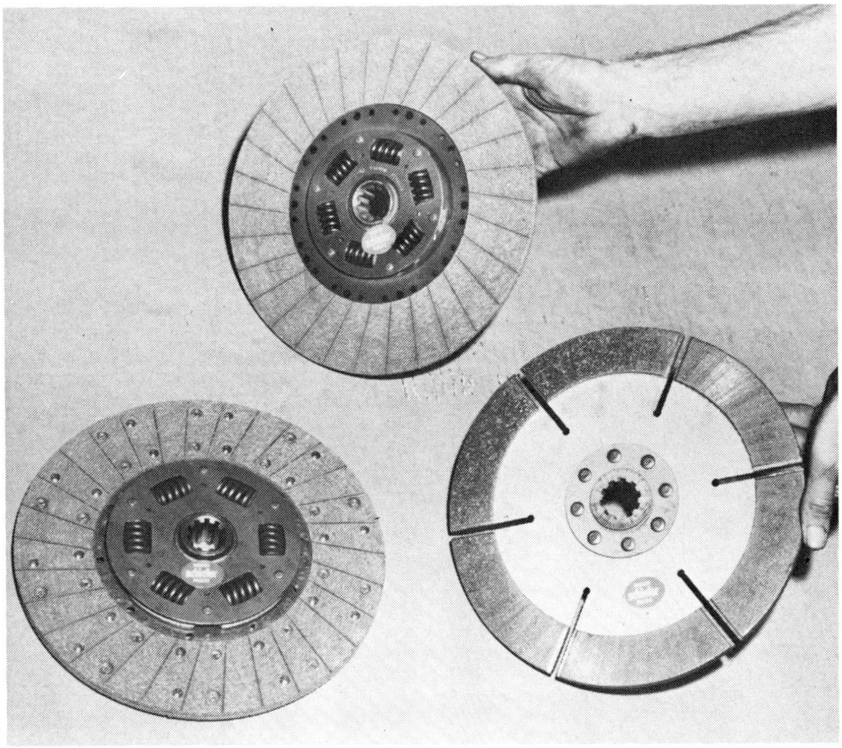

1

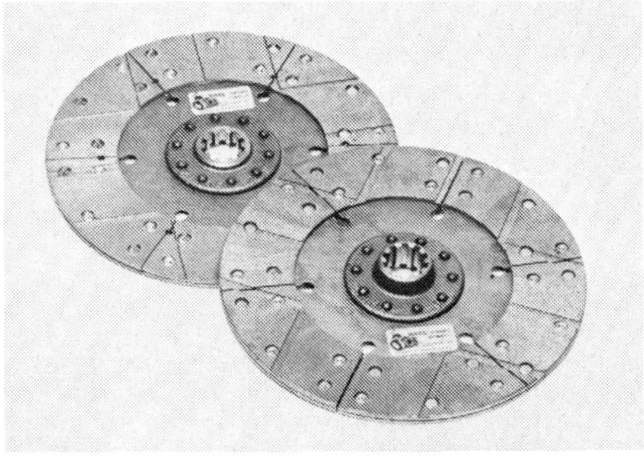

2

3

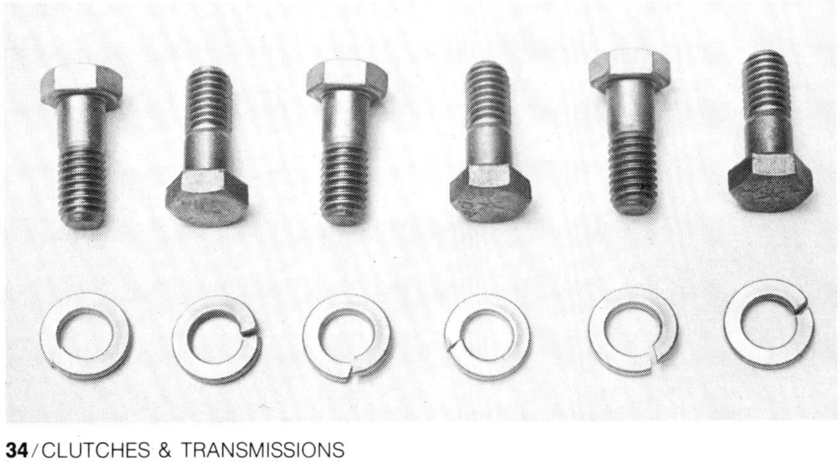

4

5

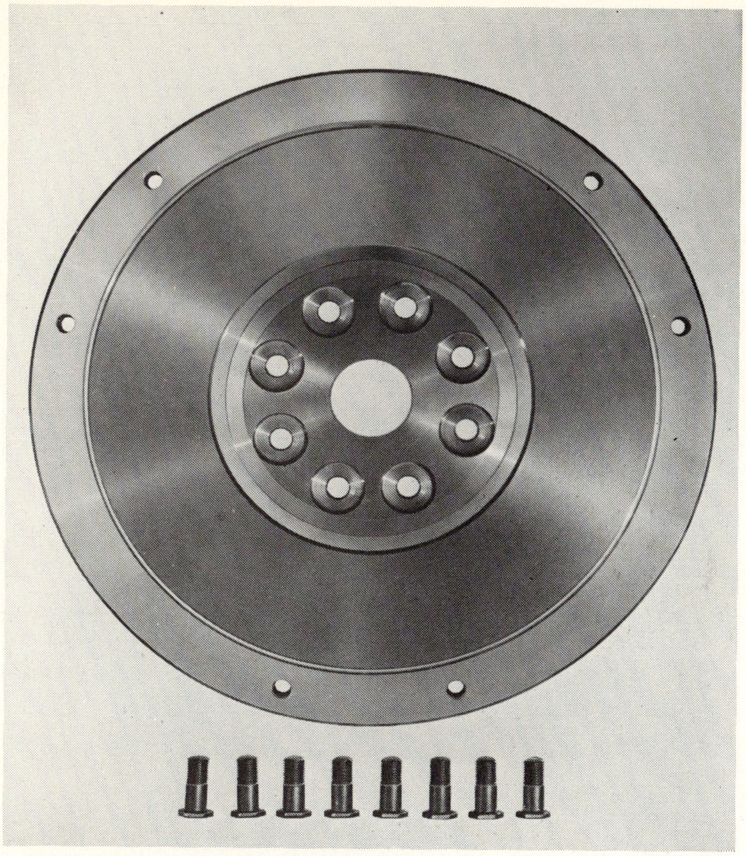

6

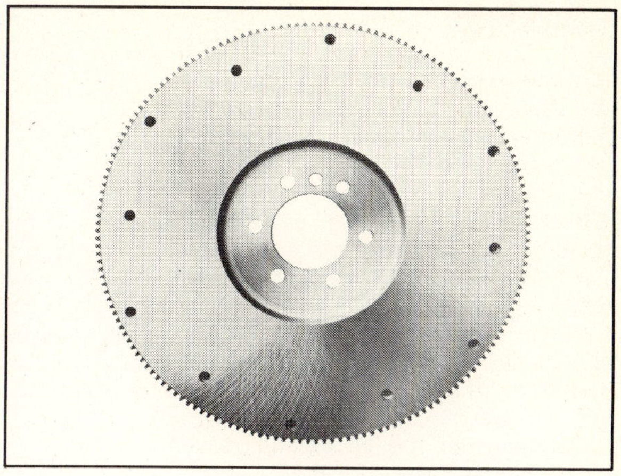

7

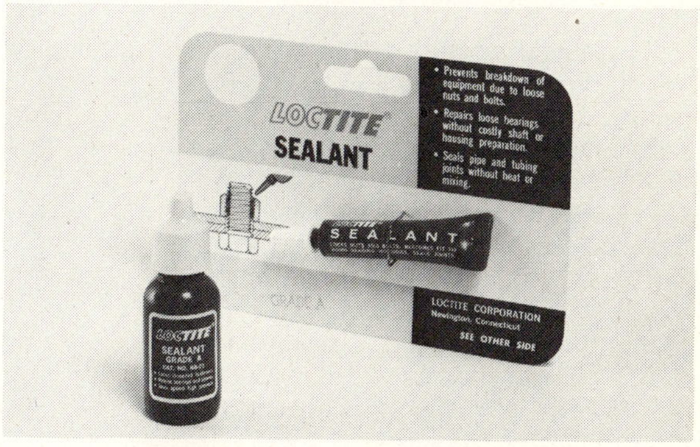

8

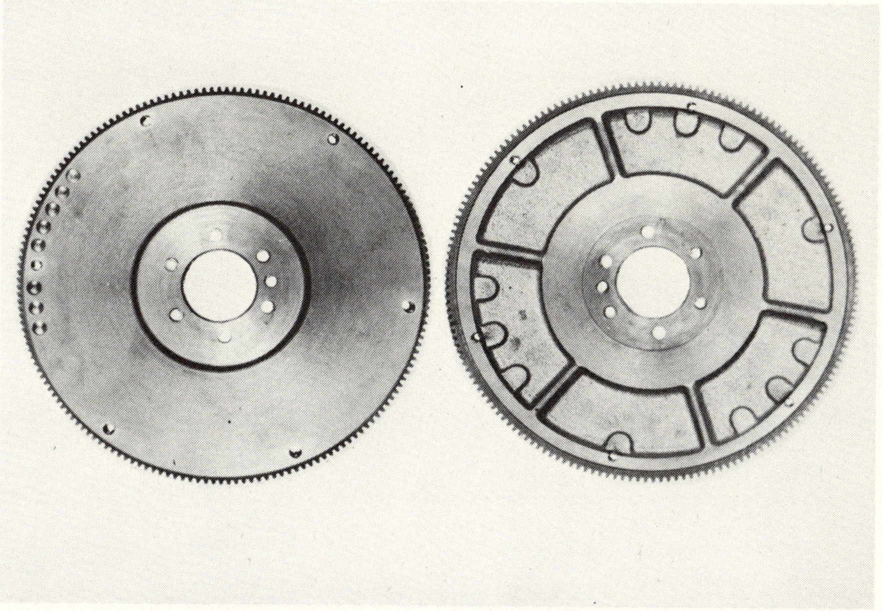

9

1. Three different Schiefer discs are the bonded lining with damped hub, good for competition or semi-competition; riveted and damped hub, good for street machines; and the sintered iron, undamped disc, for racing only.

2. Hays sintered iron, undamped discs are for all-out competition, high-gear-only cars. Facing is grooved for greater heat dissipation and improved adhesion.

3. The Hays billet steel flywheel is ideal for street and strip operation and will take a lot of punishment.

4. No clutch is stronger than the bolts which hold it together. Always use the high-strength bolts identified by the six lines on the head. These are available from all clutch specialty manufacturers.

5. Unlike the Schiefer facing which is metal sprayed, the Hays flywheels use a riveted sintered bronze friction surface.

6. The Schiefer aluminum flywheel is available with or without the ring gear, depending on method of car operation, as is the case with most other manufacturers. Flywheel bolts are also important and should be of high strength. Flywheel or pressure plate bolts that have been removed more than five times should be replaced for safety.

7. This is the Weber steel billet flywheel which is ideal for street and competition. Finish is Blanchard-ground.

8. If you're going to be doing any serious racing, you need some of this. Use it on any bolts subject to lots of vibration, such as on clutch/flywheel.

9. Weber's cast steel flywheel is good for street use where big horsepower is not used, or for low competition classes.

flywheel, pressure plate and disc combination.

This combination breeds the best of a stick transmission with an automatic. Missed shifts will be a thing of the past and there is no load on the automatic at the starting line—engine rpm can be brought up as high as needed without worrying about that ever lovin' stall speed.

Installation is very simple. If you use the transmission bellhousing the kit is an easy bolt-up job, with the main work being cutting a rectangular hole in the side of the housing for the release fork to operate. If you're going to use a combination safety shield/bellhousing, the transmission bellhousing is cut off just in front of the main case and a new adapter housing bolted on. Each type of kit comes complete with all hardware right down to a new input shaft for the clutch. They can be operated on the street as well as the drag strip.

BELLHOUSINGS

Special bellhousings are necessary if you were to install a Long-type pressure plate on Chevy or Pontiac engines equipped with a Borg & Beck or diaphragm units. The Long-type unit requires more area radially than the other types, so it's not possible to use the standard housing. When this is the case, it's best to switch to a scattershield bellhousing for safety as well as gaining the necessary room for the conversion.

Since the high-revving '55 Chevy was introduced, flywheel and clutch explosions have been a problem, and will always be a threat to driver and spectator with rpm in the six, seven and eight thousand or more bracket.

SAFETY SHIELDS

Any safety shield used should be the approved type—AHRA, NHRA or better. Naturally, safety shields are required at these sanctioned events and they should be used on all hot street machinery too—explosions are not limited to the drag strip alone. Donovan and Lakewood Industries make excellent products which are hydroformed from solid, 1-piece sheet stock. In addition to this type of shield most pro drivers are using an inner lining of heat treated alloy steel or aluminum that fits snugly inside the bellhousing and is only about ⅛-in. thick. Liners are made by Lakewood Industries and Don Long Chassis. The liner gives that extra ounce

1

2

3

4

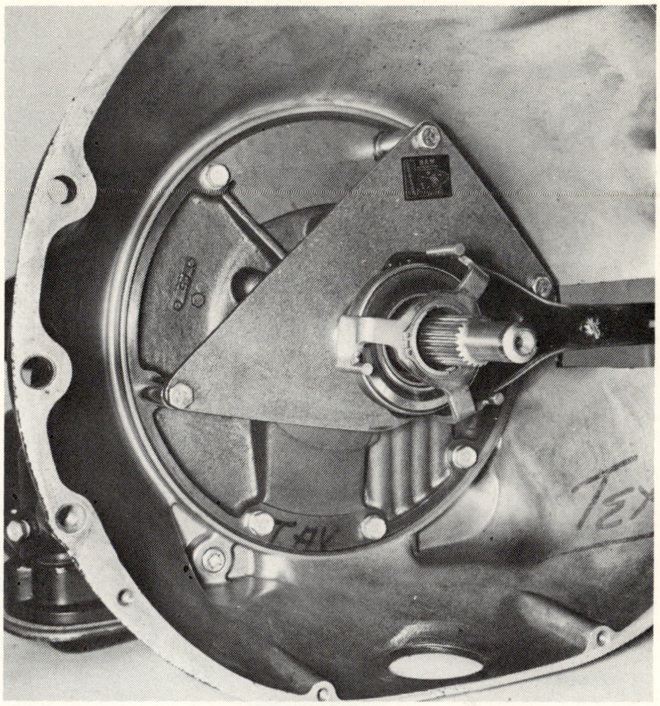

5

1. The Weber forged aluminum flywheels run a steel facing that is riveted in place and then Blanchard-ground for finish and parallelism.

2. A recent addition to the clutch-driven automatics is the Turbo-Clutch for most of the GM cars.

3. The B&M clutch-driven automatics are popular items for both the street-and-strip ranks. The Clutch-Flite shown in its two forms is modified from the Chrysler TorqueFlite transmission. The unit on the left is mainly for the street, as it uses the stock bellhousing. The competition version (at right) has the bellhousing machined off and special adapter housing installed in its place. This will hook up to most safety shield housings.

4. A close-up of the Clutch-Flite shows adapter plate, new input shaft, throw-out bearing and pump drive spider.

5. On a stock setup, an adapter kit is used for the conversion which holds throwout bearing collar and provides a pivot point for the release fork.

6. The kit can be installed at home without special tools; no machine work is necessary with the exception of cutting a slot in the case for the release fork. But this can be an easy job with a drill, hole saw and a file. Complete instructions and template are included.

7. The competition setup is about the same with the exception of the release fork and support plate. Aluminum front plate is for coupling with bellhousing.

8. The do-it-yourself kit includes the following parts from left to right: throwout bearing support plate, pilot bearing, pump drive tube (includes roller bearing and seal inside), mounting bolts and spacers, input shaft, throwout bearing, pump drive spider, teflon thrust washers and snap ring, converter valve plug, charge ball and return plug, and throwout fork with pivot ball.

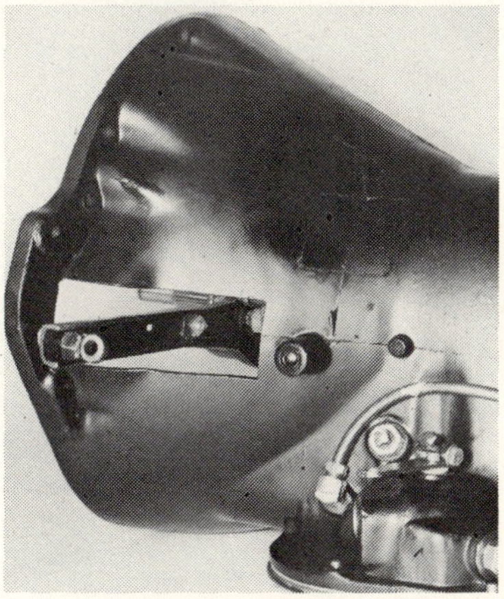

6

7

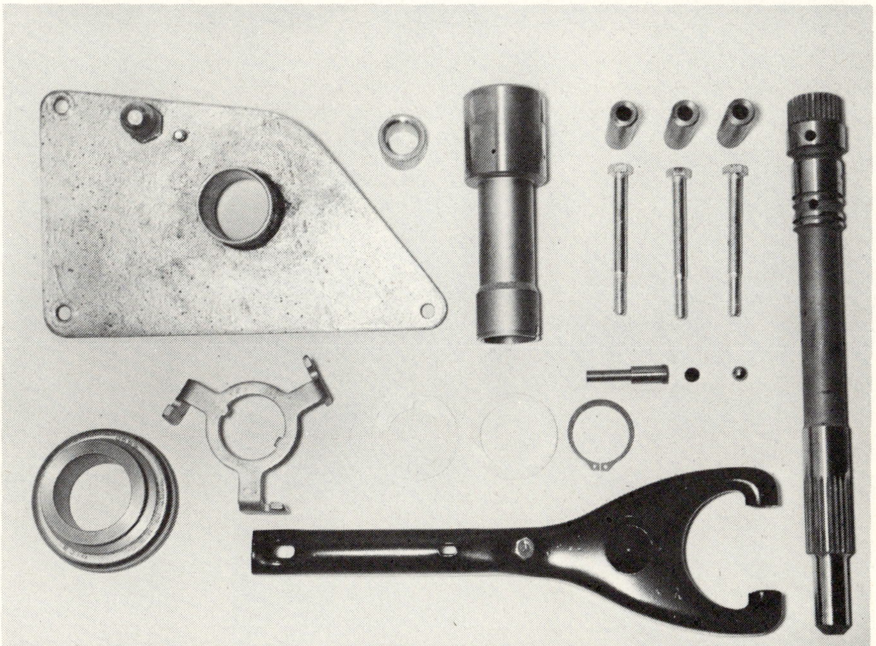

8

Clutches

of protection and strength to the bellhousing if a red hot clutch is loose inside. Another safety item that has rapidly become favored is the *safety blanket.* The safety blanket consists of 14 layers of heavy nylon shaped to conform to the bellhousing. Several big heavy-duty nylon straps and buckles similar to seat belt material hold the blanket securely to the bellhousing. Tests have shown them to withstand clutch disintegration, and when I visited Simpson Safety Equipment they invited me to try to penetrate the nylon cloth they use. With the pointed end of a pair of big scissors I stabbed the cloth as hard as I could and only penetrated four layers—and at that, only with the very point. Besides being made for competition bellhousings, Simpson makes them for automatic transmissions as well. They weigh only 10 lbs. and are worth every ounce.

By now you should have a keen idea of how clutches work and the vital link they play in the performance of a car—street or strip. It's amazing that the clutch and its components can take the pounding they do and yet still perform under the conditions they face. That "clutch" is truly a wonderful little part!

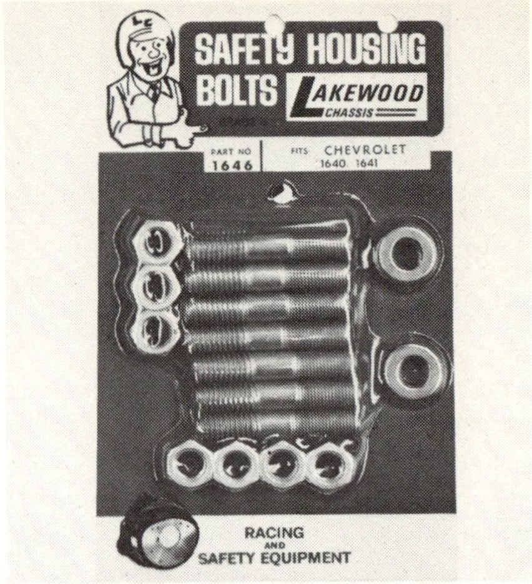

1

2

3

4

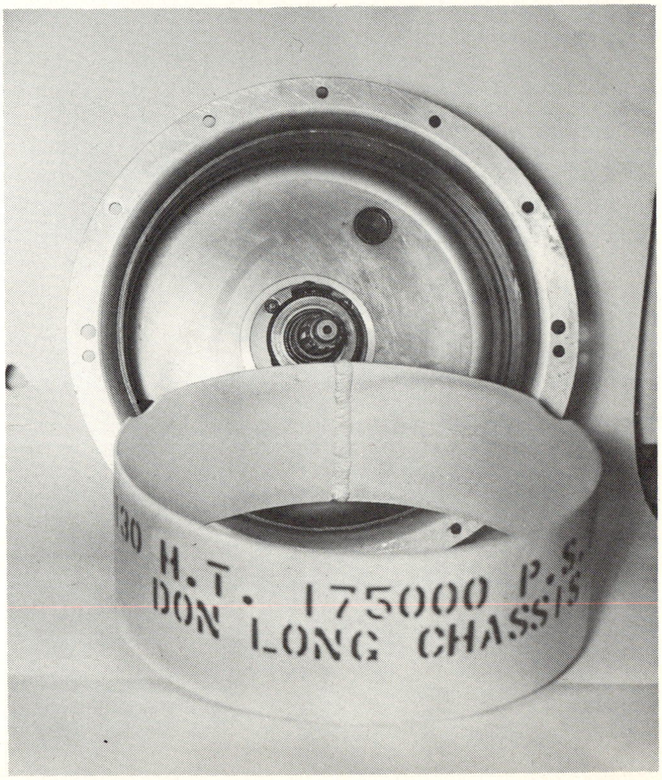

5

6

7

8

1. Special safety housing studs and nuts are a must for those hot setups.

2. To meet rigid SEMA specs clutch/flywheel/bellhousing manufacturers run test units to death. This is one bellhousing that failed to withstand blow-up.

3. This bellhousing held up after flywheel disintegrated.

4. All high-performance cars should have engine and drivetrain in perfect alignment. When installing a special bell-housing or safety shield this is a must. It's easy for the manufacturer to be just slightly off, which can cause serious damage to other components. If more than .005-in. out of concentricity, the housing should be returned for another.

5. The Donovan hydro-formed safety shield is popular among dragster drivers and all-out competition machines. Don Long makes an inner shield which fits snugly inside the housing for that extra margin of safety.

6. Lakewood also makes a rolled steel liner for the direct drive housing. Liner is held in place by two locating bolts.

7. Safety blankets consist of 14 layers of this tightly woven nylon. The resulting strength is almost unbelievable.

8. Nylon straps and steel buckles secure safety blanket to automatic transmission as shown. The complete Simpson blanket weighs only 10 pounds.

9. This safety blanket, also by Simpson, is for direct drive housings. Most pro drivers are using the complete "load"—safety housing, inner lining and blanket.

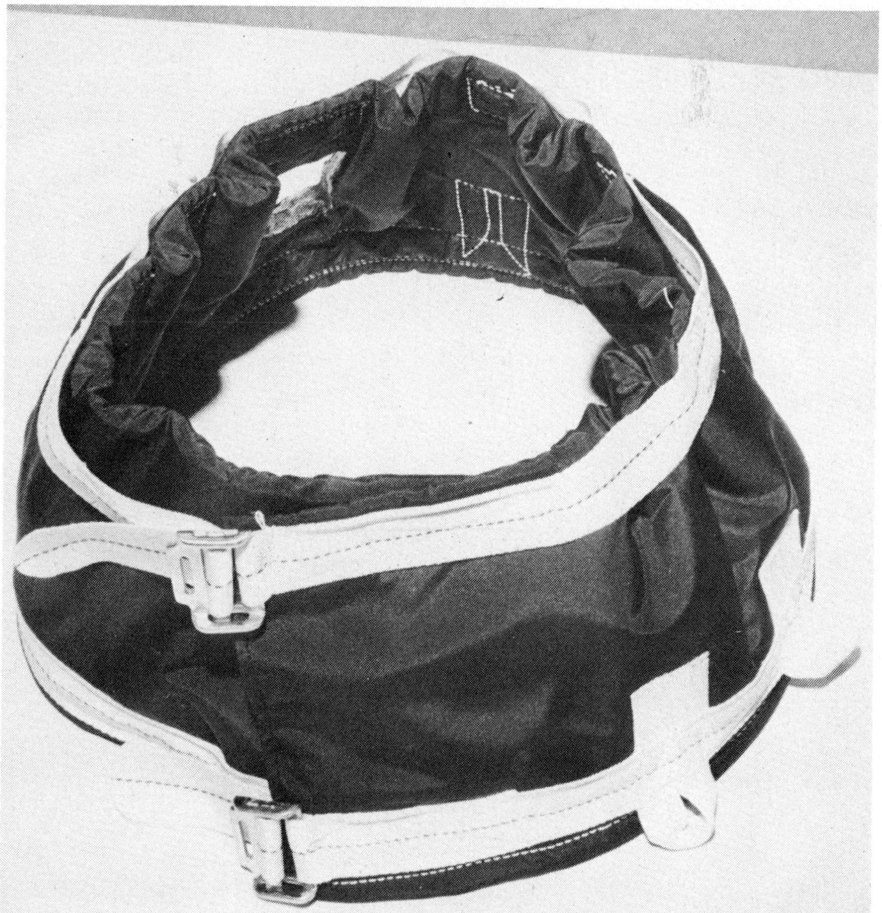

9

What Goes on Inside the Gearbox

In the 1968 Indianapolis 500 Bobby Unser found out the hard way that a car really needs a transmission. His shifter was stuck in high gear, which meant he had to slip the clutch every time he left the pits. That's hard on the engine and even harder on the clutch, but it was the only way he could get moving. Those clutch-slipping starts were slow, too, and Unser lost valuable time getting up to speed. An internal combustion engine just can't do the job efficiently without a transmission.

If Unser had a steam engine or electric motor in his car, then a transmission would have been unnecessary. Both steam and electric powerplants will put out lots of torque at low rpm. They will even put out torque at zero rpm, which is impossible with most internal combustion engines. The turbine is an exception, but that's another story.

Unser went on to win the race, which proves that if you can keep a car going fast enough at all times, you don't need a transmission. But when you slow down, there's big trouble brewing.

Some engineers explain this need for a transmission on an internal combustion engine by saying that the internal combustion engine is not flexible enough. What this inflexibility label means is that the gasoline engine operates only within a limited rpm range. If the rpm is too low, it develops so little power that there isn't enough to keep itself running. If the rpm is too high, the engine will scatter pieces all over the landscape. A steam engine or electric motor does not need a clutch or transmission. Both the electric motor and the steam engine will produce torque when they are not moving, because the power comes from outside the engine. If you had a steam engine in your car, you could run up against a brick wall and apply steam to the engine, and the car would push on the wall. The engine would not be moving and the car would not be moving, but even so a certain amount of torque would be put out by the engine in pushing on the wall.

The electric motor works the same way. If it is trying to do some work that it is not capable of doing, it will still exert a certain amount of torque or push against the work that it is not able to do. In the same situation the internal combustion engine will stall itself out.

Everybody knows the bitter experience of getting a little anxious in traffic and letting the clutch engage too fast so that it kills the engine. It's a mistake that is easy to make because of the limitations of the gasoline engine, but explain that to the guy behind you who's honking his horn.

With a gasoline engine the only way we can exert any push on a brick wall is by using a clutch and slipping it so that we apply a certain amount of torque to the wheels of the car and allow the rest to be used up in heat expended through the friction of the clutch.

A fluid coupling would work in the same way and last a lot longer than a clutch would. The fluid coupling idea has been tried on several cars, the most famous being Chrysler products of World War II vintage. But even with the fluid coupling giving some flexibility to the internal combustion engine, it was still necessary to use the transmission. With the fluid coupling alone a car is a sluggish beast that won't even get out of its own way.

Another reason for a transmission is that a 4-cycle internal combustion engine will not reverse the way a steam or electric motor will. They both have reversing mechanisms built in, but on an internal combustion engine this can't be done without a tremendous amount of extra gadgets built onto the engine that would be much more expensive than an ordinary transmission.

1

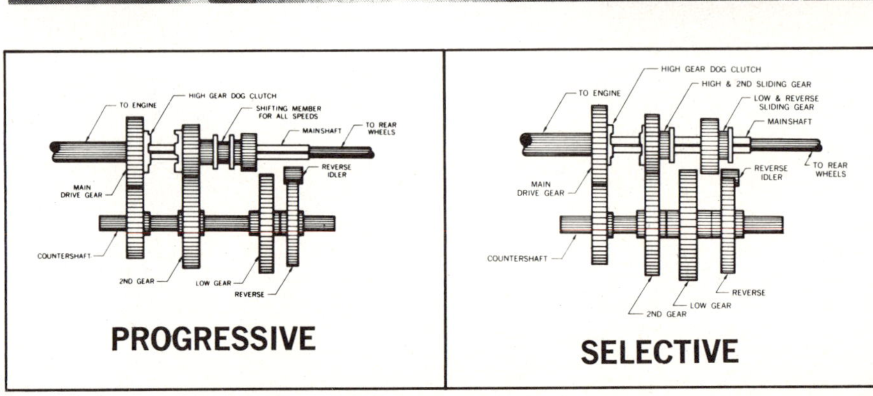

2

So no matter what we do we are stuck with the need for a transmission of some kind. To better understand the need for a transmission you might try pushing on a bicycle spoke right down near the hub and trying to roll the bicycle forward. It will be difficult, but if you move your hand up on the spoke until it is right next to the rim near the tire, then you will find you can roll the bicycle forward without any difficulty at all. You have more leverage up near the wheel than you have down near the hub, and this is what we do for the internal combustion engine when we provide a transmission. We give the engine more leverage so that it can do the job easier. Remember that it has to do the job easier or it will be outside the range where it produces enough twisting power.

Lugging an engine, which means running it too slowly, results in extreme pressures being built up inside which will wear it out very quickly. Over-revving an engine or running it too fast will have pretty much the same effect.

Through gearing, the transmission keeps the engine operating at the best speed where it puts out its most torque and can work the easiest, providing that the driver knows when to change the gears to correspond with road conditions. The driver who doesn't will be much better off with an automatic transmission. It will be a lot easier on the engine and on the transmission, especially if he's the type who clashes gears a lot.

GEAR RATIOS

It's a lot easier to use a transmission correctly if the driver understands what is going on inside the box. A lot of people who get paid to repair transmissions don't really know how they work. Take gear ratios, for example. Many mechanics who take transmissions apart every day do not know how to figure ratios on gears. There's really no excuse for it because it's simply a matter of counting the teeth on the gears to find out what's going on.

Suppose an engine is turning a gear with 10 teeth and this is meshed with another gear with 20 teeth, and then a shaft runs from the 20-tooth gear directly to the rear end and then to the wheels of the car. To find what gear ratio we have in the transmission we merely divide the number of teeth on the driving gear into the driven gear, or 10 into 20. The answer, of course, is 2, so we have a 2-to-1 gear ratio in our transmission. This means that the engine will turn over two revolutions for every one turn of the transmission output shaft.

Suppose we had a 10-tooth to 40-tooth relationship. Then the engine would turn over four times for every one revolution of the output shaft on the transmission. Whenever two gears are meshed, the one with the smaller number of teeth turns faster, and, of course, the one with the greater number of teeth turns slower. If you had your hands on a boulder and you could just barely move it, your hands would move the same distance that the boulder would move. But if you put a long 2x4 under the edge of the boulder, with another rock close to it to pry on (as a fulcrum point), and you got out about six or eight feet, you could move the boulder with very little pressure on the 2x4. Your hand has to travel a much greater distance in doing this than it did when you moved the boulder directly. When we hook up the engine to the small gear in the transmission we are accomplishing the same thing. It gives the engine leverage to do work, but at the same time the engine has to travel quite a bit further.

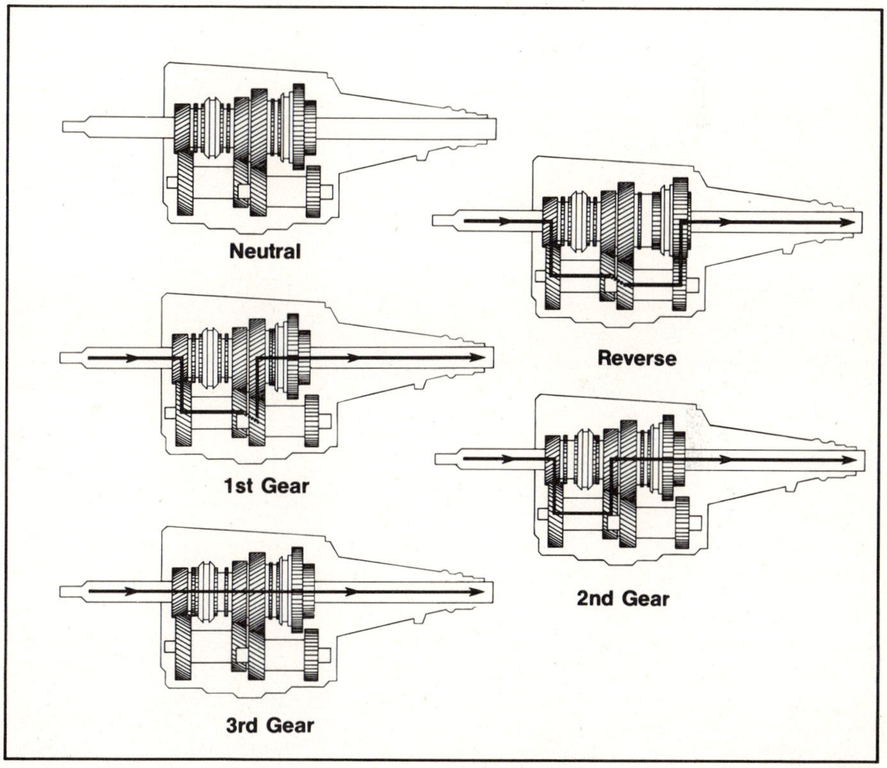

3

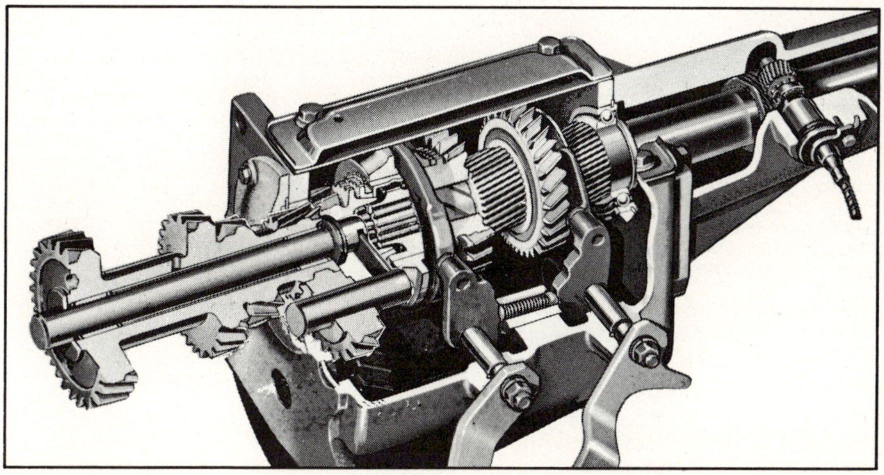
4

1. Before you do any work on or modify any transmission, you've got to know how and why it is built. The theory is the key to the application.

2. In both selective and progressive transmission, mainshaft is not joined to main drive gear until the dog clutch is engaged for high gear. In all other speeds, the mainshaft rotates free in a bearing inside the main drive gear.

3. Construction and design of 3-speed transmissions are similar. These drawings show how engine power transfers in various gear positions.

4. Cutaway of a typical 3-speed standard transmission shows how the shifting arms move gears and synchros.

Inside the Gearbox

Now to get a little bit more complicated, suppose you had 15 teeth on the engine gear and 29 teeth on the gear that meshes with it. The same arithmetic still applies. Divide the 15 into the 29 and you get 1.93 as your gear ratio. In other words, for every one revolution of the transmission output shaft, the engine will turn 1.93 times.

There isn't any transmission that only goes through two gears, however. They all go through four, but don't let this throw you. You'd simply take each pair of gears in turn and figure out the ratio of the pair. Then you multiply the ratios together to get the overall ratio of the transmission. Let's take as an example the most popular transmission ever made, the 3-speed Chevrolet. In low gear, it has two pairs of gears, meshed. The first pair has 19 and 27 teeth, the second pair 14 and 29. In each pair we divide the driving gear into the driven gear and we get ratios of 1.42 and 2.07. Multiplying those two ratios together gives us 2.939 or 2.94, which is the overall ratio of the Chevrolet transmission in low gear.

To get the overall ratio of the whole car, you would carry this a step further and multiply the transmission ratio times the rear-end ratio of 3.36 giving you 9.88 which is the number of times the engine turns for each rear wheel revolution.

With the transmission in second gear the power flow is also going through four gears or two pairs, so the same division of teeth and multiplication of ratios would give you the overall ratio in second gear. If you have a 4-speed transmission, third gear would amount to the same kind of calculations, but then when you get into high, which could be third or fourth (depending on the number of speeds your transmission has) the power flow is not going through any of the gears, so there is no multiplication. Then you are running on the ratio that is in the rear end.

If you want to figure out the ratios in reverse, it's a little bit trickier because you have five gears in mesh at one time. This is not as frightening to figure out as it sounds; it just means that the reverse idler is doing double duty and you count it twice. So you do your figuring as if you had three pairs of gears.

For example, the main drive gear meshes with the front end of the cluster gear for a 19-27 ratio. The cluster gear meshes with the reverse idler gear for a 14-15 tooth count. Last, the reverse idler meshes with low and reverse sliding gear for a 15-29 tooth count or ratio. Do your division and multiplication and you will end up with 2.9336, which is listed in the specification books as 2.94, and we won't argue for that small a difference.

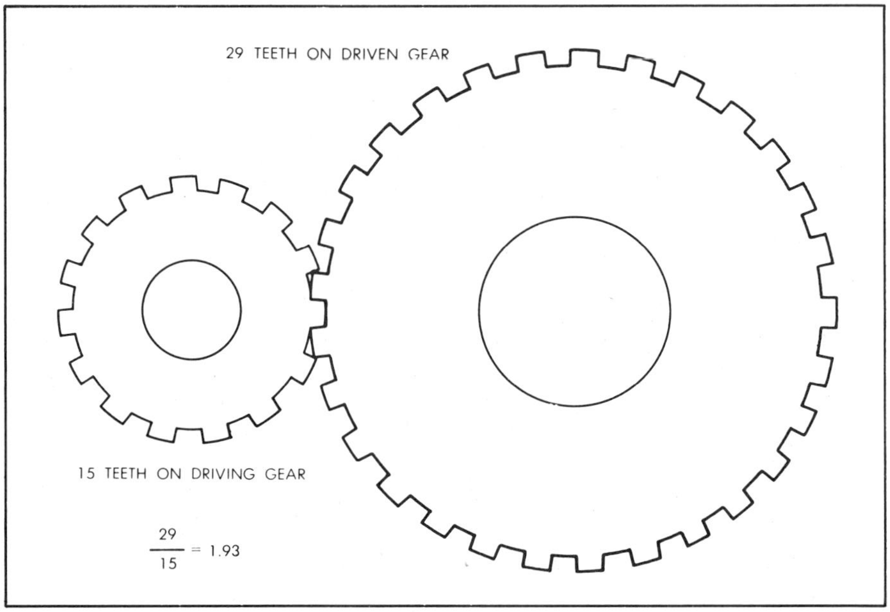

1

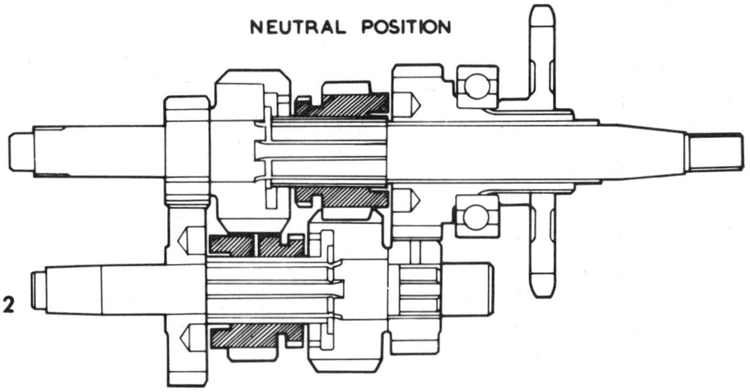

2

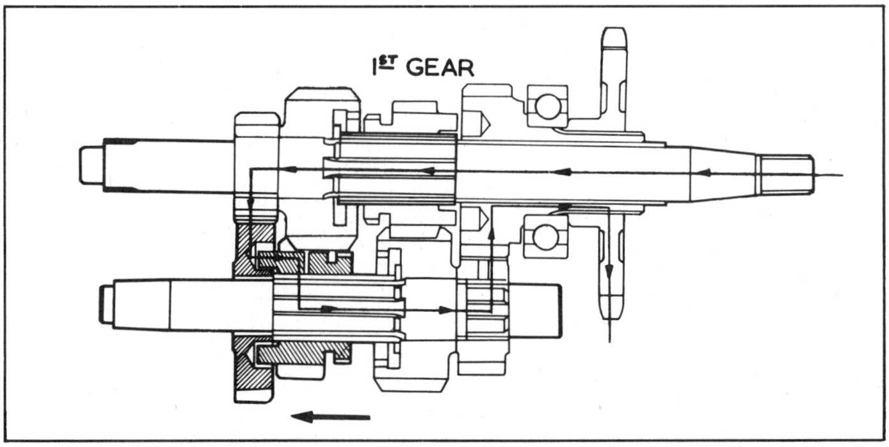

3

1. Count the teeth, do a little division and you get the ratio, but you must do the division properly according to which is the driven and which is the driving gear. In this illustration, the driving gear makes 1.93 revolutions for each single revolution of the driven gear.

2. Gear arrangement on BSA cycle transmission. Gears are constant-mesh, and the shaded portions are dog clutches that slide back and forth to engage any one of the four gears.

3. One dog clutch has been engaged, as shown by shaded portion, and the transmission is in low gear. If you are wondering about the power flow—that seems to go in and come out at the same place—it is possible because the input shaft is a running fit inside the tubular output shaft.

4. These are not all the parts in a transmission, but they are the important ones that you should know.

5. The Volkswagen is quite different from other cars. With the engine back of the rear wheels, the transmission is combined with the rear end.

We suggest counting the reverse idler twice, because it is easier to keep the relationships of the gears straight. When you do the calculation, however, you will notice that the tooth count for the reverse idler, being used both as numerator and denominator in the figuring, will cancel out. So the reverse idler can actually be ignored for ratio calculating purposes and the ratio figured as if the two gears were meshed directly. The reverse idler is actually what its name says, a gear that reverses direction and is simply an idler.

All this business of figuring gear ratios will come in handy some time if you should happen to blow a transmission and there are several different gears available for replacement. For example, on the old flathead Fords there were four different cluster gears available which, when combined with the correct main drive and second gear, would give four different ratios. Stock clusters in Fords and Mercurys were 28 and 29-tooth; 25 and 26-tooth clusters were stock in Lincoln and Lincoln Zephyr. Put Zephyr gears in your Ford and your engine had better have some excess oats or you'll have an awful time getting away from stop signals, especially if you have to start on a hill.

The advantage of the 25 and 26-tooth clusters was that they would allow an engine that could pull them to wind out, whereas the engine pulling 28's or 29's would be all done a few feet after it got started.

Gearing plays a big difference in both acceleration and horsepower. For the same reasons we mentioned before, the internal combustion engine just doesn't have the flexibility to operate efficiently over a wide rpm range. Only proper gearing will allow it to put its horsepower onto the road where it really counts.

A transmission doesn't really have to have gears in order to get the engine leverage to do its job. Fifty years ago some cars had friction wheels instead of gears. The two wheels were set at right angles to each other and when the driver stepped on a pedal, one wheel would be forced against the other and turn it. If the second wheel was near the hub of the first, it would be comparable to a low gear, but the driver also had a hand lever so he could move the second wheel back and forth across the face of the first wheel, changing the leverage ratio from low to high. He could even reverse it, by moving the hand lever so that the second wheel was over on the other side of the drivewheel.

It sounds like an excellent system but it had a lot of disadvantages. The wheels have to be very large and heavy in order to handle the horsepower, and there's always a problem of wearing out the facing so that there's slippage. Friction drive was most popular on a car called the Carter, but it soon died out and most car makers went to gear boxes.

TRANSMISSION TYPES

There are three major arrangements of gears in a transmission. They are planetary gears, sliding gears, and constant-mesh gears, all of which have been tried by different car makers throughout the history of the automobile. If you do much reading of automotive articles, you're probably tired of hearing about the Model T and its planetary gears, but this really was one of the major reasons for its tremendous sales record.

The planetary gear arrangement in the Model T transmission was really primitive and not very dependable. If the transmission bands lasted one year the driver was very lucky, and it was possible to burn them out completely on one long steep hill.

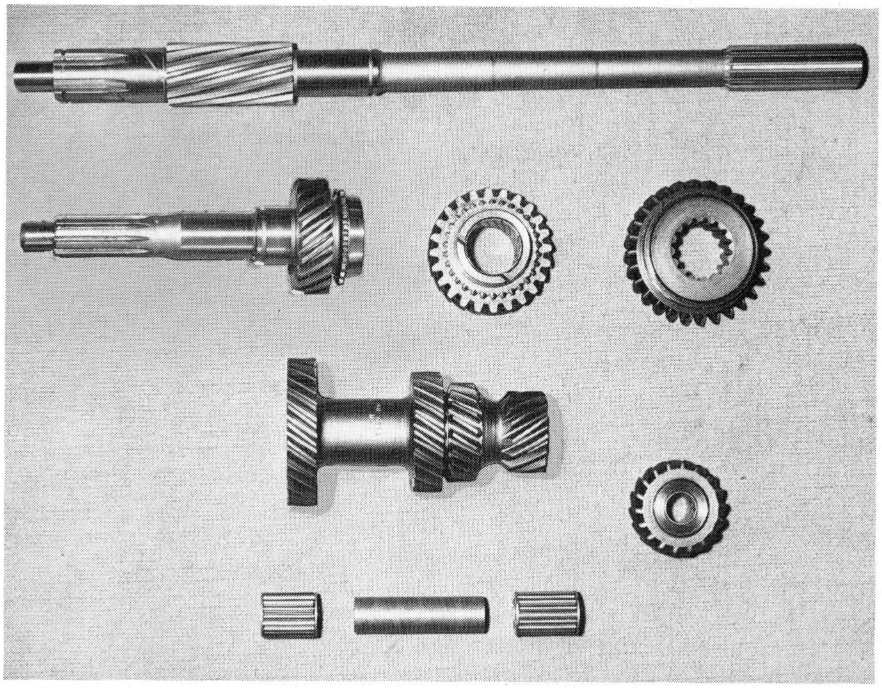

4

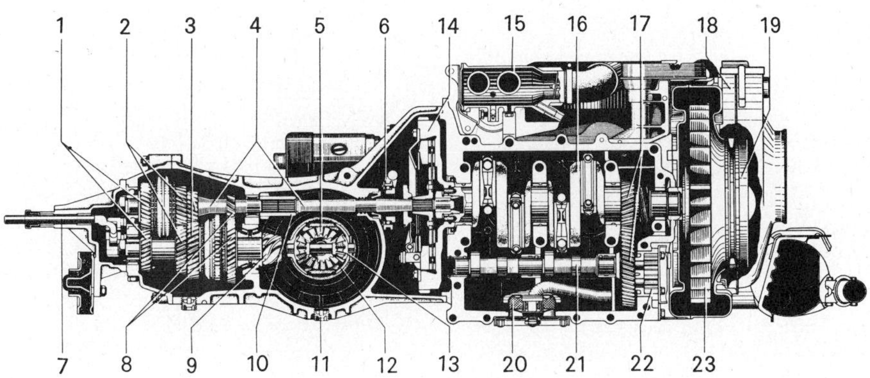

5

1 - 4th gear train
2 - 3rd gear train
3 - 2nd gear train
4 - Main drive shaft
5 - Differential side gear
6 - Clutch release bearing
7 - Transmission shift lever
8 - 1st gear train
9 - Drive pinion
10 - Reverse gear
11 - Oil drain plugs
12 - Differential housing
13 - Differential pinion
14 - Flywheel
15 - Intake air distributor
16 - Crankshaft
17 - Camshaft drive gears
18 - Fan housing
19 - Crankshaft pulley
20 - Oil strainer
21 - Camshaft
22 - Oil pump
23 - Fan

Inside the Gearbox

Sliding gears are still used today in manual transmissions. All this means is that a gear is splined onto a shaft and when the driver wants to mesh it with another gear he moves a lever that slides the gear along the splines until the position is correct.

The first sliding-gear transmissions were of the progressive type. There was no H-pattern as we know it today. All the driver had to do was to move the shift lever in one direction and he would go through all of the forward speeds. He moved the shift lever in the other direction to get into reverse. Thus all shifts were made in a straight line. It sounds like a great idea until you stop to think that when coming down to a stop he had to gear back through second and low to get to neutral. If he waited until he came to a full stop, then he had to jiggle the shift lever to mesh with second gear then wiggle it again to get into low and finally he'd make it into neutral. To get into high gear he had to go through low and second. There was no way to get around them. If you really wanted to hear some gear clashing and a few cuss words going through the hot summer air, you should have been around in those days.

The progressive transmission was really nothing more than an adaptation of the old back gears on a lathe. It was out of date before World War I when most manufactures had gone to a selective transmission. The selective transmission has shift rails and a lever which can be moved through a gate from one rail to the other so that we can select any number of gear positions from the neutral position of the lever. Most selective transmissions use the H-pattern which has stayed with us right up until today, although the gear positions on the H have changed slightly from what they were in the early part of the Twentieth Century.

The selective transmission can be either a sliding-gear or constant-mesh type or a combination of the two. In a sliding-gear type the actual gears themselves are moved back and forth to mesh with other gears. In a constant-mesh type the gears are in a fixed position and the connection between is made by dog clutches. These dog clutches usually have some type of synchronizing device which is supposed to prevent clashing of the teeth on the clutch. However, if a person is inept enough, it is possible to make a lot of noise and do a lot of damage with just about any transmission.

1

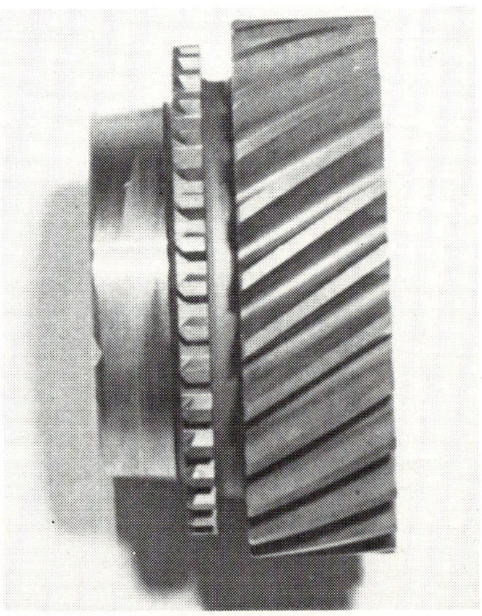

2

1. Disassembled synchro and gear it mates with. When the outer shell is moved by the shift fork, it pushes on the little spacer bars in the outer part of the hub. The bars push against the brass ring, forcing it against the gear cone, which makes the gear revolve at the same speed as the synchro. Finally, the outer shell of the synchro rides over the spacer bars and engages the teeth on the gear.

2. A constant-mesh gear. The cone is the rubbing surface for the brass synchro ring, and the small teeth engage outer shell of the synchro.

3. Heavy-duty synchro out of a Clark 280V 5-speed truck box.

4. Both of these are sliding gears, one uses a shift fork that fits into a grooved collar on the gear, the other requires a horseshoe-shaped fork that goes over the gear.

The advantage of the constant-mesh type is that it is easier to shift and if a shift is missed, the resulting clash will not be as damaging on a dog clutch as it would be on the gears themselves. With careful study of the illustrations, you will easily see how the sliding-gear and constant-mesh transmissions work. All automotive transmissions made today are of the selective type. Even motorcycles use a selective gearset but their linkage is hooked up so that it is applied in a progressive manner thus giving straight-line shifting. If you've ever ridden a motorcycle you'll know why straightline shifting is necessary. You're too busy riding to try to feel your way through any kind of a pattern and take a chance on getting into the wrong gear.

If you cringe when you hear someone clashing gears in a transmission, then you probably have some appreciation of the damage that can be done by careless shifting. Transmissions with sliding gears are called "clash boxes" because it takes a little bit of thought and care to keep from ending up with a lot of gear teeth in the bottom of the box.

There isn't much that can be done to a sliding-gear box to eliminate the clashing. The driver just has to be careful. But on a constant-mesh gearbox the designers have eliminated most of the clashing of the dog clutches by providing a little friction device that synchronizes the speed of the two gears so that their dog clutches can slip easily into mesh.

POWER FLOW

Let's go through a few terms so that you can better understand how a transmission works. Some parts of a transmission are called by different

3

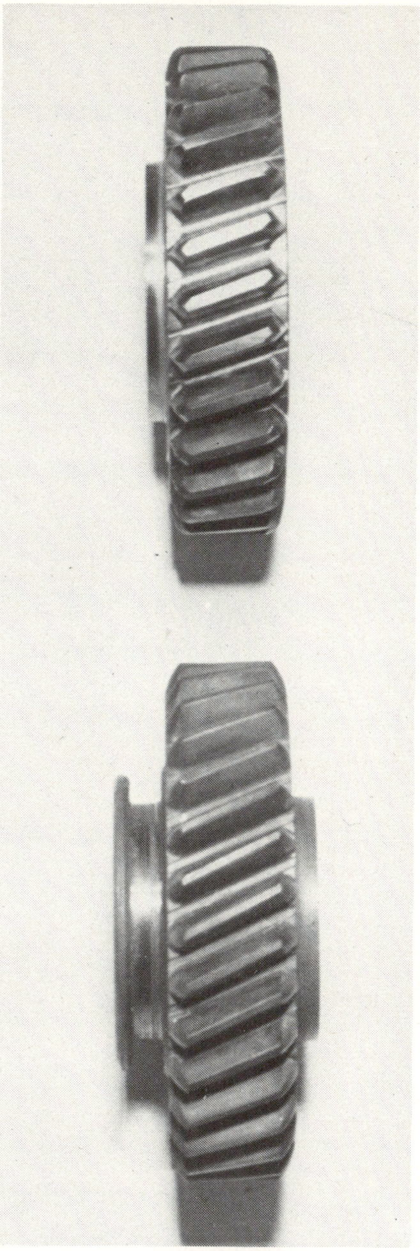

4

names that mean the same thing. The shaft that is splined to the clutch and is in one piece with the main drive gear is usually called the clutch shaft. But because it is in one piece with the main drive gear, many parts men may refer to it as the main drive gear and let it go at that. Sometimes it is even referred to as a jack shaft.

Torque from the clutch comes into the front of the transmission and rotates the main drive gear, which meshes with the cluster gear, sometimes called the countershaft gear. In England they call it the laygear because it usually lies at the bottom of the transmission case.

The cluster gear is just what the name implies—a series of gears cast in one piece so that it looks like a cluster of gears. It rotates any time the clutch is engaged to a running engine, whether the transmission is in gear or not.

If our transmission is a simple sliding-gear type, then nothing is turning inside except the main drive gear and cluster gear when the shift is in neutral. The reverse idler is turning, but it is only going along for the ride. In order to mesh the gears and get some torque applied to move the car, we slide a gear along a shaft mounted directly above the cluster. That shaft is known by mechanics as the mainshaft.

Once we have the gears meshed, we let out the clutch pedal and the torque is applied through the clutch shaft, main drive gear, cluster gear, mainshaft gear, through the mainshaft and out to the rear end. There can be several gears on the mainshaft of different diameters and teeth counts. Transmission linkage is designed so that the driver has to unmesh one gear before he can mesh another. It works fine except for the clashing of gears when the driver is trying to mesh them. If at any time you should get a transmission into two gears at once because the shifter is worn out, you have a surprise, and maybe a broken box, too. With more than two gear positions in mesh at the same time, the transmission is trying to go at two speeds at once. Of course it can't do this, so the gears will lock up solid and something may give. If it happens when the car is standing still, then no damage will be done, providing the driver doesn't try to move the car.

In a constant-mesh transmission the same arrangement of gears applies, except that all the gears on the mainshaft are in constant mesh with the cluster gears. That may not sound possible after what you were told above, but it is because the gears on the mainshaft are not splined to the shaft, but free to rotate on it. So with a constant-mesh transmission in neutral we not only have the main drive gear and cluster gear turning, but also all the mainshaft gears.

Alongside each gear on the mainshaft is a dog clutch, with a hub that is splined to the shaft and an outer ring that can slide over against the gear. The gear has some extra teeth on its side, and there are some teeth on the ring of the dog clutch. All that is necessary is to move the dog clutch against the gear and the gear will be locked solidly to the mainshaft. Then let out the clutch and we are on our way.

Dog clutches can clash and make noise just as much as the gears in a sliding-gear transmission. But the dog clutches are a little bit tougher than gear teeth, and then there is the matter of the synchros.

SYNCHRONIZERS

Somebody figured out that if a projection on the dog clutch rubs against the side of the gear as the dog clutch is moved into engagement, the projection will cause the gear to revolve at the same speed as the dog clutch. When the gear and the dog clutch are turning at the same speed, then there can't be any clashing, and the dog clutch will slip into mesh much more easily.

The gadget that speeds up or slows down the gear, so that its speed matches the dog clutch's speed, is called a synchronizer. Through the years most dog clutches on passenger car transmissions have had synchronizers built in, so the whole works has come to be known as a synchro.

The inner hub and the outer ring of the synchro are made of steel, but the part of the synchro that rubs on the gear to change its speed is usually made of brass. In early transmissions the brass was permanently attached to the outer ring of the synchro, but in most later designs the brass is a ring that floats between the synchro ring and the face of the gear. That is a good design, because the rings can be replaced when they wear out. The type with the floating rings is known as the "blocker" type of synchro.

Of course there is more to the design than just a floating ring. A spring and some little spacer blocks keep the ring in position, and the ring has teeth that match the teeth on the dog clutch. Most synchros do double duty, too. Push the synchro in one direction and it locks one gear to the mainshaft. Push it the other way and it disengages the first gear, passes through a neutral position and engages a gear on the other side.

Chrysler products and the early Cadillacs use a synchro that is slightly different, utilizing pins instead of spacer blocks to locate the brass ring. The pins have a shoulder that will not allow the dog clutch to engage until the pin lines up with a hole. It's just added insurance that the dog clutches are lined up so there won't be any clashing.

For many years most transmissions were synchronized with constant mesh gears in second and high positions, but low was a sliding gear. In 1963 a 3-speed transmission appeared on Fords that had a constant-mesh synchronized low gear. This meant that the driver didn't have to come to a dead stop in order to get into low. The convenience of it made the transmission popular, and now many car makers have transmissions that have a synchro on low, or are "fully synchronized" as they say.

DOUBLE-CLUTCHING

Drivers who know how to double-clutch have been shifting into a non-synchronized low gear without stopping for many years, but you have to have good timing to do it right. And no matter how many times you have done it, you always give a little sigh of relief that you made it without clashing the gears.

What double-clutching means is that the driver leaves the shift in neutral and removes his foot from the clutch pedal. Of course the clutch is engaged and the engine is turning the main drive and cluster gears as well as the constant-mesh gears in the transmission. However, the mainshaft and the low sliding gear are

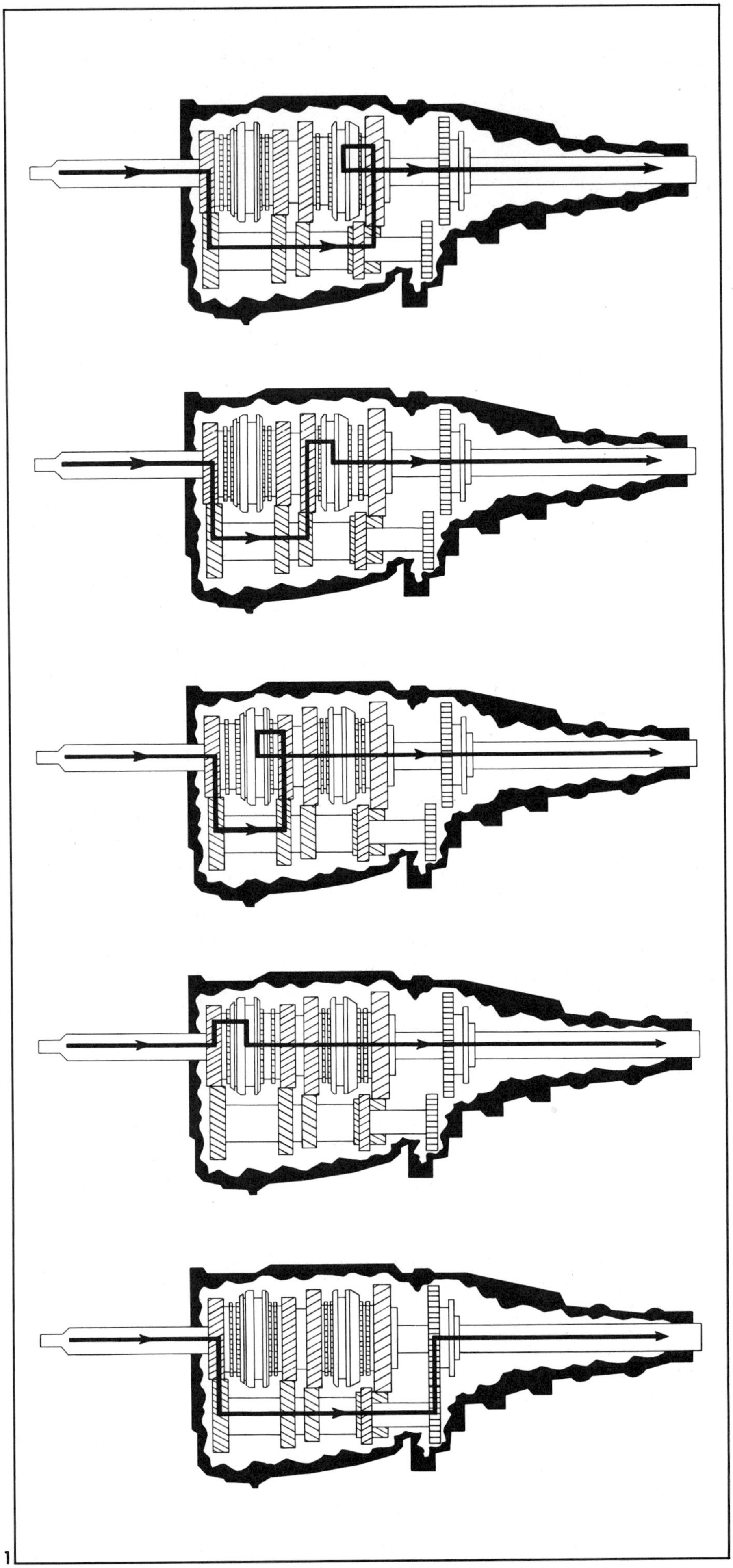

1

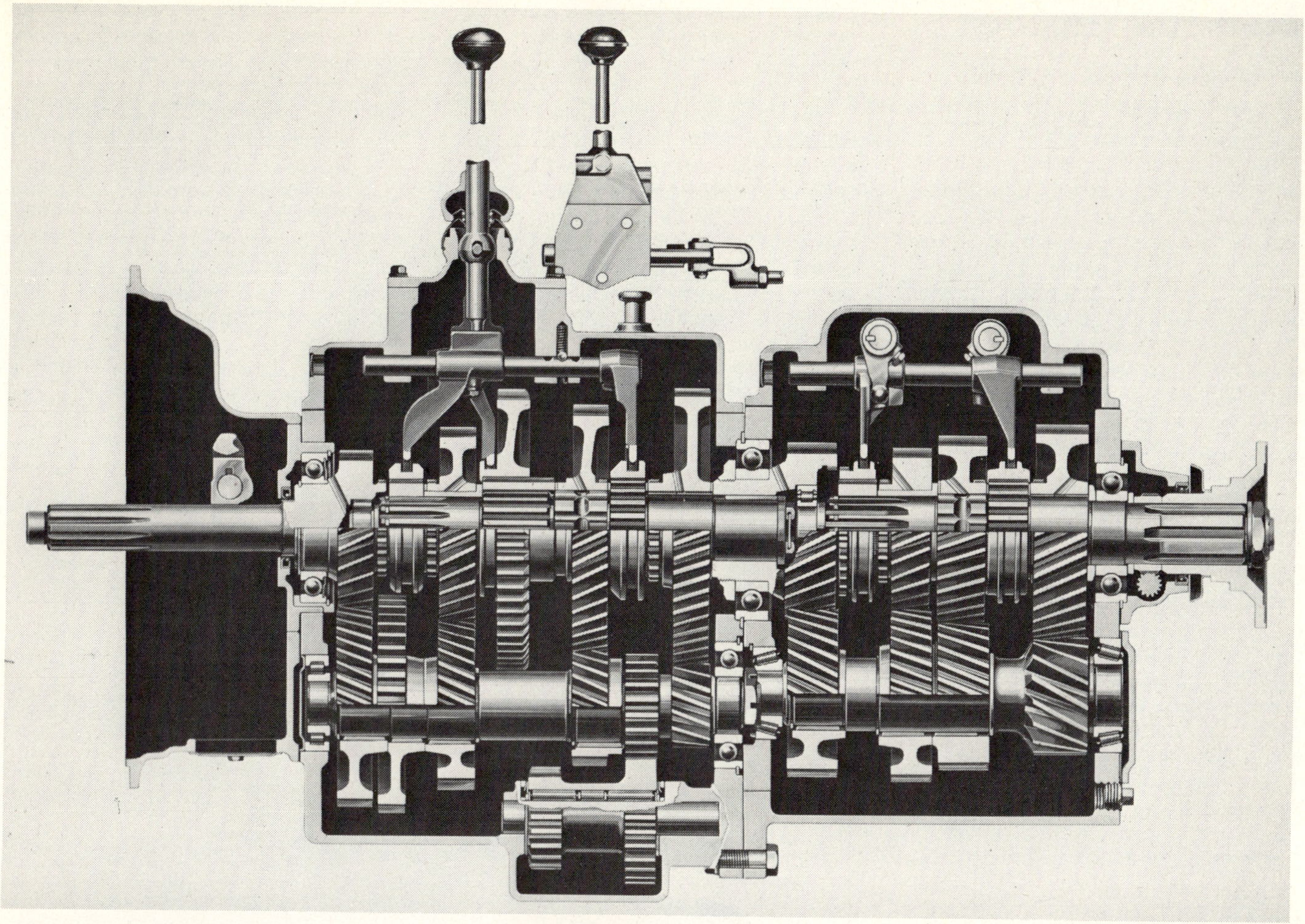

2

1. Arrowed lines represent the routing of power in typical 4-speed transmission as it progresses through gear changes. To get some kind of an idea of the gear ratios involved, you can compare various sizes of meshed gears. High, or 4th, is of course 1:1 or simply straight through.

2. They call the Spicer a compound transmission because it is actually two 4-speed boxes bolted together. Those gears look husky, and they are. This transmission will handle engines that put out 700 lbs.-ft. of torque and pull a truck weighing 80,000+pounds.

turning with the rear wheels. All the driver has to do is speed up the engine so that the main drive gear and cluster are spinning at the same speed as the low gear. When this is done he quickly disengages the clutch, pulls on the shift lever, and the trans will slip into low gear so smoothly he can do it with two fingers. If the driver is really good, he will match the speeds of the gears so perfectly that he can engage the gears without disengaging the clutch. Don't try this yourself, unless you want to spend next weekend putting new gears in your trans.

A shift into reverse is always done with the car standing still. Do it any other way and you have bought yourself a transmission.

In all this discussion of gear trains in a transmission, the result of any gear position has been to hook up the main drive gear with the mainshaft, with the cluster and other gears actually making the connection. When we put the transmission into high gear we are essentially doing the same thing, but the power flow is not through any of the gears. Of course the cluster is still engaged, so it is turning, but it is only going along for the ride. The same applies to any constant-mesh gears on the mainshaft. They are meshed with the cluster, but are just idling.

In high gear we have no gear reduction in the transmission. A dog clutch on the end of the mainshaft slides over and connects the end of the shaft with the main drive gear. In effect we have one solid shaft going through the transmission, with a lot of gears going along for a free ride. It doesn't have to be this way. There are transmissions that send the power flow through gears in the high gear position, but these are usually truck transmissions or special applications that are rarely seen.

When the power flow in high does go through gears, it is usually for the purpose of providing a less than 1-to-1 ratio. Remember that in all gear positions up to high we were turning the engine faster than the output shaft of the transmission. In high the power flow was straight through so the engine and transmission output shaft were turning at the same speed. With a less than 1-to-1 ratio, the engine will turn slower than the output shaft of the transmission. The term "overdrive" is used to describe any gear position that results in the engine turning slower than the output shaft of the transmission. An overdrive can be an auxiliary unit that bolts to the back of the regular trans and provides an extra overdrive ratio for high-speed cruising.

In order to get reverse gear in a transmission, the designers have simply added what is known as a reverse idler. It is an idler because it is always meshed with the cluster gear, and goes along for a free ride in all the other gear positions.

When the driver wants reverse, he slides a gear attached to the mainshaft until it meshes with the reverse idler. The power flow then comes from the main drive gear to the front cluster gear, off the rear cluster gear to the reverse idler, then to the mainshaft gear and out the tailshaft of the transmission. In an automobile transmission only one speed is provided in reverse, but truck transmissions provide several speeds in reverse.

Inside the Gearbox

TRUCK TRANSMISSIONS

Stand near the highway in a small quiet town on a calm night, and listen to the trucks starting up from a roadside diner. The big powerful diesel engines only go up a few hundred rpm before the driver shifts to the next gear. The engine rpm drops, climbs quickly to the peak of that gear and then the driver shifts again. Time after time he works the shift stick, throttle, and clutch until at last he is up to speed and can settle down for a long pull on the highway.

In the old days it took a lot of doing to make these shifts. There were two sticks in the cab, one for the main transmission and one for the auxiliary. The main transmission might have five speeds, and the auxiliary three. That gave 15 speeds forward, but some trucks also had a 2-speed rear end, which gave them 30 speeds forward and 6 in reverse. Not all the speeds were usable, however, because the ratios worked out so that some were too much the same and others were so far apart as to be totally useless.

The usual practice was to use all five speeds in the transmission, but only the top two speeds in the auxiliary, reserving the lowest speed in the auxiliary as a "granny" for pulling at extremely slow speeds under heavy loads. If the auxiliary transmission were left in the granny gear, then all five speeds of the main transmission might be used to keep from having to crawl along too slowly.

Auxiliary transmissions were known as "Brownies" in those days because the most popular one was made by the Brown-Lipe Corp. Shifting a Brownie was easy if you moved each stick individually, but the real pros put one arm down through the steering wheel and shifted both sticks at the same time.

With a loaded truck the driver started out in first in the main transmission, and the middle (second) position in the Brownie. His first shift was to the No. 3 or "over" position in the Brownie. On the next shift things got complicated, because he had to go to second in the main and back to the middle position in the Brownie. After that the shift was to the over position in the Brownie, and so on until he got to the fifth over position. Then, if he had a 2-speed rear end he might flip the switch to put it into high.

Double-clutching was an art in those days, and the driver had to do it at every shift. If he paid attention to what he was doing, the gear would slip into position without so much as the tick of a tooth. But if he missed a shift and ended up with either of the sticks in neutral, he was in trouble. A big truck slows down amazingly fast on the level or uphill. Going downhill it will pick up speed quickly. If the driver missed a shift, it wasn't just a matter of revving up the engine and dropping the trans into the gear he had missed. The truck was probably going faster or slower than when he missed the shift. There might be three or four gears in the middle of the range to choose from. He had to make the right decision quickly and get into the correct gear as fast as he could. For a lot of drivers it was too much. If they missed a shift they simply stopped and started all over.

Anytime the Brownie was shifted it was called "splitting a shift." The Brownie actually split the ratio so that there wasn't such a big jump between gear ratios in the main transmission. Going up through the gears

1

SPECIFICATIONS

Tractor	IH CO-4000
GCW	73,280 lbs
Engine	250-hp diesel
Transmission	Fuller RT-915 ROADRANGER
Axle	4.44:1
Tires	10.00 x 20
Tractor	Kenworth K923
GCW	76,000 lbs
Engine	280-hp
Transmission	Fuller RTO-913 ROADRANGER
Axle	4.62:1
Tires	10.00 x 20

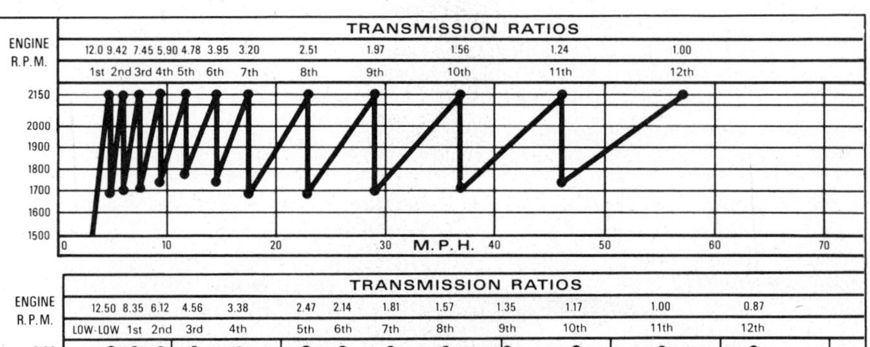

2

was comparatively easy. Most drivers shifted one stick at a time, revving the engine on the second double-clutch when the Brownie was shifted.

Gearing down was a different matter, especially if the hill was steep. There wasn't time to shift one stick at a time. If you wanted to keep your truck under control you had to spear one arm through the steering wheel and make both shifts at once. Miss a shift while gearing down on a hill and it was possible to have a runaway truck on your hands. When one of those loaded behemoths gets galloping down a hill, it isn't a simple matter of putting the brakes on and coming to a stop. Their brakes are excellent, but once they get rolling 70 or 80 mph, with many tons of load pushing down a steep hill, there isn't

1. Spicer 16-speed compound transmission. Each stick has four gear positions, with the main stick including a reverse. Depending on the model, gear ratios run from a 15.20 Lo-Lo to a .805 overdrive. It has four speeds in reverse, too.

2. How'd you like to shift every 250 rpm? In the upper ranges of a Fuller RTO-913, it's necessary if your truck is fully loaded. Diesel engines put out usable power over a very narrow rpm band. The right transmission can make the difference between a powerful truck and one that pulls like a sick horse.

3. One stick controls this Fuller Roadranger RTO-913. Both the range section and the overdrive splitter are operated by compressed air.

4. Fuller T-905 Twin Countershaft transmission. The RTO-913 has similar construction. Twin countershafts are used to divide the torque for longer gear life. Dog clutches use conical teeth for easier engagement, however there are no synchronizers.

5. Clark 280V 5-speed box. It has synchronizers on the top four speeds. Note the bolted cover on the power takeoff opening.

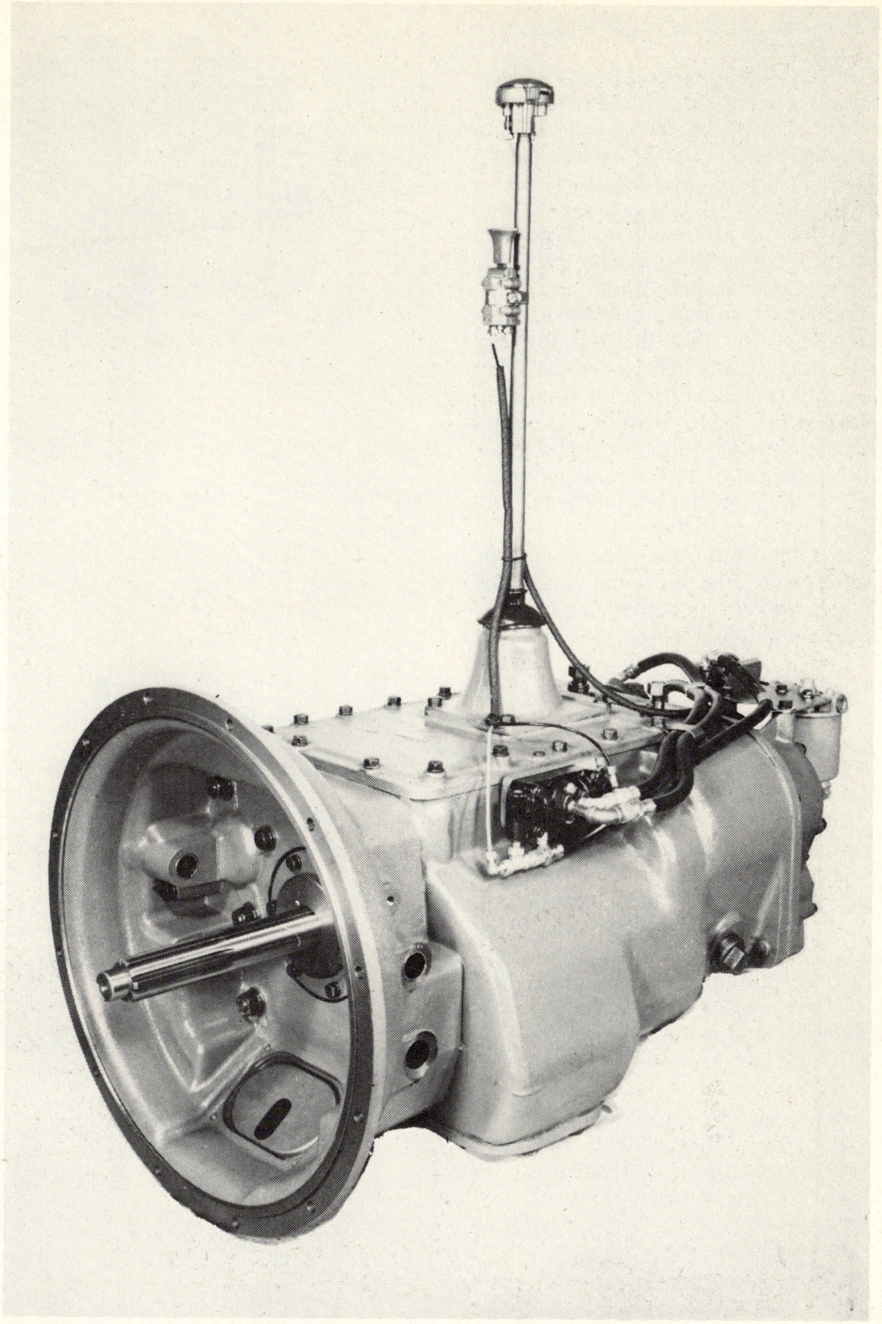

3

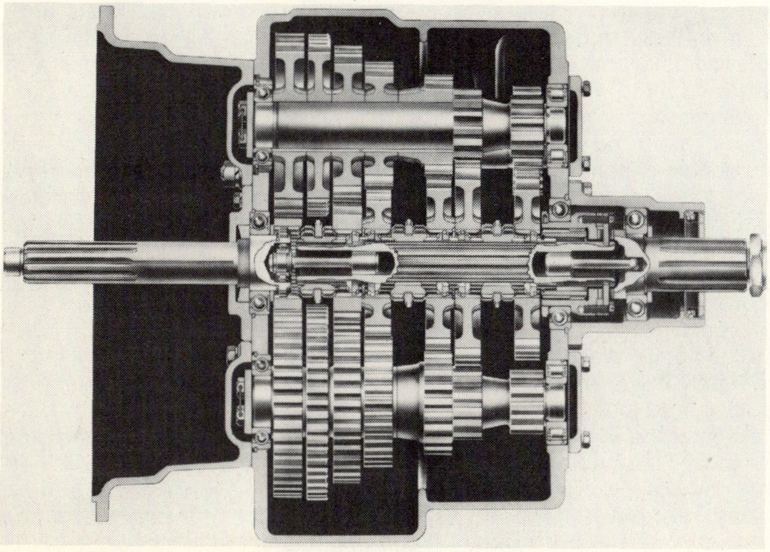

4

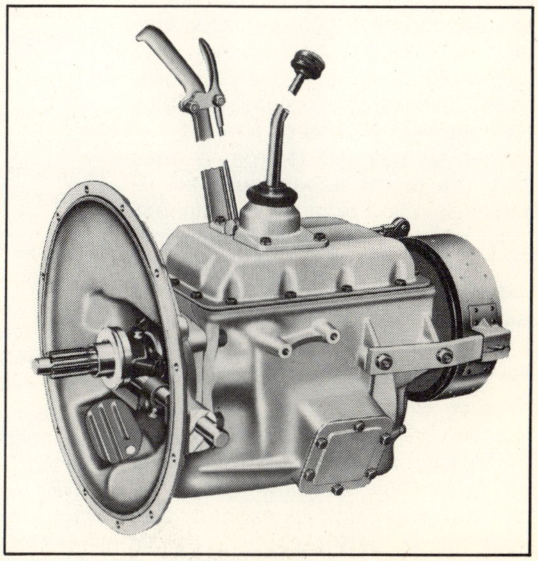

5

Inside the Gearbox

any brake that will stop them. Several highways have sand traps at the bottom of steep hills so that a truck can turn into the sand and come to a safe stop. If you ever see the track left by a big truck as it plows through deep sand for several hundred feet, you will realize that no brake in the world could ever stop such a juggernaut.

Modern transmissions do not require the skill that the old Brownie did. Most trucks use air brakes, so you will find air-operated transmissions on a lot of trucks. In some models one stick does all the shifting, with a button or small lever mounted on the stick so that all shifts are controlled with one hand.

Let's take the Fuller Roadranger as an example and see what has to be done to go through the gears. Fuller makes lots of Roadrangers, but the RTO-913 is a typical one for big trucks. It has 4 speeds, with a granny, a dual range, and a splitter overdrive, all built into one box only 34-1/16 inches long and weighing 710 pounds.

The granny is used only for starting from a dead stop, and then the driver goes through the familiar H pattern of four gears. His next step is to pull up a knob on the side of the shift handle to put the transmission in high range. Now he can go through the same four gears again, but he also has the splitter, a little lever at the top of the shift stick that brings in the overdrive. In each of the four gear positions he uses the splitter, so he has eight usable ratios. That, combined with the four speeds in low range and the granny, gives a total of 13 speeds forward. The splitter can be used only in the four high range ratios.

1

2

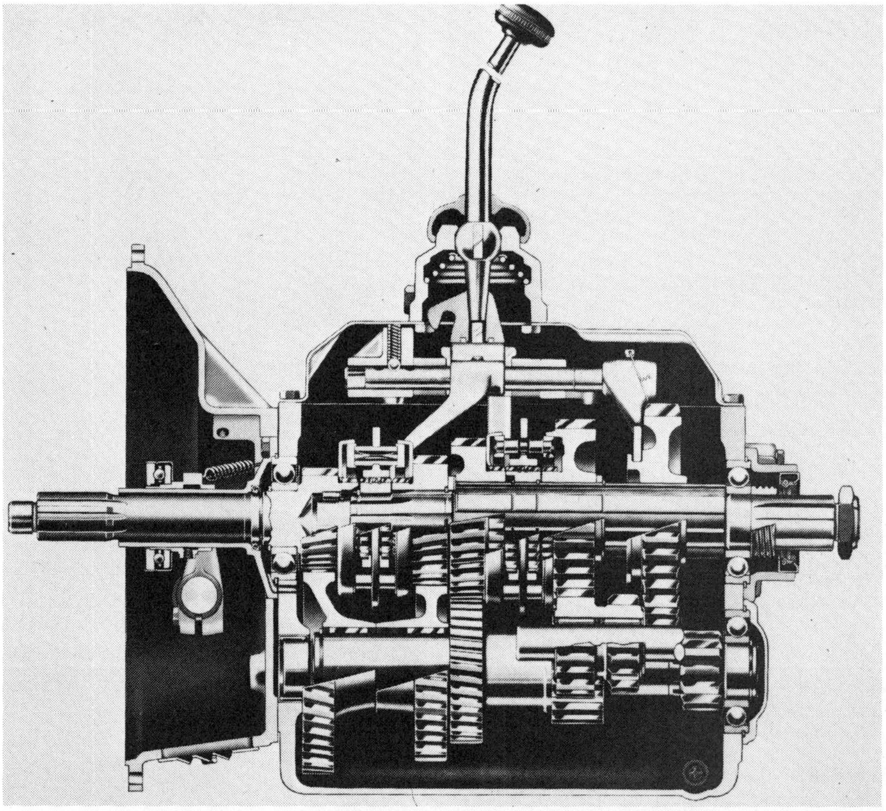
3

Two-stick transmissions are still around, too. They are much improved from the old days. The sloppy linkage and wiggling sticks that would flop all over the cab have been replaced with positive linkage, short throws, and easy-shifting gears. Dana Corp. makes Spicer transmissions, and they have one with 16 speeds forward.

When a truck starts from a dead stop, the driver doesn't use all 16 speeds in the Spicer. The main box has 4 speeds and reverse. The auxiliary box has Low-Low, Under, Direct, and Over. Low-Low is the granny, used only for extremely heavy pulling. Bottom gear in it has a 15.20-to-1 ratio in one model.

Starts are usually made with the main box in first, and the auxiliary in Under. Then the driver goes to Direct, and Over. On the following shift he puts the main box in second, and drops the auxiliary back to Under. In each of the four gears of the main box he goes through the three gears in the auxiliary, for a total of 12 speeds forward. If the truck is lightly loaded, the driver can skip gears if he wants to, selecting any gear that the engine is capable of pulling. Gears are constant-mesh, but no synchronization is used on the dog clutches, so double-clutching is still necessary.

POWER TAKE-OFFS

Some truck transmissions come with a plate on the side or the bottom that can be removed and a power take-off installed in its place. Power take-offs are handy if you want to run a saw for cutting up timber or power a posthole digger. The old Model T driver would scoff if he could see all the modern conveniences on transmissions today. His power take-off consisted of jacking up one rear wheel of his T, running a belt over the tire, and powering everything from a saw to a water pump.

The modern power take-off is a lot more dependable. All it consists of is a gear in a housing that meshes with one of the gears in the transmission that is exposed by removing the cover plate. A shaft comes out the end of the power take-off and anything that needs to be run is hooked to the shaft. A transmission must be designed to accept the power take-off. If the cover plate is not present, then a power take-off cannot be used.

IN-AND-OUT BOXES

Circle track racing cars, such as sprint cars and midgets, do not have any need for a clutch or transmission because their speeds are never outside the range where their engine produces usable torque. Race car builders found out long ago that by

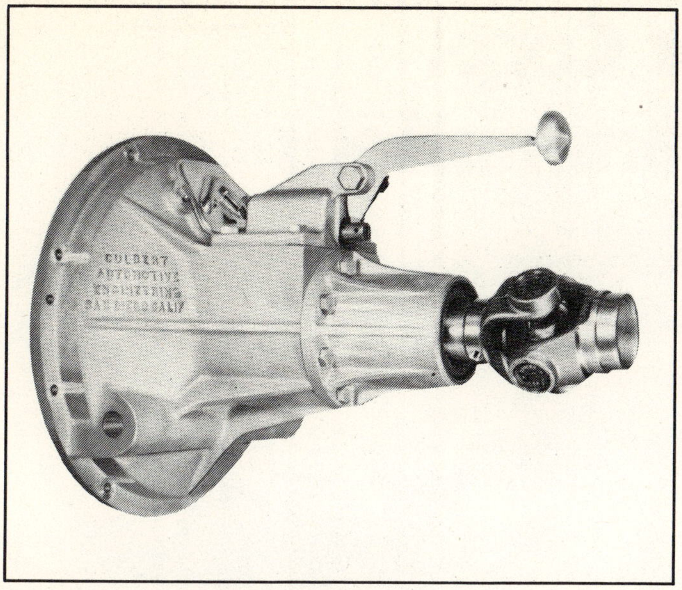

4

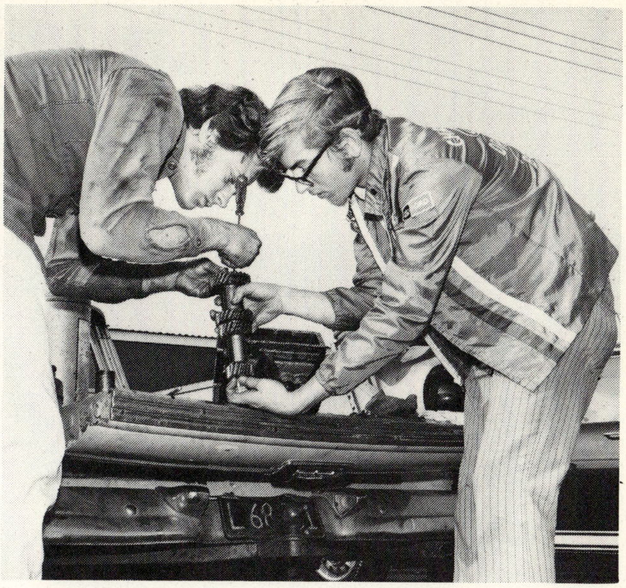

5

1. If your truck needs more gears, it may be easier to put an auxiliary transmission behind the main box rather than change the main box for a bigger trans. This is a remote-controlled Spicer 3-speed auxillary, with power takeoff opening on the side.

2. Spicer 8-bolt power takeoff. It has a shifter with a neutral position so it doesn't have to run all the time.

3. Inside the Clark 280V. Gears are all constant-mesh except for low and reverse.

4. CAE race car in-and-out box. CAE makes boxes for just about any engine and driveline combination a race car owner can think up.

5. Despite their complexity, transmissions are forgiving enough that "field stripping" should not be shunned.

6. Four-wheeldrive grows up with the introduction of both Warner Gear (on Jeeps) and New Process (on Chevy, Dodge, Plymouth and Ford) full-time systems. Secret is a 3rd differential in conjunction with either a standard or automatic trans which can be locked-up for hard going without front lock hubs.

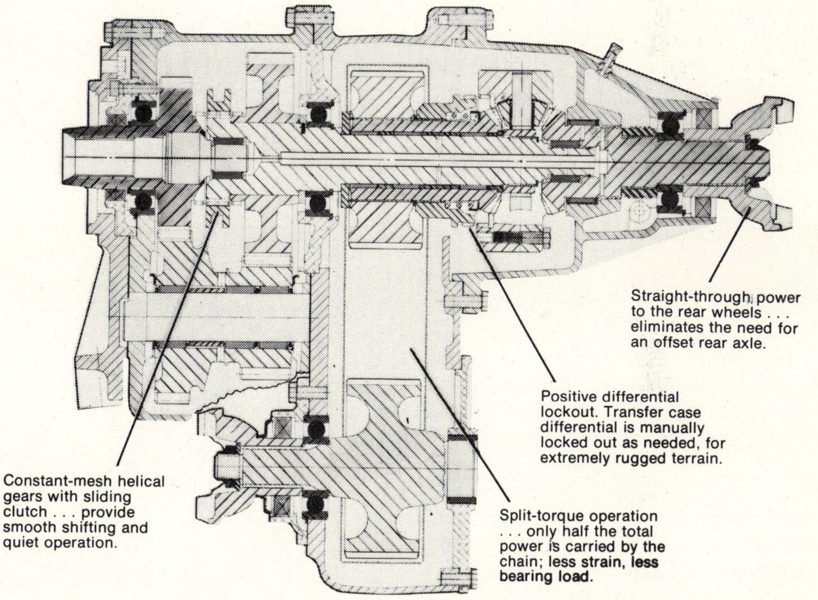

6

eliminating the clutch and transmission a lot of excess weight was pared off and they could go faster.

They still had to have some way to disconnect the engine from the wheels so that the engine could be warmed up with the car standing still, and also some way to make stopping easier. The in-and-out box was just what they needed, and it has been used for many years in all kinds of race cars. It is not used in championship cars or at Indianapolis, however, because those cars are required to have a reverse gear for backing up in emergencies. They also have two or more speeds forward in their transmission, besides reverse.

In-and-out boxes have one lever coming out the top of the box between the driver's legs. When he wants to engage the engine with the wheels, he pushes the lever down, or pulls up to disengage. He engages the engine only when the car is standing still and the engine is dead, unless he likes to hear gear teeth chewing away at each other. Disengagement can be done at any time, but if he wants to re-engage he has to come to a complete stop, shut the engine off, and then call for the push truck to start him again.

Some in-and-out boxes have been made with changeable gears. In that case it is still an in-and-out box, but also a 1-speed transmission. To change gear ratios for different size race tracks, the box is unbolted from the back of the engine and slid back far enough to reach in and change the gears. That type of box was not too popular. Most car builders preferred to use a plain in-and-out box with a quick-change rear end.

As we said in the beginning, an internal combustion engine has to have a transmission, but it is more of a blessing than a liability. A good transmission allows the power of the engine to be used to greatest advantage. In following chapters we will take a look inside the different makes of automobile transmissions.

Chrysler Corp. Standard Transmissions

Chrysler Corp. uses two 3-speeds and a 4-speed, all of their own manufacture. Revisions in transmission design and gear ratios necessitated by engine changes to control emissions have resulted in new 3-speeds designated A-250 and A-230 to replace the previous A-903 and A-745 models, but the 4-speed remains the time-honored A-833.

1

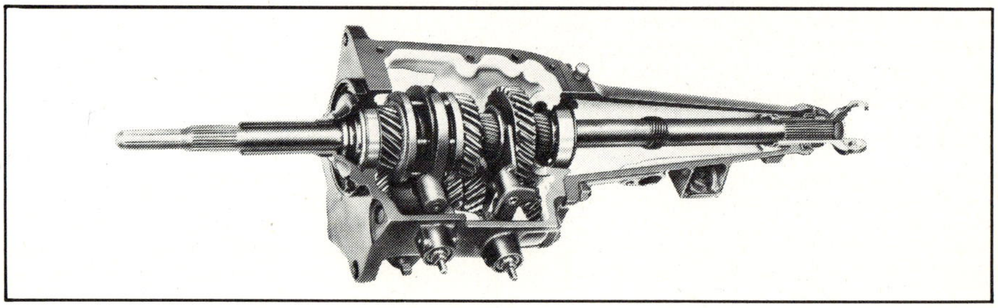

2

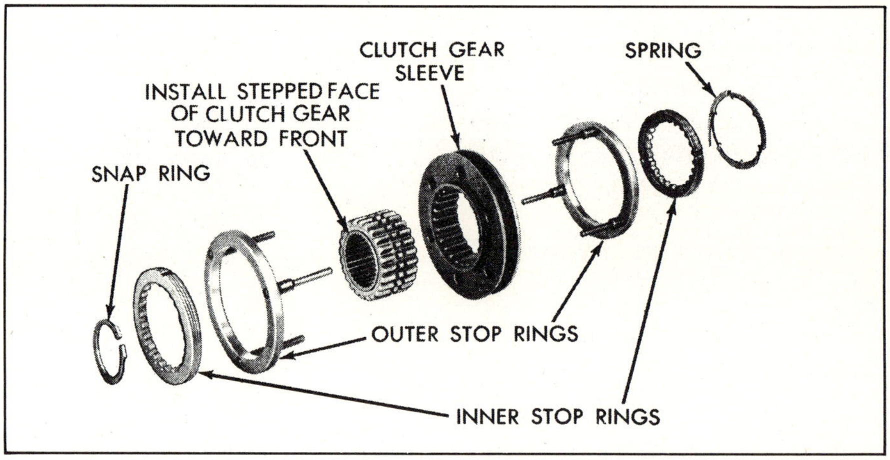

3

A-250 3-SPEED

An updated version of the previous A-903, the Model A-250 is of the synchromesh type with helical gears to provide silent operation. The countershaft is in constant mesh and is supported by roller bearings at each end.

The mainshaft front end is piloted in roller bearings in the end of the main drive pinion, and is supported by a ball bearing in the rear of the case. The rear end of the mainshaft is supported by the sliding yoke and the bushing in the extension housing.

Used as standard equipment in the 6-cyl. compact versions of Chrysler Corp. cars, the A-250 is supplied with a low gear ratio of 3.18:1, a second gear of 1.83:1, a direct high gear, and a reverse of 3.97:1.

The transmission case is of the closed case type of construction which places the access cover on the top of the case rather than on the side. This type of case enables the use of rugged pivot supports for the shifter shafts in the case casting itself rather than the more common side cover type. In addition, the thickness of the casting can be reduced with a saving in total weight.

A-230 3-SPEED

Chrysler's heavy-duty 3-speed is used in both 6-cyl. and V-8 powered cars and is found in Chrysler, Dodge, Plymouth and Valiant models as a standard equipment item. When coupled with the 198 or 225-in. 6-cyl., or 318-in. V-8, it has a low gear of 3.08:1, a second gear of 1.90:1, a direct high, and a reverse of 2.90:1. The 360-in. V-8 has a higher low gear at 2.55:1, a higher second of 1.49:1, the same direct high, and a reverse of 3.34:1. As a standard item, these ratios will be found only in the larger Dodge and Plymouth; the Chrysler of this size is fitted with a TorqueFlite automatic.

While the A-745 which it replaces was of the same closed case construction as the lighter duty A-250, the A-230 uses the side cover design in which shift forks and detent cams can be replaced without pulling the transmission from the car. It should continue the reputation of the A-745 as one of the most efficient and reliable transmissions available. Chrysler has replaced the pin-type gear synchronizer (once considered the best for competition) with the blocker type, common in the industry.

A-833 4-SPEED

Built in 1964 to replace the Warner T-10 then being used, the A-833 was designed with the knowledge that bigger and more powerful V-8's were to come in the years ahead, and with the reduced output of today's crop of engines, it has all the guts necessary to remain around for another decade.

The four forward speeds are synchromesh and consist of a series of helical gears designed for high torque capacity with teeth so proportioned as to operate at high speeds without excessive friction or heat generation.

Like the A-230 3-speed transmission, the A-833 has two separate sets of ratios for use with different engines. The 318-cu.-in. V-8 and earlier 6-cyl. engines (the 4-speed is no longer available with a 6) have a low gear ratio of 3.09:1, a second gear of 1.92:1, a third gear of 1.40:1, a direct

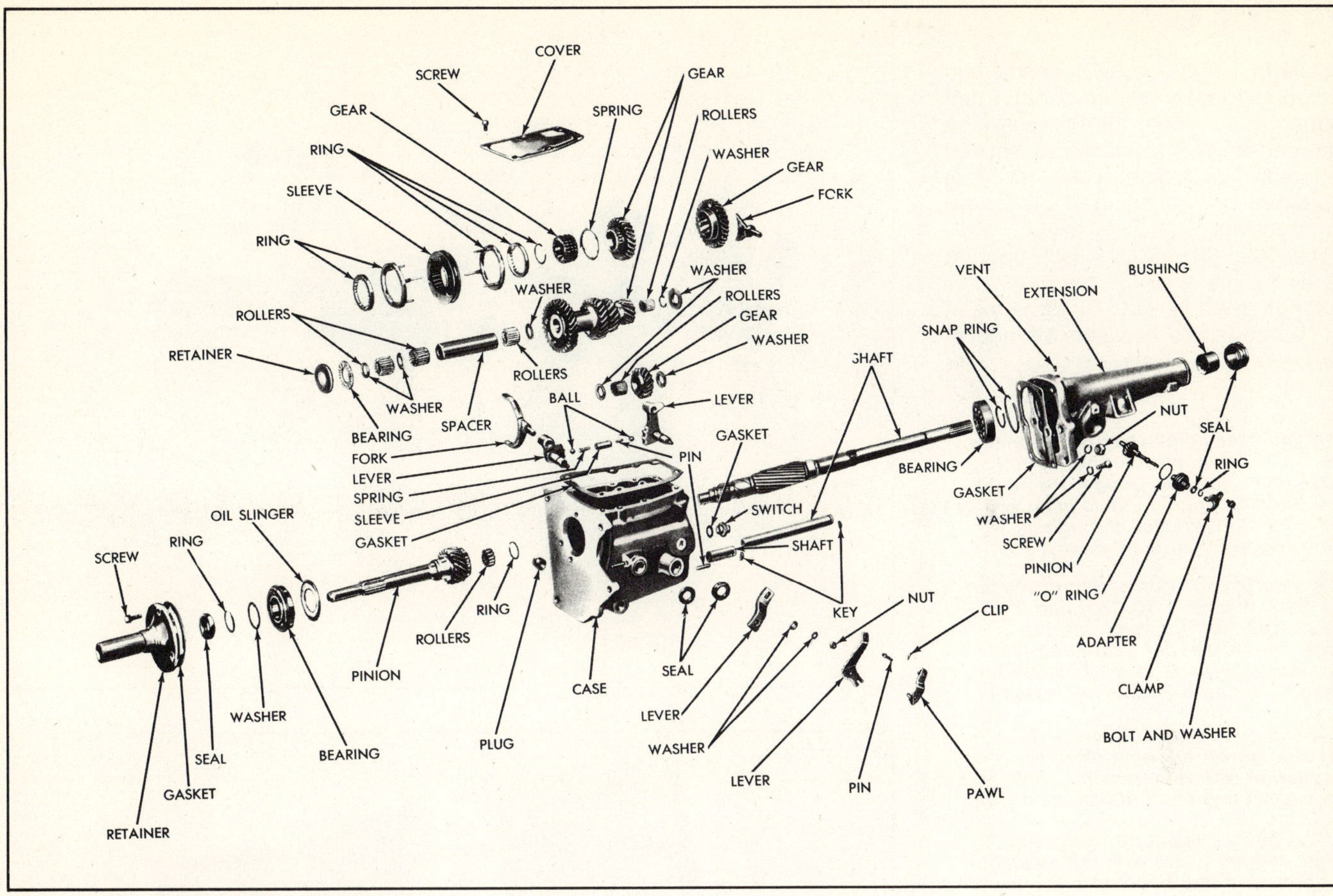

4

1. Chrysler Corp. continues to offer three standard transmissions in their '74 car lineup; two 3-speeds and one 4-speed. However, both the 3-speeds are all-new (A-250 and A-230) while the 4-speed (A-833) remains unchanged.

2. Cutaway view of "closed case" type of construction shows heavy bosses around shift lever pivots and the thin casting made possible with this design.

3. Synchronizer is important part of any manual transmission. By closely matching speed of the gears, it provides clash-free gear changes. This is from the light-duty A-250 3-speed.

4. The light-duty A-250 is used behind the 6-cylinder engines in Chrysler's compact cars, replacing the older A-903, and features rugged construction.

5. The 2nd and 3rd synchronizer from the heavier duty A-230 3-speed. It has larger clutch gear sleeve and stronger stop rings to handle extra loads.

6. Cross section exposes the needle bearings used in the center of the countershaft on the A-833 4-speed.

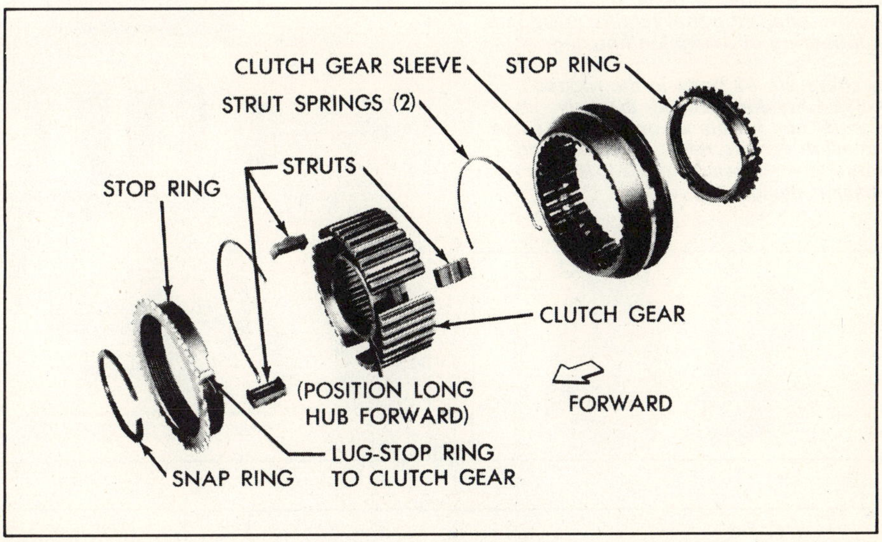

5

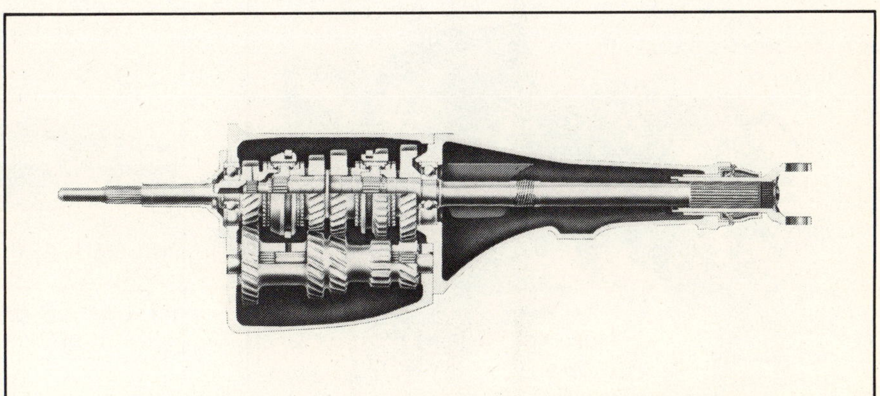
6

1.00:1 high gear, and a reverse of 3.00:1. The larger 360 and 400-in. engines come equipped with a low gear of 2.47:1, a second of 1.77:1, a third of 1.34:1, a direct high gear, and a reverse of 2.48:1.

The main drive gear is supported by a heavy-duty ball bearing at the transmission case, and an Oilite bushing pressed into the end of the crankshaft. The mainshaft front end is

Chrysler Corp.

piloted in needle roller bearings and is supported by a heavy-duty ball bearing in the center of the case. The rear end of the mainshaft is splined to the sliding U-joint yoke which is supported by a bushing in the extension housing.

The countershaft gear is supported by a double row of needle roller bearings at each end. Thrust is taken on thrustwashers between the end of the gear and the case.

1. If you ever completely disassemble the A-230, standard on both 6- and 8-cylinder Chrysler, Dodge, Plymouth and Valiant models, this is what you will find. Construction throughout is more rugged than the lighter A-250.

2. A side by side comparison between the Borg-Warner T-10 (left) and the A-833 Chrysler. Chrysler unit is larger due to beefier gears and the fact that reverse is in the box itself, while T-10 has it in tailshaft housing.

3. A-833 case has shift levers in left side cover, allowing internal inspection without removing trans from car. Long shaft is box mainshaft.

4. The beefy Chrysler 4-speed has large, rugged gears with the mainshaft supported in ball bearings.

5. Pliers are shown being used to remove snap ring that retains mainshaft ball bearing in extension housing.

6. There are 92 parts in the rugged A-833 Chrysler 4-speed, the only manual box to remain unchanged since emission requirements have dropped horsepower levels requiring new transmission design.

1
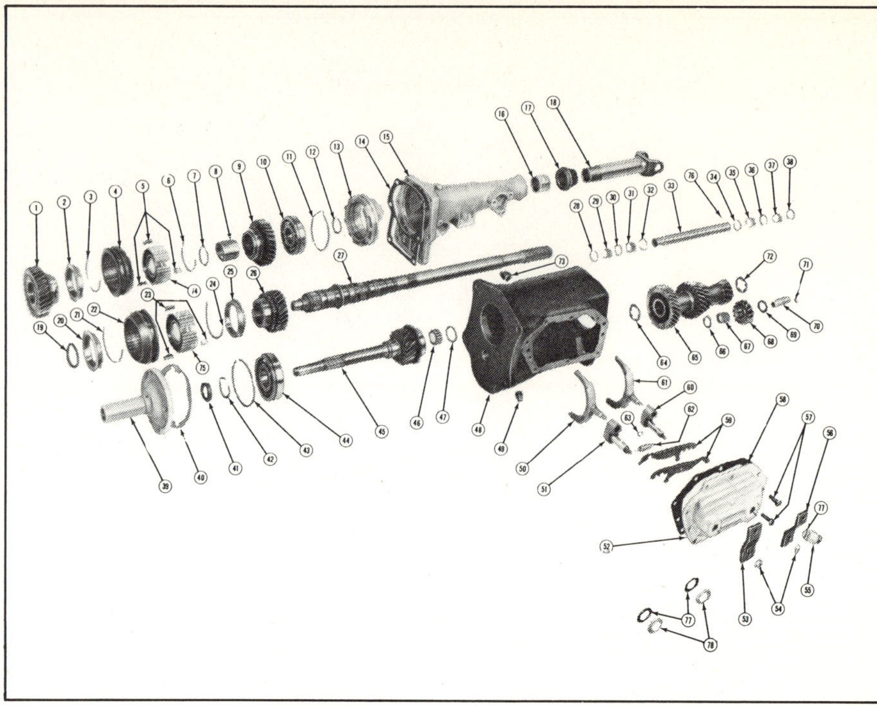

2
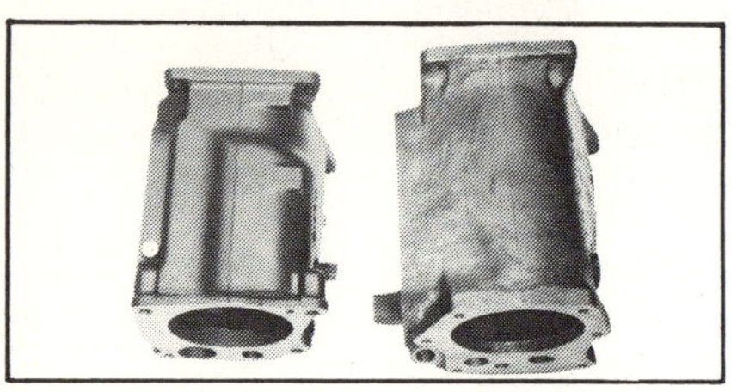

3
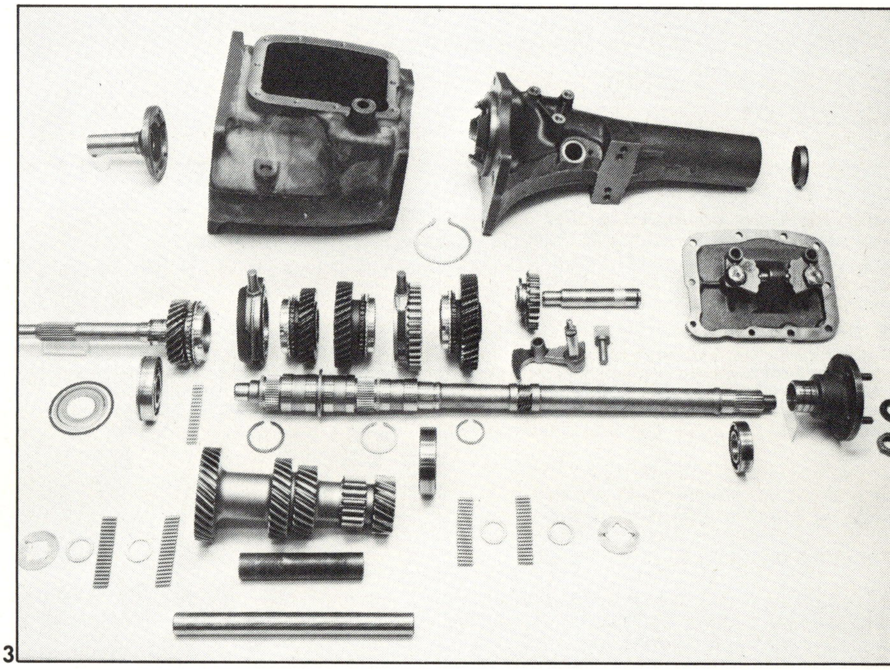

4
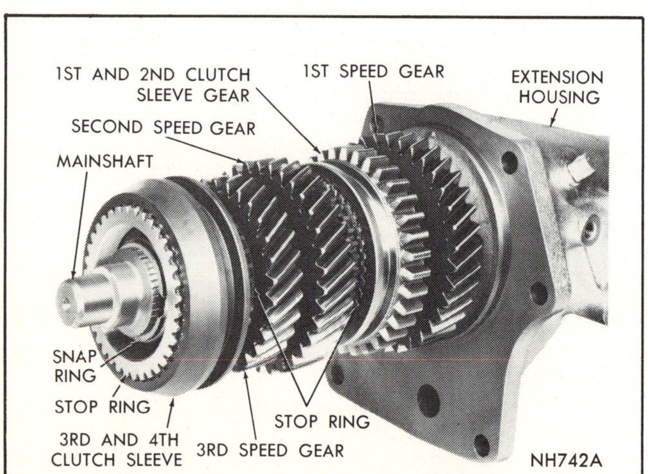

5
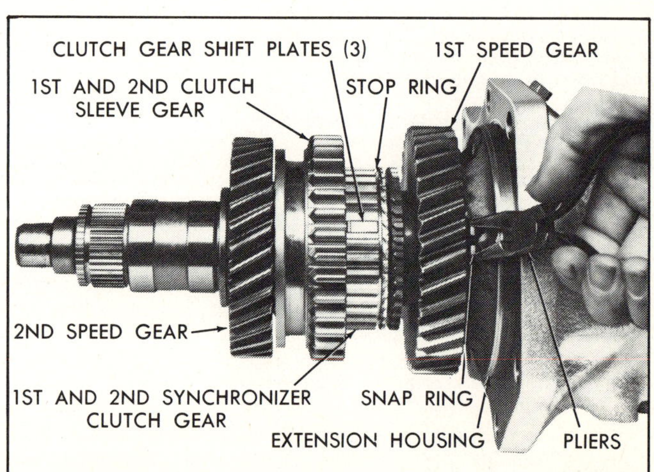

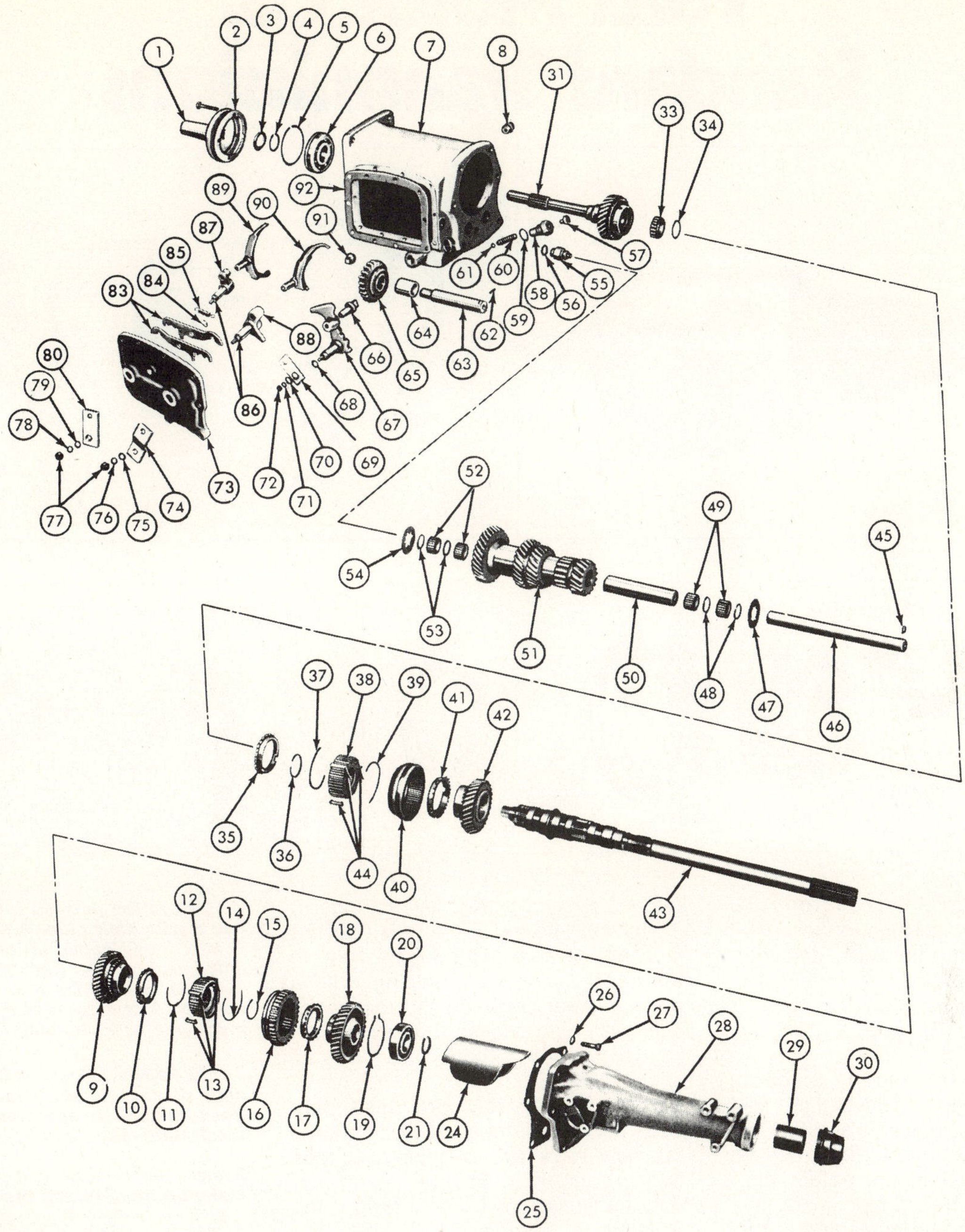

1. Bearing Retainer
2. Bearing Retainer Gasket
3. Bearing Retainer Oil Seal
4. Snap Ring, Bearing (Inner)
5. Snap Ring, Bearing (Outer)
6. Pinion Bearing
7. Transmission Case
8. Filler Plug
9. Gear, 2nd Speed
10. Snap Ring
11. Shift Strut Springs
12. Clutch Gear
13. Shift Struts (3)
14. Shift Strut Spring
15. Snap Ring
16. 1st and 2nd Clutch Sleeve Gear
17. Stop Ring
18. 1st Speed Gear
19. Bearing Retainer Ring
20. Rear Bearing
21. Snap Ring
24. Baffle
25. Gasket, Case to Extension Housing
26. Lockwasher
27. Bolt
28. Extension Housing
29. Mainshaft Yoke Bushing
30. Oil Seal
31. Main Drive Pinion
33. Needle Bearing Rollers
34. Snap Ring
35. Stop Ring
36. Snap Ring
37. Shift Strut Spring
38. Clutch Gear
39. Shift Strut Spring
40. Clutch Sleeve
41. Stop Ring
42. 3rd Speed Gear
43. Mainshaft (Output)
44. Shift Struts (3)
45. Woodruff Key
46. Countershaft
47. Thrustwasher, Gear (1)
48. Spacer Ring Needle Roller Bearing
49. Needle Bearing Rollers
50. Bearing Spacer
51. Countershaft Gear (Cluster)
52. Needle Bearing Rollers
53. Spacer Ring Needle Roller Bearing
54. Thrustwasher, Gear (1)
55. Backup Light Switch
56. Backup Light Switch Gasket
57. Plug
58. Retainer, Reverse Detent Ball Spring
59. Gasket
60. Spring, Reverse Detent Ball
61. Ball, Reverse Detent
62. Woodruff Key
63. Reverse Idler Gear Shaft
64. Bushing, Reverse Idler Gear
65. Gear, Reverse Idler
66. Fork, Reverse Shifter
67. Reverse Lever
68. Oil Seal, Reverse Lever Shaft
69. Reverse Operating Lever
70. Flatwasher
71. Lockwasher
72. Nut
73. Gearshift Control Housing
74. 1st and 2nd Operating Lever
75. Flatwasher
76. Lockwasher Lever
77. Nut, Lever
78. Lockwasher, Lever
79. Flatwasher, Lever
80. 3rd and 4th Operating Lever
83. Interlock Lever (2)
84. "E" Ring
85. Spring
86. Oil Seal (2)
87. 3rd and 4th Lever
88. 1st and 2nd Lever
89. 3rd and 4th Speed Fork
90. 1st and 2nd Speed Fork
91. Drain Plug
92. Gasket, Shift Control Housing

6

Ford Motor Co. Standard Transmissions

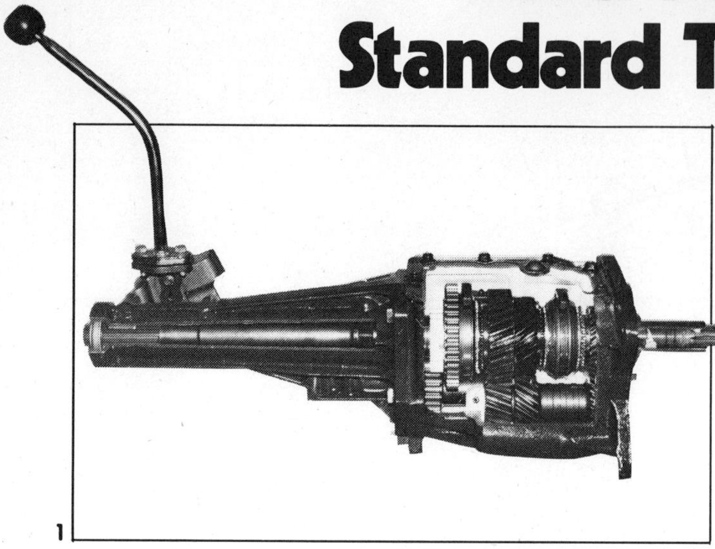

1

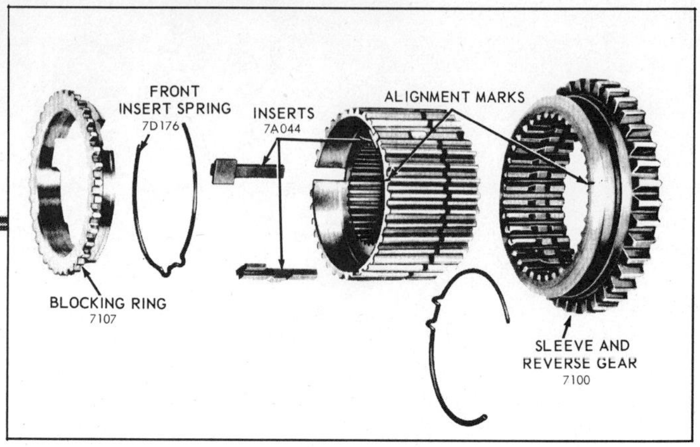

2

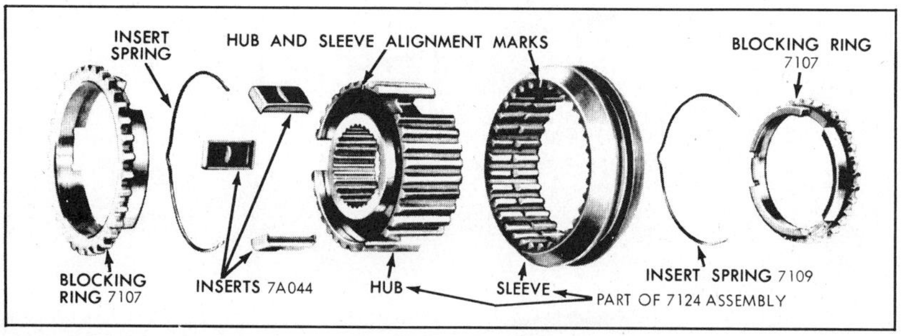

3

4

FoMoCo currently uses one 3-speed transmission in their passenger cars, the 3.03, and four 4-speed boxes, two of which are manufactured in Europe.

3.03 3-SPEED

The 3.03 RAN model 3-speed transmission is used with all Ford vehicles equipped with 200 and 250 6-cyl. engines, and the 302-in. V-8. The designation 3.03 is the actual distance between the centerline of the countershaft and the centerline of the input shaft. The RAN model has a low gear ratio of 2.99:1, a second gear of 1.75:1, a direct third, and a reverse of 3.17:1.

Until Ford discontinued its 390-in. V-8, a RAT version of this transmission was available. The RAT used a low gear of 2.42:1, a second gear of 1.61:1, a direct third, and a reverse ratio of 2.33:1.

The RAN transmission is of the fully synchronized type, with all gears except reverse being in constant mesh. The forward-speed gears are helical cut, with those in the reverse gear train spur-cut and not synchronized.

Ball bearings support the input shaft and gear and the center of the output shaft. Roller bearings in the input shaft bore support the front of the output shaft. The countershaft runs on two rows of roller bearings.

The shift forks, shift rails, detent mechanism and related parts are in the transmission case casting, with a pressed steel cover on the top of the case providing access to the gears.

4-SPEED

For many years, Ford used only Borg-Warner 4-speed transmissions, but now have two of their own plus two produced in Europe. A heavy-duty transmission of Ford design, the RUG model is equipped with the 351-in. V-8. Used in Cougar, Torino and Montego cars, it has a low gear of 2.78:1, second gear is 1.93:1, third is 1.36:1, a direct high, and reverse of 2.78:1.

The RAD version used in Mustang and Mustang II has a low gear of 3.50:1, a second of 2.31:1, a third of 1.43:1, a direct high gear and 3.38:1 reverse, all contained in an aluminum die cast housing.

The 72WG and 72WT are both of European manufacture and are essentially of the same design except for the countershaft gear bearings. The 72WG contains 38 countershaft roller bearings and a single thrust washer at the rear, while the 72WT has 42 roller bearings and four rear thrust washers.

The 72WG was coupled with all 1.6

1. Back in the mid-'60's, Ford had had it with their older 3-speed. A "better idea" came out of it; the fully synchronized (including low) 3-speed, termed the 3.03. It proved so good that some GM divisions even made use of it on some hi-perf models.

2. The RAN's 1st and reverse synchronizer disassembled. If you get your trans completely apart, don't mix up the 1st-Reverse, 2nd-3rd synchro parts.

3. Here are the bits and pieces that make up the 2nd and 3rd gear synchro.

4. The 2nd and 3rd gear shift rails are rotated to gain access to a set screw in the shifting fork during teardown or rebuild of the RAN trans.

5. Here's a RUG 4-speed going back together after disassembly. The 3rd and 4th gear shifting forks are being re-installed in the box.

6. Exploded view of the RUG model 4-speed, available in '74 behind the 351-in. V-8 in Cougar, Montego and Torino cars, showing primarily the shift rails and forks. Note location of seat belt sensor switch; yes, Maud, restraint system interlocks extend to even the lowly transmission.

7. Power flow of the RAN 3-speed looks like this. The 3.03 designation comes from the distance between the center-lines of the input and counter-shafts.

8. The RUG box was designed to live for at least 100,000 miles and is one of the most trouble-free transmissions.

5

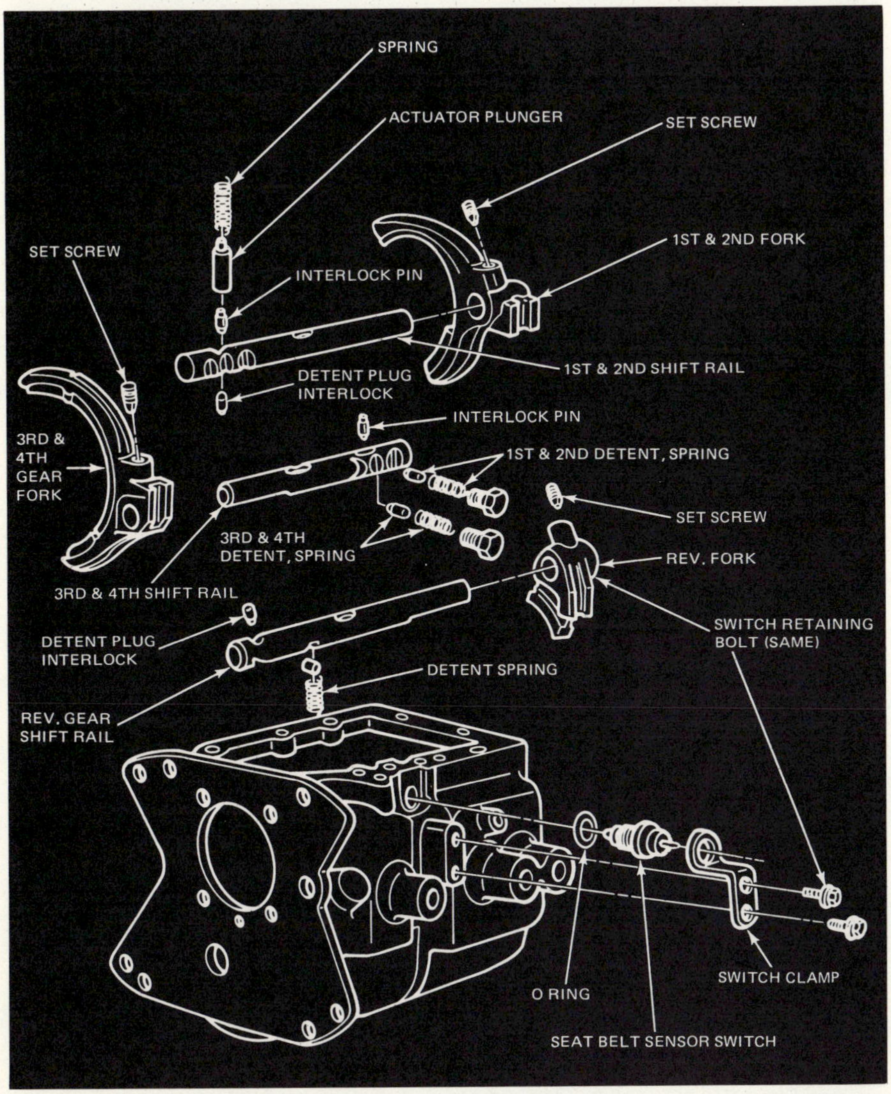

6

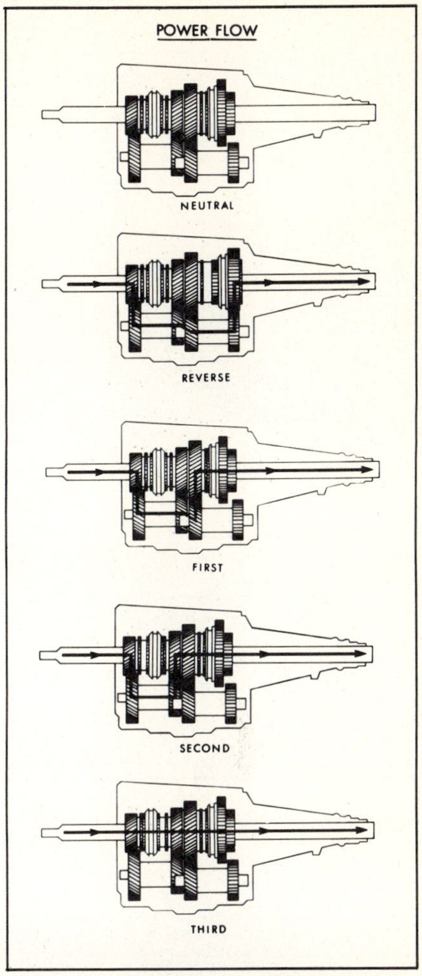

7

8

and 2.0 liter Pinto engines through 1973 and has a low gear of 2.78:1; second is 1.69:1, third is 1.29:1, with a direct high gear and reverse of 2.32:1. The 72WT is fitted to the new 2.3 liter 4-cyl. engine and used in Pintos for 1974. First gear has been strengthened to handle the new engine's extra torque. Sedan application has a 3.65:1 low gear, second is 1.97:1, third is 1.37:1, high is direct and reverse is 3.66:1. When used in the Pinto wagon, the gear ratio is slightly different, with a low of 3.36:1, second of 1.81:1, third of 1.26:1, a direct high and 3.36:1 reverse gear.

All transmissions are fully synchronized in the four forward speeds with the gears in constant mesh. Reverse on the RUG is engaged by a sliding gear and on the other three by a meshing of the reverse idle gear and reverse gear on the countershaft gear and output shaft.

The RUG shift linkage is mounted directly on the transmission housing and enters the driver's compartment through an opening in the floorpan. A flexible rubber boot is provided to seal the driver's compartment from the exterior. Shift linkage on the 72WG and 74WT is completely enclosed within the transmission; the shift lever is connected to a boss at the rear of the extension housing. As the shift mechanism is completely enclosed, it is also completely lubricated at all times.

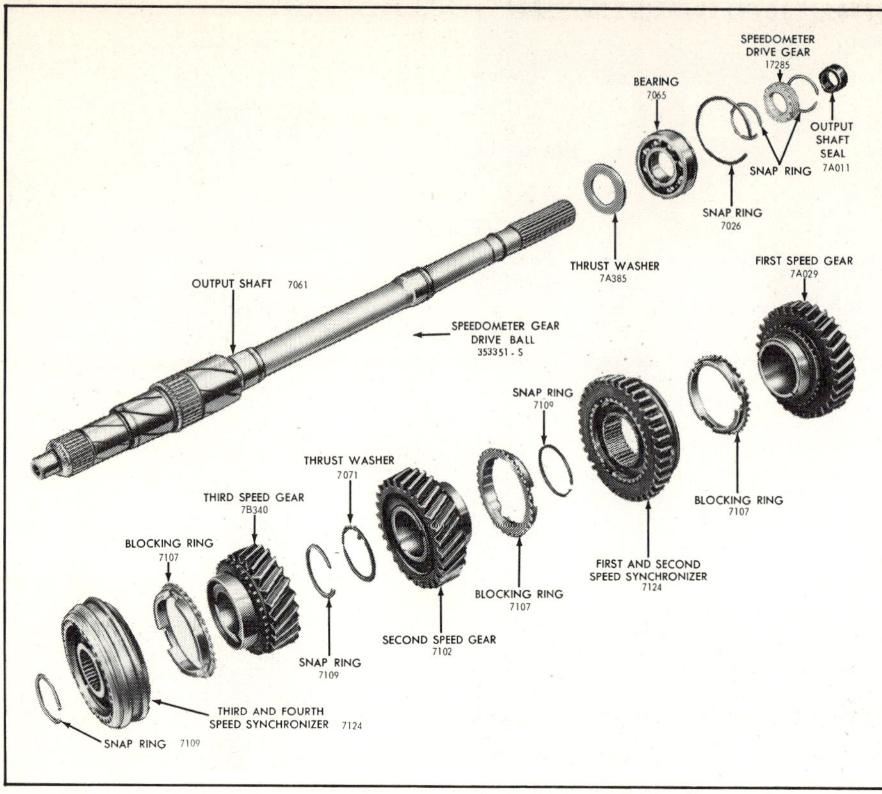

1

1. More innards of the 4-speed RUG, one of FoMoCo's two domestic boxes, one of a total of four 4-speeds used in the company's car lines. This is a look at the disassembled output shaft. The shaft (and tailshaft housing) will vary in length depending upon what model the RUG is used in; end-play is controlled by shims.

2. Extension housing with gear-laden mainshaft comes off back of transmission case as a unit.

3. This is the foreign-built 72-WG 4-speed. Main difference between it and 72-WT is in the number of countershaft roller bearings. Picture shows removal of lengthy shift rail, necessary since small engine and small trans are so far ahead of driver position, and his reach to rear-mounted gear lever must be accommodated.

4. Some Ford-driving speed shifters may tell you to take teeth off the bronze synchros to make shifts easier, faster; don't do it. If the thing doesn't shift properly, it's doubtful trouble lies inside the gearbox itself.

5. Disassembly drawing of Ford's RAD 4-speed shows all component parts and their relationship. Transmission case is aluminum die cast and is used in earlier Mustangs and Mustang II.

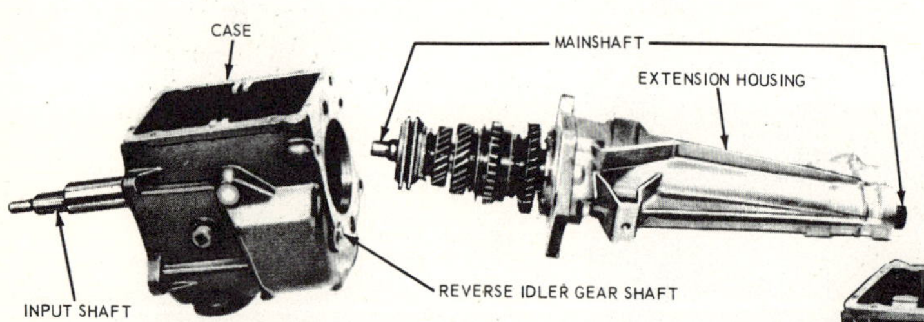

2

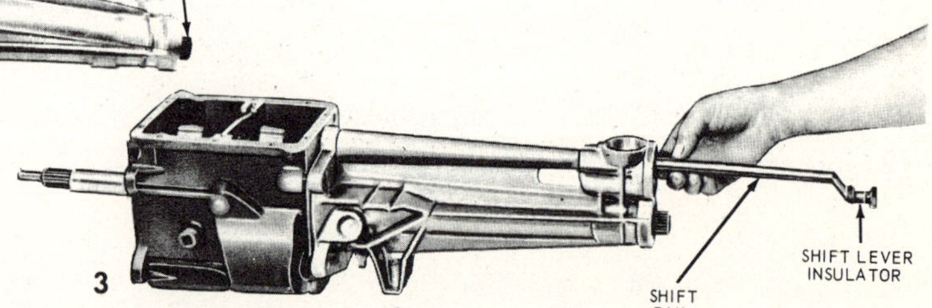

3

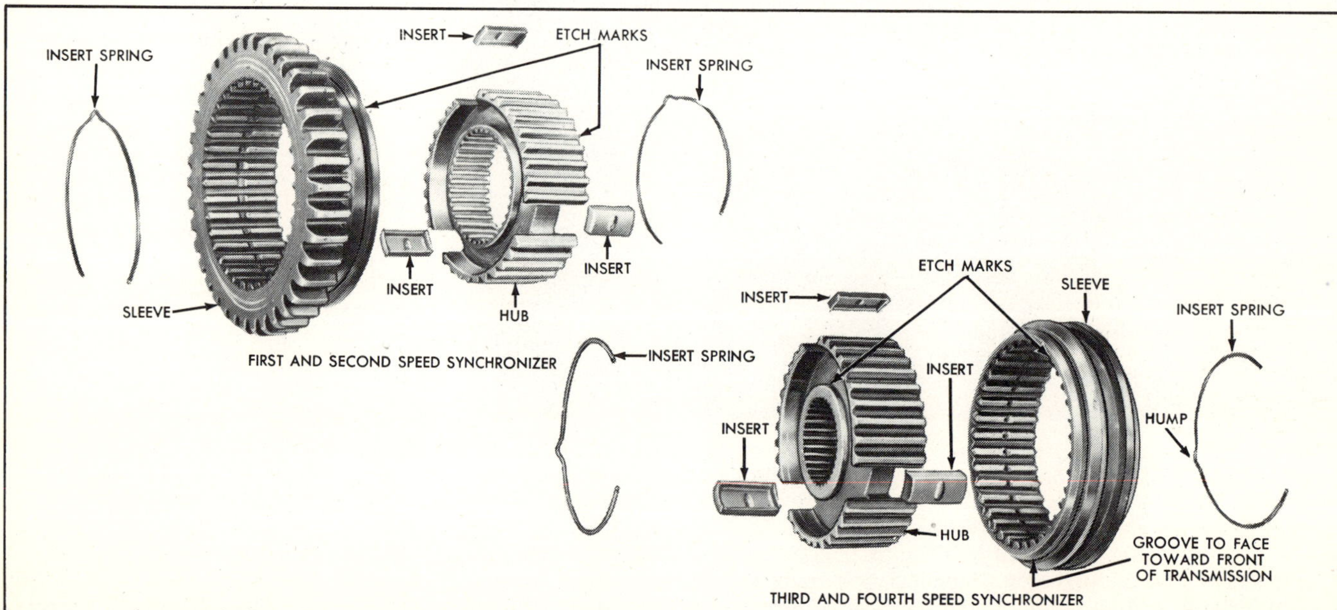

4

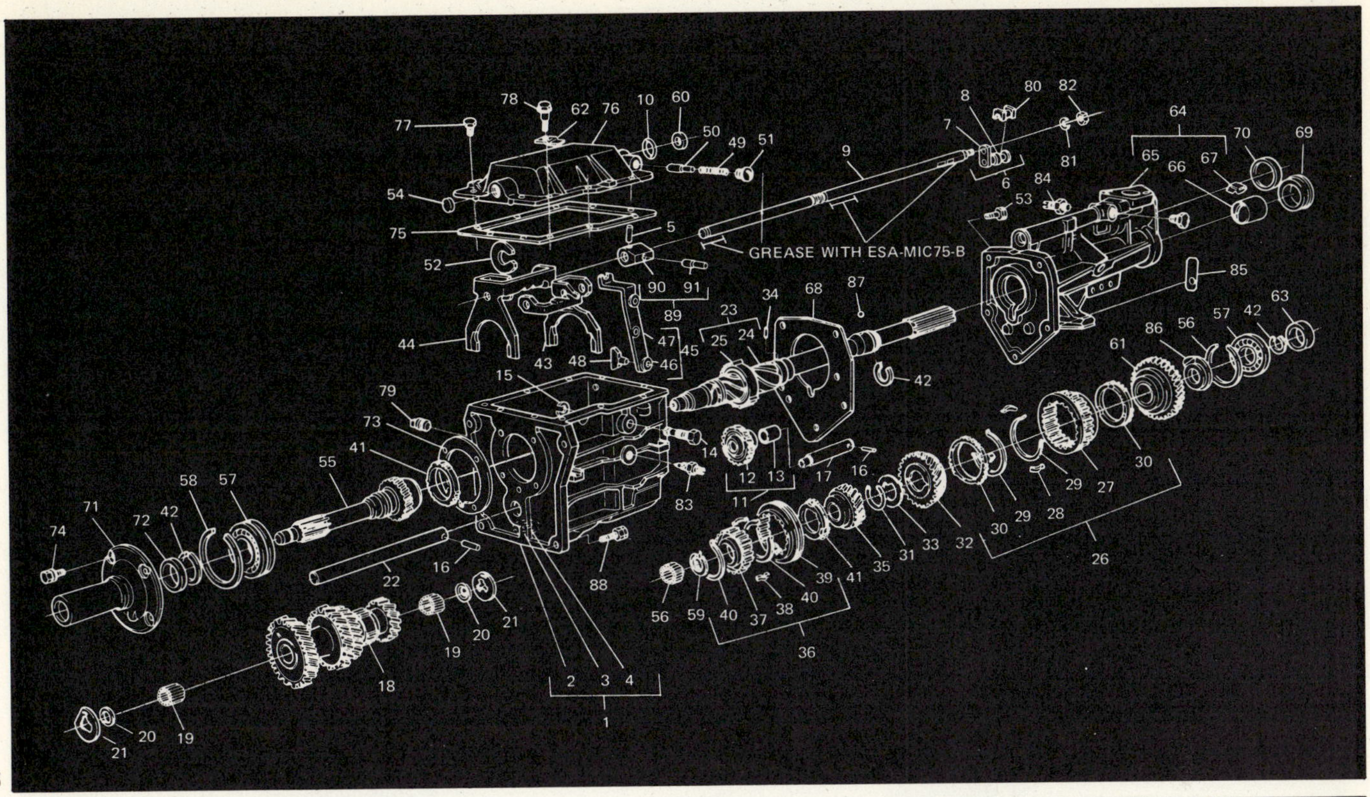

5

ITEM	PART NAME	ITEM	PART NAME
1.	CASE ASSY—TRANS.	47.	LEVER-TRANS REV. GEAR SH. RELAY
2.	CASE—TRANS.	48.	FORK-TRANS. REVERSE GEAR SHIFT
3.	MAGNET-TRANS. CASE CHIP	49.	SPRING-TRANS. SHIFTER INTERLOCK
4.	NUT. SPRING 9/64	50.	PLUNGER-TRANS. MESHLOCK
5.	PIN-3/16 DIA. X 13/16 ROLLED SPRING	51.	SCREW-M12 X 10 ROUND HEAD FLAT
6.	LEVER ASSY.-TRANS. GEARSHIFT SHAFT OFFSET	52.	PLATE-TRANS. GEAR SELECTOR INTERLOCK
7.	LEVER-TRANS. GEARSHIFT SHAFT OFFSET	53.	SCREW & WASHER ASSY. M10 X 30 HEX. HEAD
8.	PIN-TRANS. GEAR SHIFT SHAFT OFFSET LEVER	54.	PLUG-3/4 DIA. WELSH TYPE
9.	SHAFT-TRANS. SHIFTER	55.	SHAFT-TRANS. INPUT
10.	SEAL-O-RING	56.	ROLLER-TRANS. M'SHAFT BEARING
11.	GEAR & BUSH. ASSY.-TRANS. REV. IDLER SLIDING	57.	BEARING ASSY.-TRANS. M/D GEAR BALL
12.	GEAR-TRANS. REV. IDLER SLIDING	58.	RING-M/D GEAR BRG. RETAINING SNAP
13.	BUSHING-TRANS. REV. IDLER GEAR	59.	RING-1.00 RETAINING
14.	PIN-TRANS. REV. GEAR SELECTOR FORK PIVOT	60.	SEAL-TRANS. SHIFT SHAFT
15.	RING-7/16 RETAINING	61.	GEAR-TRANS. 1ST SPEED
16.	PIN-¼ X 1 SPRING	62.	CLIP-SPARK CONTROL SWITCH WIRE RET.
17.	SHAFT-TRANS. REVERSE IDLER GEAR	63.	GEAR-SPEEDOMETER DRIVE
18.	GEAR-TRANS. COUNTERSHAFT	64.	EXTENSION ASSY.-TRANS.
19.	ROLLER-TRANS. C'SHAFT GEAR BEARING	65.	EXTENSION-TRANS.
20.	WASHER-208/.918 FLAT	66.	BUSHING-TRANS. EXTENSION
21.	WASHER-TRANS. C'SHAFT GEAR THRUST	67.	STOP-TRANS. EAR SHIFT LEVER REVERSE
22.	COUNTERSHAFT-TRANSMISSION	68.	GASKET-TRANS. EXTENSION
23.	SHAFT ASSY.-TRANS. OUTPUT	69.	SEAL ASSY.—TRANS. EXTENSION OIL
24.	SHAFT-TRANS. OUTPUT	70.	PLUG-TRANS. EXTENSION
25.	HUB-TRANS. SYN. 1ST & 2ND GEAR CL.	71.	RETAINER-TRANS. INPUT SHAFT GEAR BRG.
26.	SHAFT & GEAR ASSY.-TRANS. OUTPUT	72.	SEAL ASSY.-TRANS. INPUT SHAFT OIL
27.	GEAR-TRANS. REVERSE SLIDING	73.	GASKET-TRANS. INPUT SHAFT BRG. RETAINER
28.	INSERT-TRANS. SYNCHRO. HUB	74.	BOLT M8 X 20 HEX HEAD-LOCK
29.	SPRING-TRANS. SYNCHRO. RETAINING	75.	GASKET-TRANS. CASE COVER
30.	RING-TRANS. SYNCHRO BLOCKING	76.	COVER-TRANS. CASE
31.	RING-TRANS-2ND SPEED GEAR RET. SNAP	77.	SCREW-M6 X 20 HEX. HEAD
32.	GEAR-TRANS. 2ND SPEED	78.	BOLT-M6 X 32 HEX. WASHER HD. SHOULDER
33.	WASHER-TRANS. 2ND SPEED GEAR THRUST	79.	PLUG-1/2-14 PIPE (FILLER)
34.	PIN-1/8 x 3/8 ROLLED SPRING	80.	BUSHING-TRANS. GEAR SHIFT DAMPER
35.	GEAR-TRANS. 3RD SPEED	81.	WASHER-SPRING LOCK
36.	SYNCHRONIZER ASSY.-3RD & 4th SPEED	82.	NUT-HEXAGON
37.	HUB-TRANS. SYNC. 3PD 84TH GEAR CLUTCH	83.	SWITCH ASS'Y.-BACK-UP LAMP
38.	INSERT-TRANS. SYNC. HUB	84.	SWITCH ASSY.-TRANS. SEAT BELT WARNING SENSOR
39.	SLEEVE-TRANS. 3RD & 4TH GEAR CLUTCH HUB	85.	TAG-TRANS. SERVICE IDENTIFICATION
40.	SPRING-TRANS. SYNCHRO. RETAINING	86.	WASHER-TRANS. 1ST GEAR THRUST
41.	RING-TRANS. SYNCHRO BLOCKING	87.	BALL-.25 DIA.
42.	RING-TRANS. M/D GEAR BRG. SHAFT-SNAP	88.	SCREW & LOCKWASHER ASSY. M12 X 40
43.	FORD-TRANS. 1ST & 2ND GEAR SHIFT	89.	ARM ASSY-TRANS. CONTROL SELECTOR
44.	FORK-TRANS. 3RD & 4TH GEAR SHIFT	90.	ARM-TRANS. CONTROL SELECTOR
45.	LEVER ASSY.-TRANS. REV. GEAR SH. RELAY	91.	PIN-TRANS. GEARSHIFT
46.	RING.-TRANS. REV. GEAR SH. RELAY LEVER		

General Motors Standard Transmissions

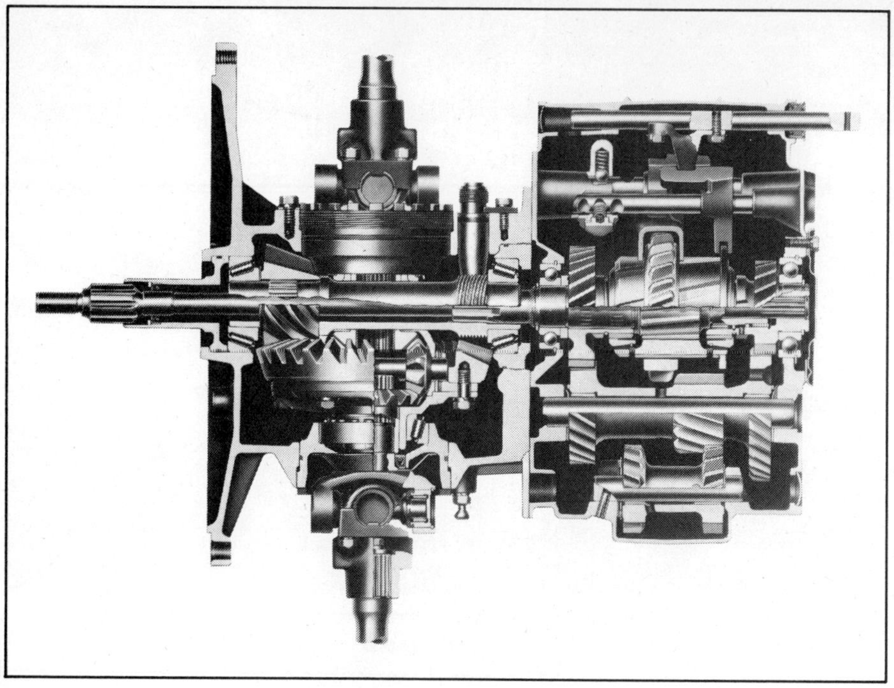

1

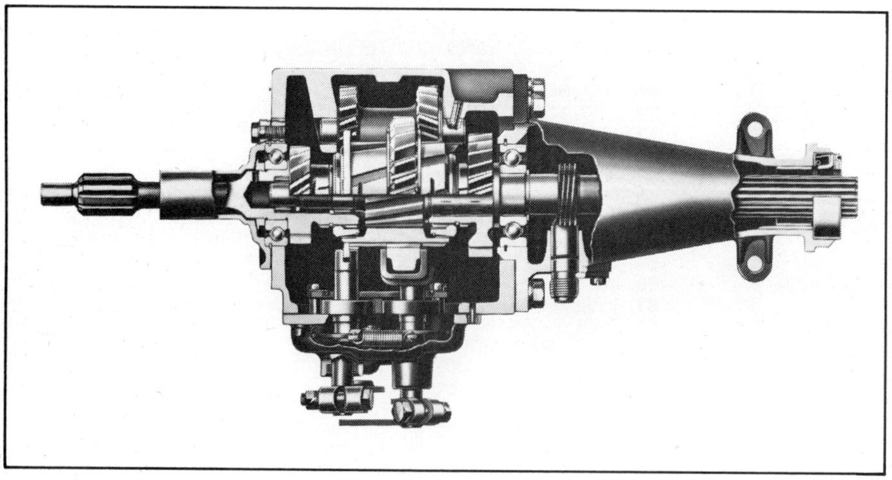

2

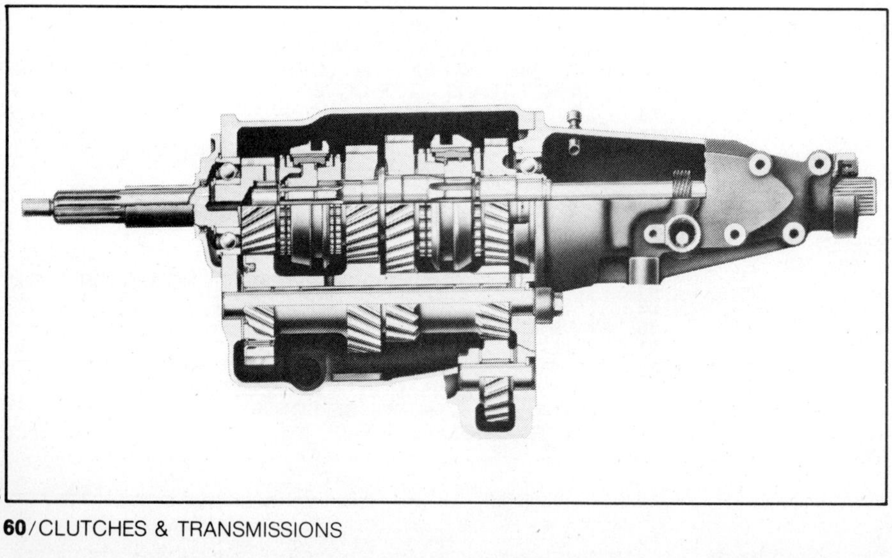

3

The present General Motors line of standard transmissions includes the Saginaw 3- and 4-speed and the Muncie 3- and 4-speed.

SAGINAW 3-SPEED

Provided as standard equipment in many General Motors products, the fully synchronized Saginaw 3-speed is manufactured with various gear ratios depending upon application. When used in the Vega, low gear is 3.11:1, second is 1.84:1, high is direct and reverse is 3.22:1.

For 6-cyl.-equipped vehicles such as Firebird, Nova and Camaro, the low gear ratio is 2.85:1, second is 1.68:1, third is direct and reverse is 2.95:1. On V-8's such as the LeMans and Grand Am, Camaro and Chevelle, a 2.54:1 low gear is used, with a second of 1.50:1, a direct third and a reverse of 2.63:1.

These transmissions incorporate all helical gears machined from drop-forged steel gear blanks, heat-treated and shot-peened for strength and long life. The shafts are machined from high-grade steel, heat-treated and ground to close limits. An anti-lash plate assembly at the front face of the countergear puts constant spring tension between the countergear and the main drive gear to reduce torsional vibrations.

The input shaft is supported by a heavy-duty ball bearing at the front of the transmission case piloted in the

1. Here's a cutaway view of a '60 Chevy Corvair 3-speed transaxle package. Mainshaft is hollow, permits passage of clutch shaft to front of trans then on to clutch gear. Clutch gear then drives a countergear with the balance of the power flow being the same as standard 3-speed manual.

2. The '58 3-speed has a smaller pilot bearing shaft when used with the 6. Tailshaft is also shorter than the one used on the earlier Chevy 3-speed.

3. The other 3-speed manual trans that Chevrolet introduced in mid-'65 was this heavy-duty model available for use with 396, 427-in. V-8's. Gear ratios were 2.41:1, 1st; 1.57:1, 2nd; 1.00:1, 3rd; 2.41:1, reverse.

4. This is the synchronizer assembly on the Saginaw 3-speed trans.

5. The Saginaw 3-speed manual has this anti-rattle plate mounted on the front of the countergear.

6. This is an exploded view of a late '50's, 3-speed Chevrolet trans used on their 6-cylinder models.

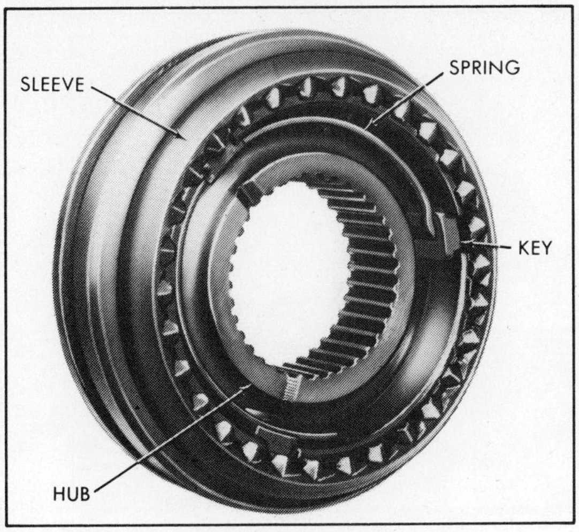

4

5

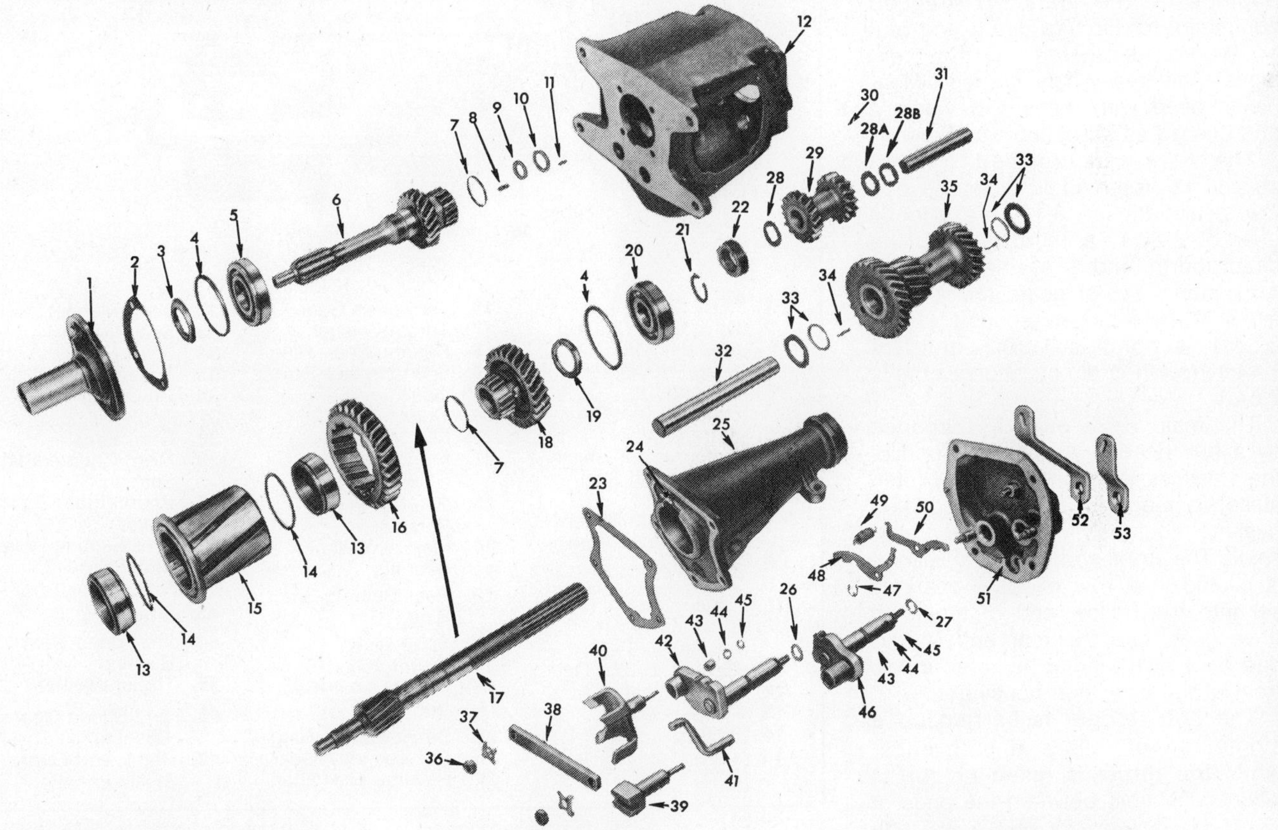

1. Clutch Gear Bearing Retainer
2. Bearing Retainer Gasket
3. Bearing Nut and Oil Slinger
4. Bearing Snap Ring
5. Clutch Gear Bearing
6. Clutch Gear
7. Energizing Spring
8. Front Pilot Bearing Roller
9. Thrust Washer
10. Thrust Washer
11. Rear Pilot Bearing Rollers
12. Transmission Case
13. Synchronizer Ring
14. Snap Ring
15. Second and Third Speed Clutch
16. First and Reverse Sliding Gear
17. Mainshaft
18. Second Speed Gear
19. Thrust Washer
20. Mainshaft Rear Bearing
21. Snap Ring
22. Speedometer Drive Gear
23. Extension Gasket
24. Rear Bearing Snap Ring
25. Extension
26. First and Reverse Shifter Shaft "O" Ring
27. Second and Third Shifter Shaft "O" Ring
28. Thrust Washer

28a. Thrust Bearing

28b. Thrust Bearing Washer

29. Reverse Idler Gear
30. Reverse Idler Shaft Pin
31. Reverse Idler Shaft
32. Countershaft
33. Countergear and Roller Thrust Washers
34. Bearing Roller
35. Countergear
36. Shifter Interlock Retainer Stud Nut
37. Shifter Interlock Retainer Stud Nut Lock
38. Shifter Interlock Retainer
39. Second and Third Shifter Fork
40. First and Reverse Shifter Fork
41. Shifter Interlock
42. First and Reverse Shifter Shaft and Plate Ass'y
43. Shifter Fork Roller
44. Shifter Fork Washer
45. Shifter Fork Retainer
46. Second and Third Shifter Shaft and Plate Ass'y
47. Detent Cam Retainer
48. First and Reverse Detent Cam
49. Detent Cam Spring
50. Second and Third Detent Cam
51. Side Cover
52. First and Reverse Shifter Lever (Outer)
53. Second and Third Shifter Lever (Outer)

6

General Motors

flywheel by an oil-impregnated bushing on the 6-cyl. version and by a roller bearing on the V-8's.

MUNCIE 3-SPEED

The 3-speed Muncie transmission differs from the Saginaw design primarily in the extra ¼-in. center distance between the mainshaft and countershaft, and the use of larger bearings, gears, input shaft and main shaft; but power flow is essentially the same for both transmissions. Gear ratios provided are 2.42:1 in low, 1.58:1 in second, a direct high and a reverse of 2.41:1.

SAGINAW 4-SPEED

A medium-duty transmission, the Saginaw is the standard 4-speed equipment for Firebird, Nova and other GM models ordered with 8-cyl. engines. The heavy-duty 4-speed Muncie is used with some V-8 versions and offered as an option with most.

The fully synchronized forward speeds in Vega applications have a low gear ratio of 3.11:1, a second gear of 2.20:1, a third of 1.47:1, a direct fourth, and a reverse of 3.11:1. An optional set of gears for V-8 engines is more closely spaced; low is 2.54:1, second is 1.80:1, third is 1.44:1, fourth is direct, and reverse is 2.54:1.

The main drive gear is supported by a ball bearing at the front end of the transmission case and is piloted at its front end in an oil-impregnated bushing mounted in the engine crankshaft. The front end of the mainshaft is piloted in a row of roller bearings set into the hollow end of the main drive gear, and the rear end is carried by a ball bearing mounted in the front of the extension housing.

The countergear is carried on a single row of rollers at both ends, while the thrust is taken on thrust washers located between the ends of

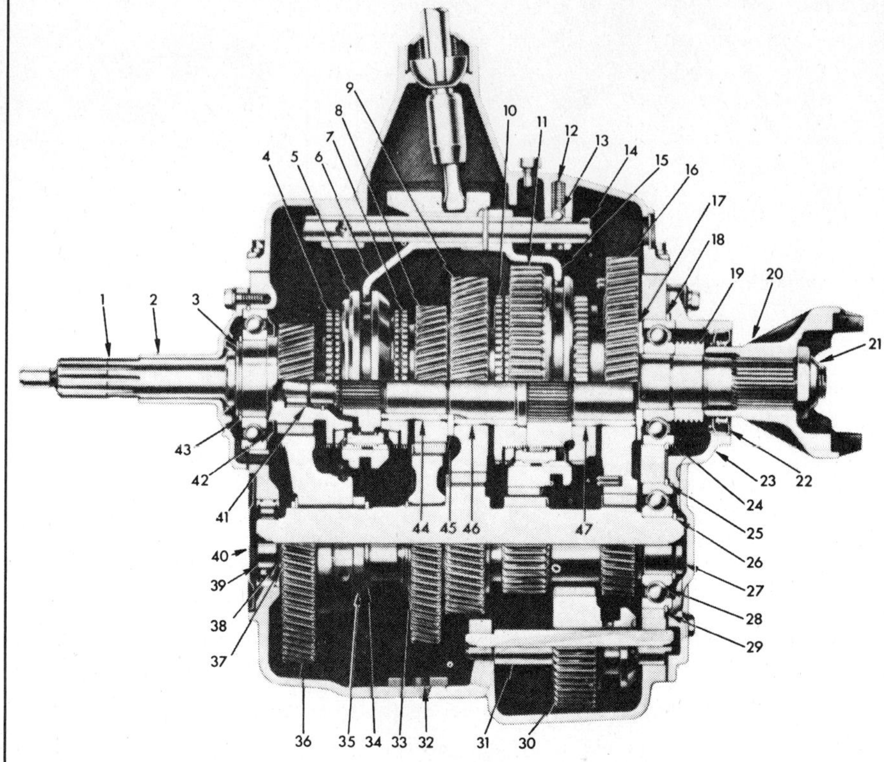

1. Main Drive Gear
2. Drive Gear Bearing Retainer
3. Snap Ring-Outer
4. 3rd and 4th Synchronizer Ring
5. 3rd and 4th Synchronizer Collar
6. 3rd and 4th Shift Fork
7. 3rd and 4th Speed Synchronizer Ring
8. 3rd Speed Gear
9. 2nd Speed Gear
10. 1st and 2nd Synchronizer Assembly
11. Reverse Driven Gear
12. Poppet Spring
13. Poppet Ball
14. Shift Rail
15. 1st and 2nd Shift Fork
16. 1st Speed Gear
17. Thrust Washer
18. Bearing Snap Ring
19. Speedometer Drive Gear
20. Output Yoke
21. Flange Nut
22. Rear Bearing Retainer Oil Seal
23. Rear Bearing Retainer
24. Mainshaft Rear Bearing
25. Rear Bearing Snap Ring
26. Snap Ring
27. Countershaft
28. Countershaft Rear Bearing
29. Bearing Snap Ring
30. Reverse Idler Gear
31. Reverse Idler Shaft
32. Case Magnet
33. Snap Ring
34. Snap Ring
35. Spacer
36. Countergear
37. Thrust Washer
38. Snap Ring
39. Front Countershaft Bearing
40. Countergear Front Cover
41. Pilot Bearing Rollers
42. Clutch Gear Oil Slinger
43. Snap Ring
44. 3rd Speed Gear Bushing
45. Thrust Washer
46. 2nd Speed Gear Bushing
47. 1st Speed Gear Bushing

1

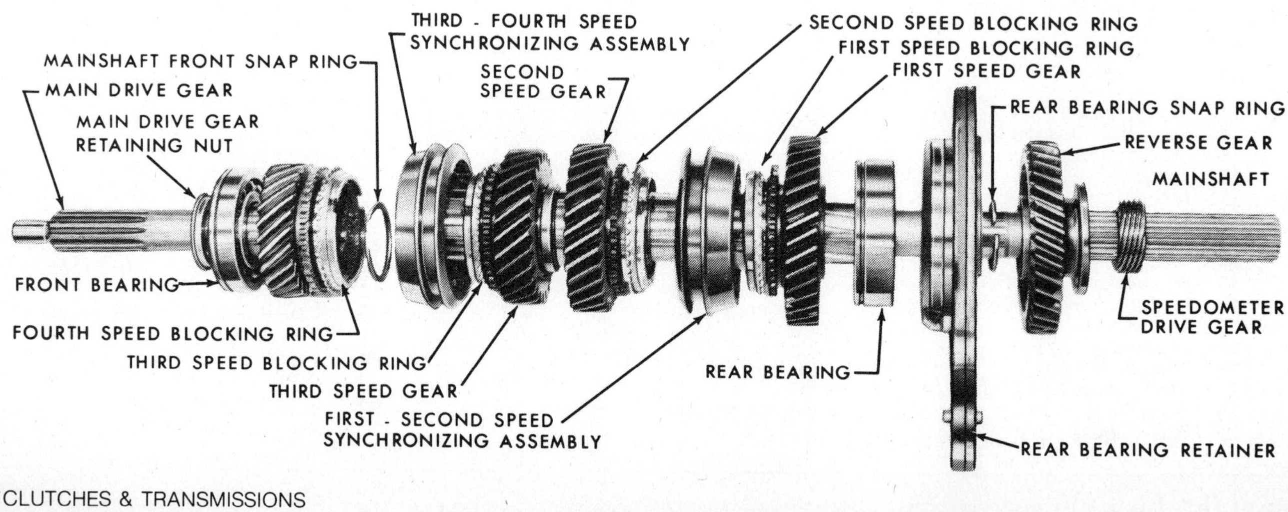

2

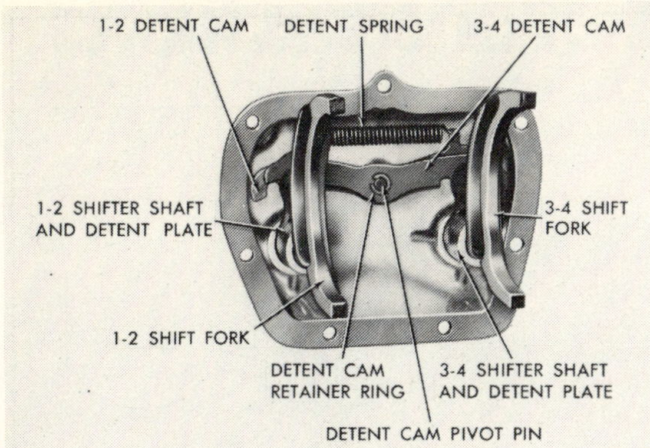

3

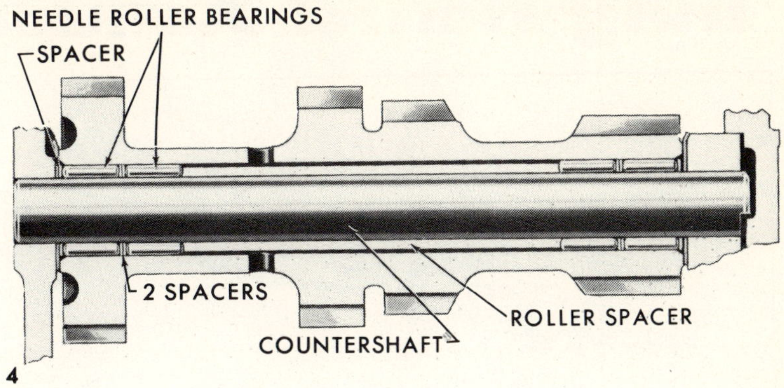

4

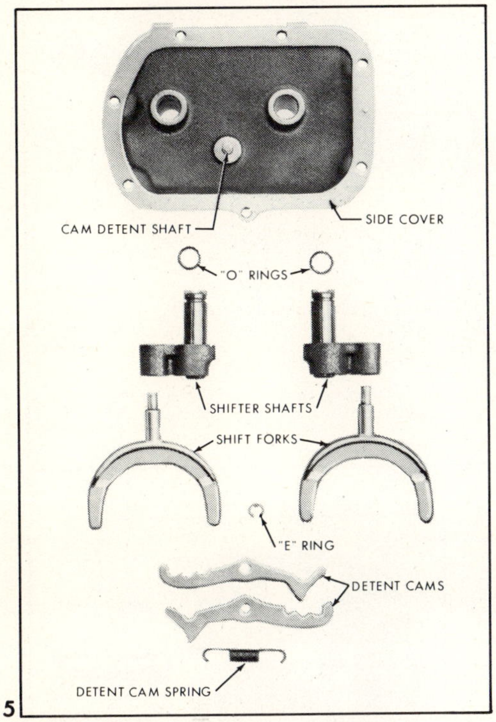

5

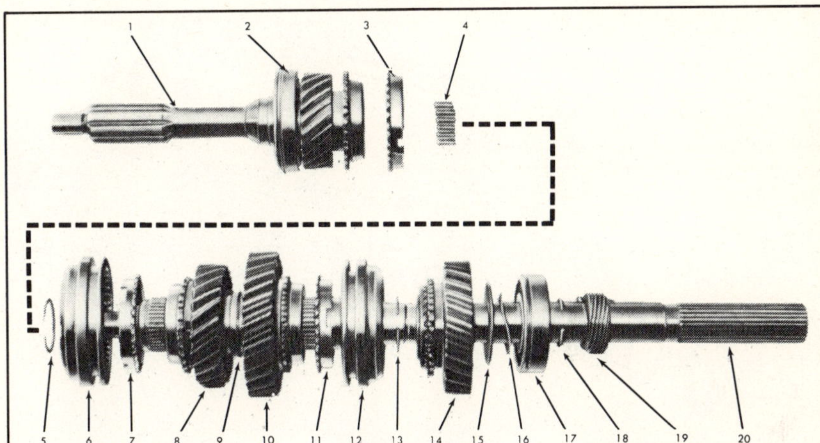

1. Drive Gear
2. Drive Gear Bearing
3. 3rd Speed Blocker Ring
4. Mainshaft Pilot Bearings (14)
5. Snap Ring
6. 2-3 Synchronizer Assembly
7. 2nd Speed blocker Ring
8. 2nd Speed Gear
9. Shoulder (Part of Main Shaft)
10. 1st Speed Gear
11. 1st Speed Blocker Ring
12. 1st Speed Synchronizer Assembly
13. Snap Ring
14. Reverse Gear
15. Reverse Gear Thrust Washer
16. Spring Washer
17. Rear Bearing
18. Snap Ring
19. Speedo Drive Gear and Clip
20. Mainshaft

6

1. The 4-speed Muncie transmission is a big, hefty unit. Hefty enough, in fact, that it's used in Chevy, GMC 4-wheel drive models. When it is, rear yoke (#20 in illustration) is eliminated and shorter output shaft substituted, so transfer case can be bolted right up to the case.

2. This exploded view of a Muncie mainshaft assembly gives the reader an insight into the complexity of a 4-speed transmission. Muncies are quite bulletproof, provide close-ratio gearing for minimum rev loss during upshifting of trans.

3. The Muncie side cover assembly looks like this. If shifting forks, cams, springs or plates are damaged, cover can be pulled without having to remove the trans from the car.

4. Long, needle roller bearings are used to support the countershaft.

5. Exploded view of the side cover assembly on a 3-speed Saginaw shows relation of the shifting mechanism.

6. Here's an exploded-view look at the clutch gear and mainshaft assembly of the 3-speed Saginaw transmission.

the gear and the thrust bosses in the case.

The first-second gear synchronizer sleeve has gear teeth on the outside diameter to let it act as a reverse gear. To shift into reverse, the sliding reverse idler and reverse gear on the countergear must be engaged.

The cover is located on the left-hand side of the transmission case and can be easily distinguished from the Muncie 4-speed by the reverse lever mounted on the side cover rather than in the extension housing.

MUNCIE 4-SPEED

The Muncie 4-speed manual transmission consists of two basic sections; the transmission case, or forward section, and the case extension, or rear section. The forward section contains the four forward speed gear assemblies and the synchronizing mechanisms; the rear section conains the reverse gear assembly. All forward speeds are provided with synchronizing clutches which can be engaged while the car is in motion.

In 8-cyl. Pontiacs, Chevrolets and Oldsmobiles equipped with the 4-speed box, low gear is 2.52:1, second is 1.88:1, third is 1.46:1, high is direct, and reverse is 2.59:1. An additional set of ratios is supplied in cars such as the Camaro Z28 and Corvette, with a low gear of 2.20:1, a second of 1.64:1, a third of 1.27:1, a direct high, and a reverse of 2.26:1.

The front end of the mainshaft is piloted in roller bearings set into the hollow end of the main drive gear. The rear end is carried by a heavy-duty ball bearing located at the rear end of the transmission case.

The countergear is carried on a double row of roller bearings positioned at both ends of the gear, with thrust taken on thrust washers at the front and rear of the gear.

The sliding reverse gear is a spur instead of helical design, located behind the rear bearing and splined to the mainshaft. When reverse gear slides forward for engagement, it is driven by the reverse idler gear, in constant mesh with the countergear.

Borg-Warner Standard Transmissions

1

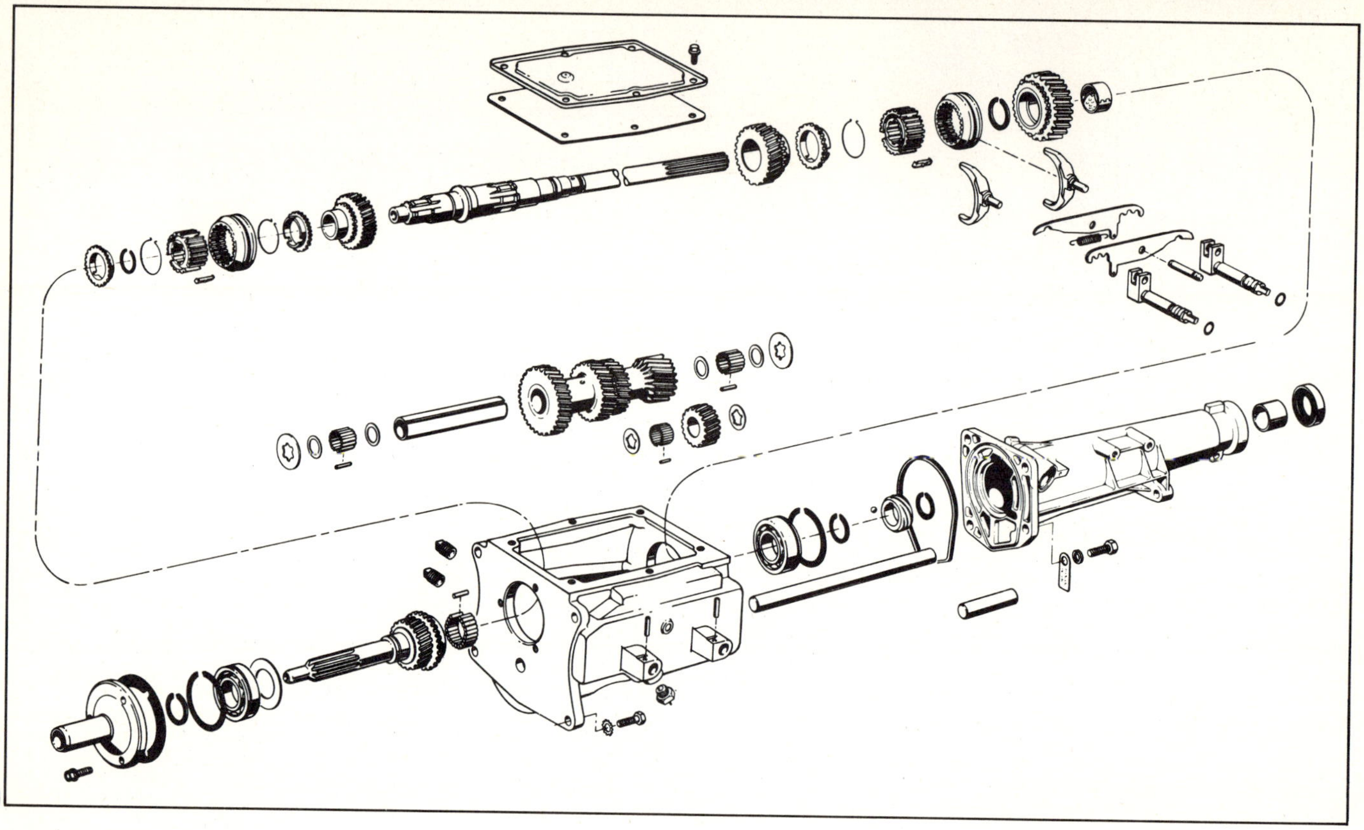

A long-time supplier of standard transmissions to the automotive industry, Borg-Warner's manual boxes are now used only in American Motors cars. The recent automakers' trend toward manufacture of their own has reduced the size of its market, but over the years, Borg-Warner has made innumerable types of transmissions for mass production and special purposes alike. The following were selected for description, as their current production and use is testimony to their quality and efficiency.

T-14 3-SPEED

Used in AMC series 01, 10, 40, and 70 cars, this medium-duty 3-speed has all forward gears synchronized. Gear shifting is accomplished through shifter rods and floor or column mounted shift levers. The low gear ratio is 2.64:1, second is 1.60:1, high is a direct 1.00:1, and reverse is 2.64:1.

Ball bearings support the input shaft and gear and the center of the output shaft. Roller bearings in the input shaft bore support the front of the output shaft. The countershaft runs on two rows of roller bearings.

The shift forks, shift rails, detent mechanism and related parts are in the transmission case casting, with a pressed steel cover on the top of the case providing access to the gears.

T-10 4-SPEED

Introduced in 1957 for use in the Corvette, the T-10 was an instant success. Having been used at one time or another by all of the Big Three automakers when a heavy-duty 4-speed box was required, it is basically a modified T-85 3-speed with reverse gear moved into the extension housing to make room for the fourth gearset. Its longevity in use attests to the quality of the basic T-10 design.

When used with Ford products, there were three different sets of ratios. In the Fairlanes, Falcons, and Meteors, low gear is 2.73:1, second is 2.06:1, third is 1.62:1, high is direct and reverse is 2.80:1. The Ford had a low of 2.36:1, a second of 1.76:1, a third of 1.41:1, a direct high, and a reverse of 2.42:1. In the Galaxie with the 427-cu.-in. engine and the Shelby Cobra, the low gear is 2.36:1, second is 1.62:1, third is 1.20:1, high is 1.00:1, and reverse is 2.42:1.

Chrysler Corp. have their own 4-speed but have used the T-10 in a great number of applications over the years. Most common usage called for a low gear of 2.54:1, a second of 1.89:1, a third of 1.51:1, a direct high, and a reverse of 2.61:1. But there was also a set of ratios that came to be known as "close ratio" with a low of 2.20:1, a second of

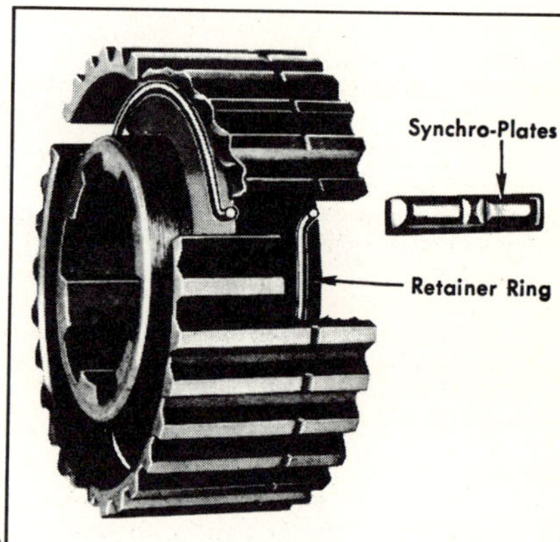

2

1. Schematic drawing of disassembled B-W T-14 3-speed, showing relationship of components. The once widely used gearbox is now found only in American Motors products.

2. The T-14's synchro-plate and location of the retainer ring.

3. Countergear is supported by needle bearings at both ends. The plate on the large gear houses anti-rattle springs that take up backlash in the gears when they are turning without a load. This feature reduces noise level.

4. Plan view of T-10 4-speed shows the placement of the forward speeds in the main case and the reverse gear in the extension housing.

5. The T-10 4-speed is a complex unit when disassembled, as revealed in this schematic breakdown.

3

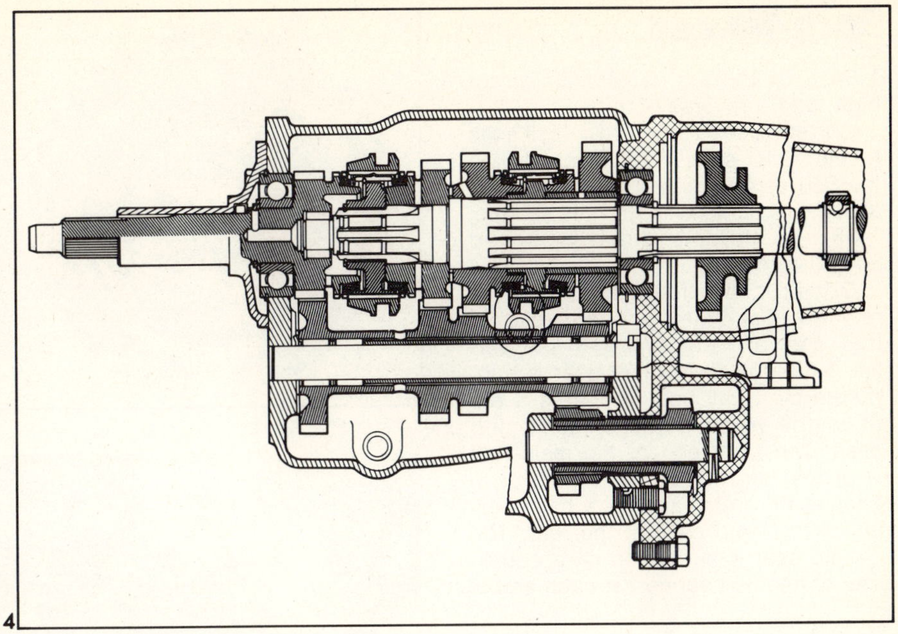

4

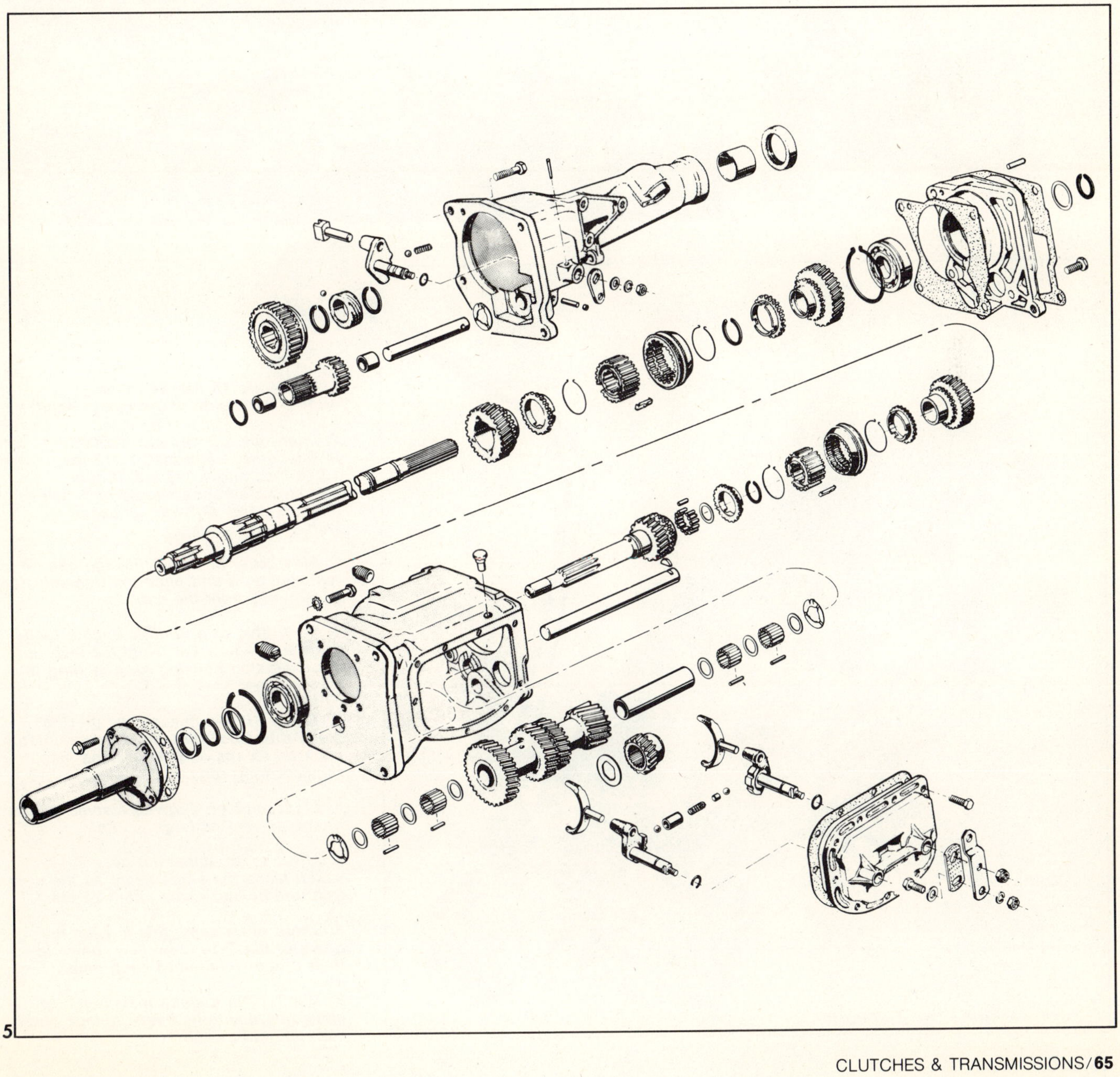

5

Borg-Warner

1.64:1, a third of 1.31:1, a direct high, and a reverse of 2.26:1.

Current use of the T-10 is limited to the AMC series 70 equipped with the 360 or 401-in. V-8. Its gear ratios are 2.23:1 in low, 1.77:1 in second, 1.35:1 in third, a direct high and 2.16:1 in reverse.

Construction features have remained essentially unchanged over the years and are similar to the earlier T-85—the main drive gear is supported by a ball bearing that is a slip fit in the front of the transmission case. The front end of the mainshaft is piloted on rollers in the main drive gear at the front, and by a ball bearing in the main case at the rear. The countergear is supported on a double row of needle bearings at each end.

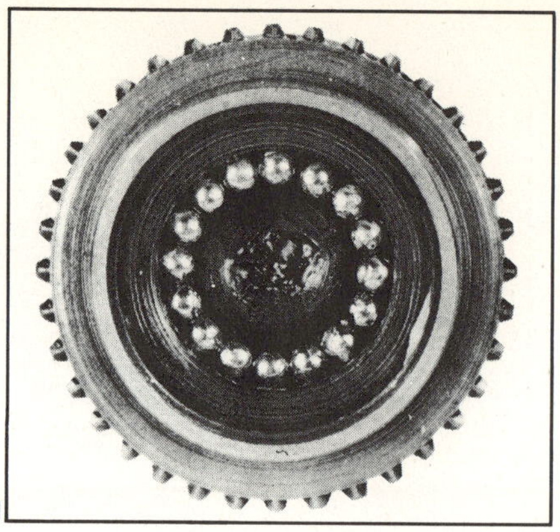
1

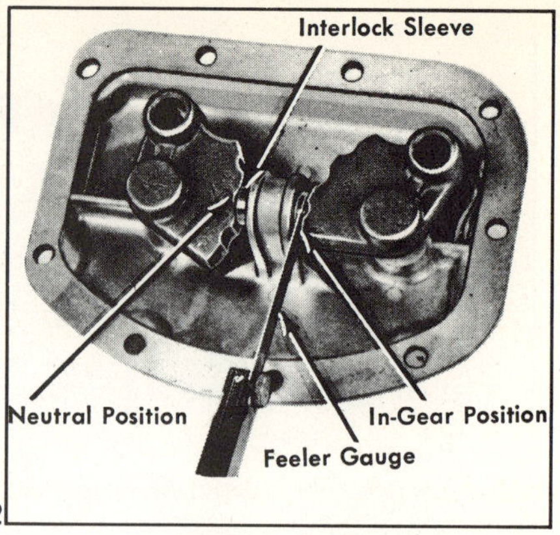

2

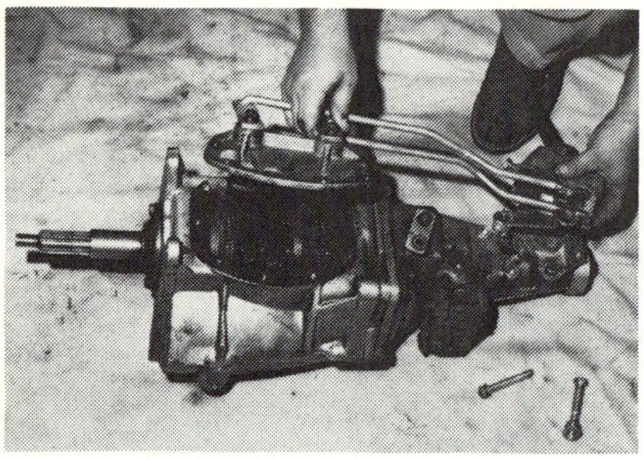
3

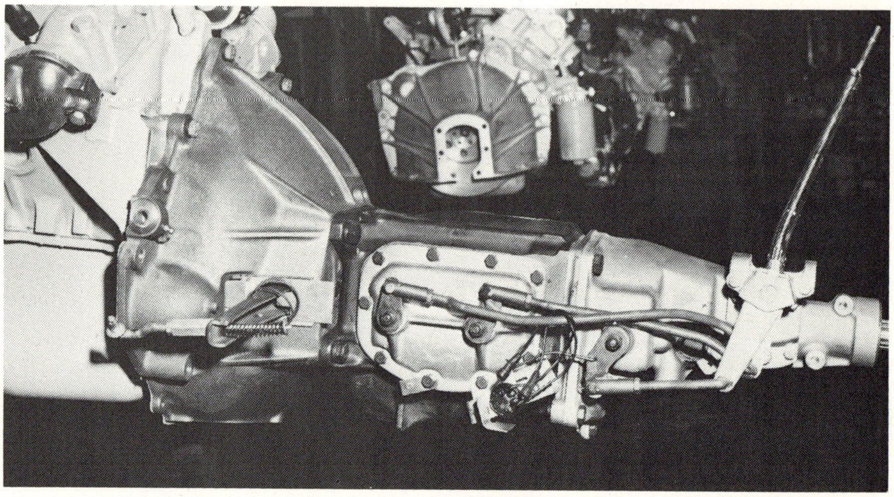
4

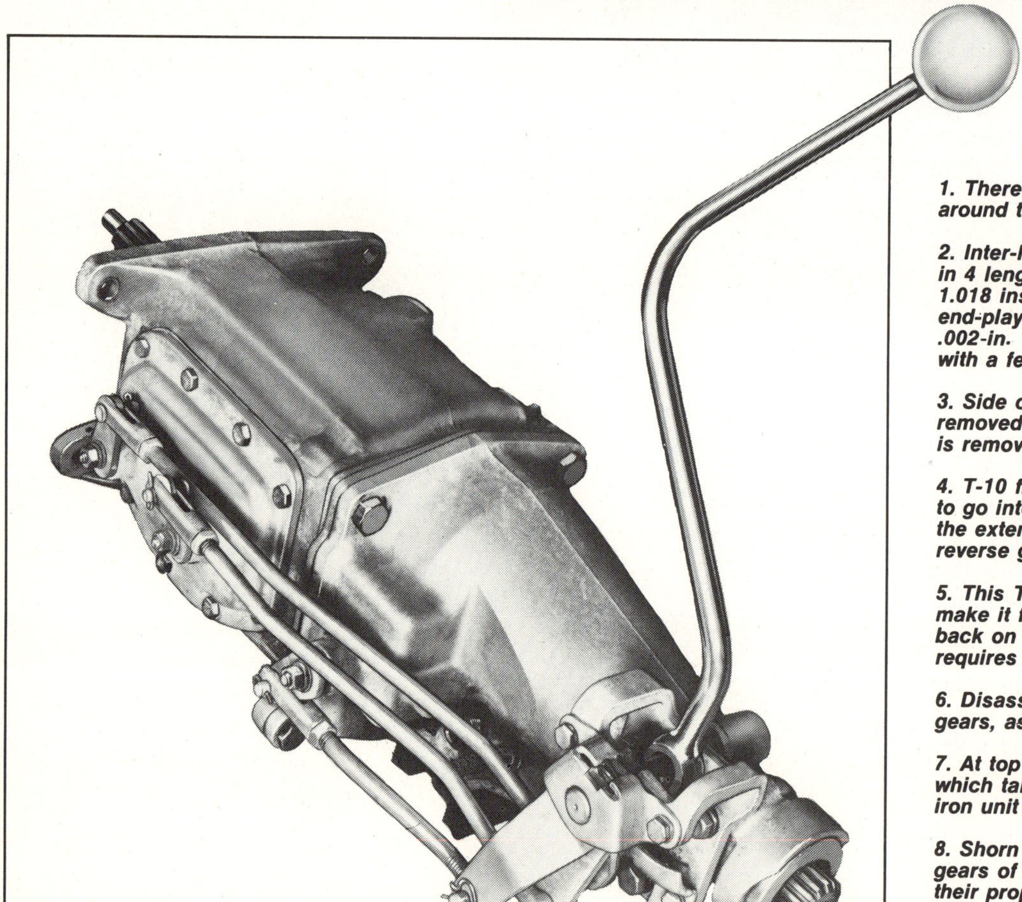
5

1. There are 16 needle rollers around the inside of the clutch shaft.

2. Inter-lock sleeves are available in 4 lengths; 1.008 ins., 1.013 ins., 1.018 ins., and 1.023 ins. Proper end-play clearance should be between .002-in. and .008-in., checked with a feeler gauge.

3. Side cover and floor linkage can be removed as a unit once the transmission is removed from the car.

4. T-10 fitted to a Pontiac engine ready to go into a new car. Note the lever on the extension housing for engaging the reverse gear.

5. This T-10 has linkage required to make it fit Chevrolet. Lever is mounted back on extension housing and also requires long rods.

6. Disassembled 4-speed reveals rugged gears, as well as large shafts.

7. At top is an aluminum-case T-10, which takes the same gears as the cast iron unit below, saves about 25 lbs.

8. Shorn of its case, this is what the gears of the T-10 look like, shown in their proper relation to each other.

9. The T-10 in a stock installation is shown here with its beefy linkage and backup light switch.

6

7

8

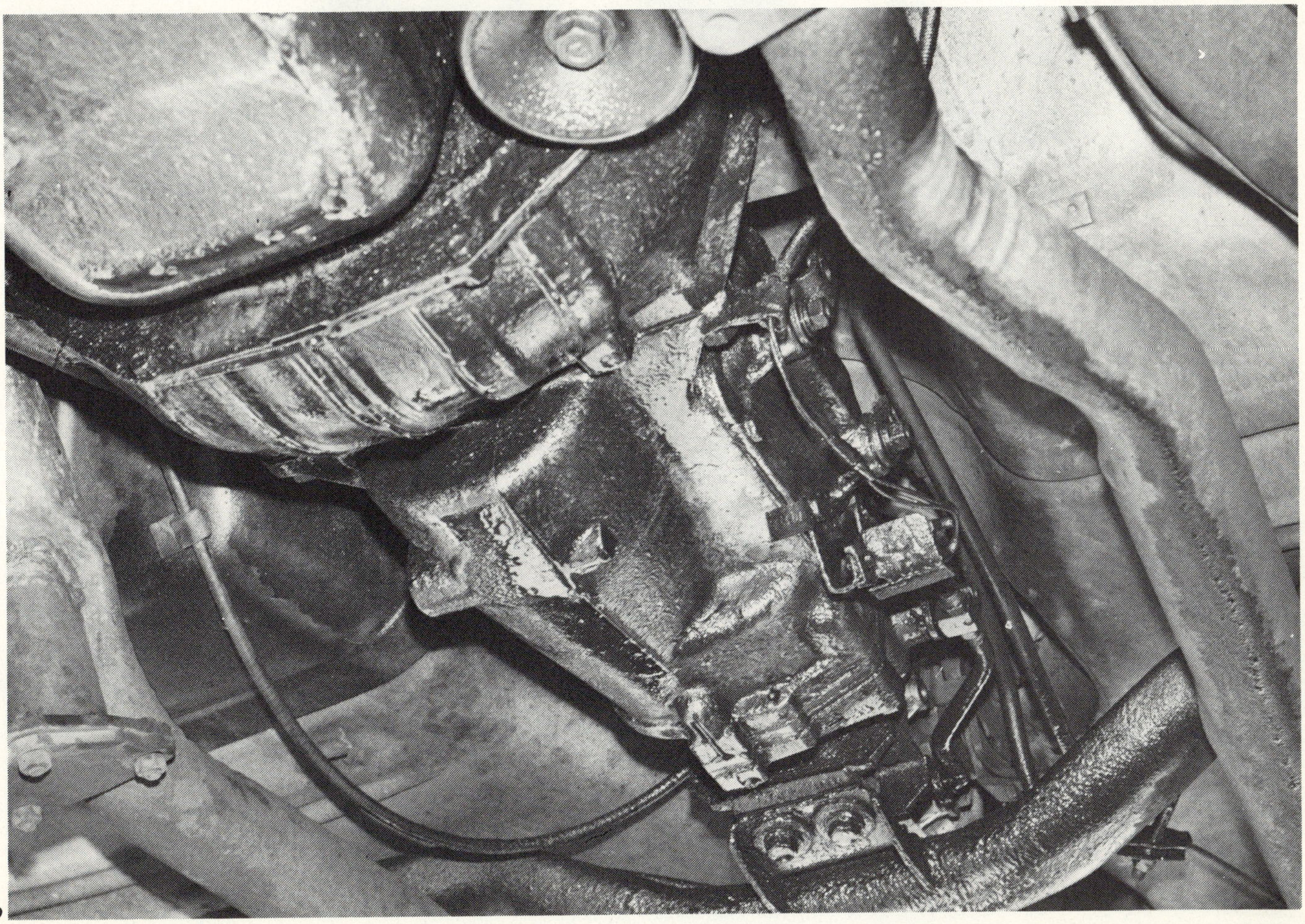

9

Removal and Repair of Standard Transmissions

New car customers have shown a sufficient resurgence of interest in the standard box during the past couple of years to bring forth several new ones since our last edition. Now, with the new emphasis on smaller cars and economical operation brought about by the current energy crunch (which will most likely be with us for some time to come), you can look for even more interest, particularly in the Pinto/Vega class.

While manual transmission design has not changed materially over the past few years, manufacturers have made their standard boxes more reliable and it's not uncommon these days to find mechanics whose sole experience with stick shifts has been that of playing with an old T-10 or Muncie 4-speed in the jalopy they're building at home.

Manual transmission work is not the cleanest job in the world, but it's interesting and really quite easy, as it's straightforward. The hardest part is getting the transmission out and back in the car. In between, it's pretty much a case of muscling the gearbox over to the bench, tearing it down and getting it back into the car.

Most manual transmission malfunctions are obvious. If a gear has lost a tooth or two, you replace it with a new gear. If there's too much play in the synchros, you replace the brass rings. A constant speed gear that's sloppy on its shaft either needs a new bushing or a replacement. Any slop between the cluster gear and the case is solved with new thrust washers. A careful inspection will tell you what needs to be replaced and what is okay as is.

HOW-TO: Removal Procedures

1. While draining transmission lubricant, chalk mark the U-joint as shown to help you get it back in its original position. Inspect U-joint after removal.

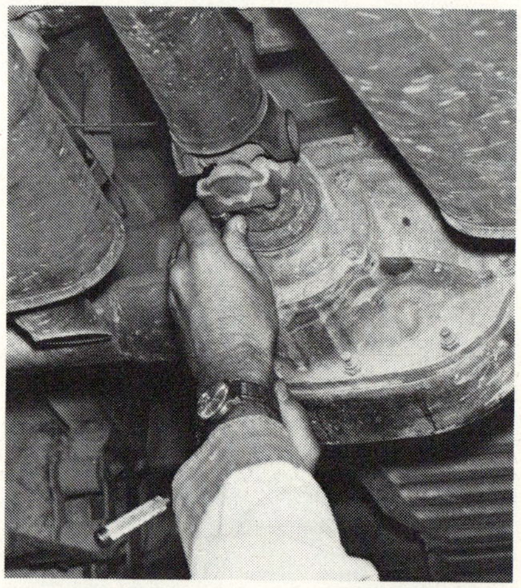

2. Four nuts and two U-bolts complete removal of rear U-joint assembly. Be careful not to drop the bearing cups—no point in goofing up this soon.

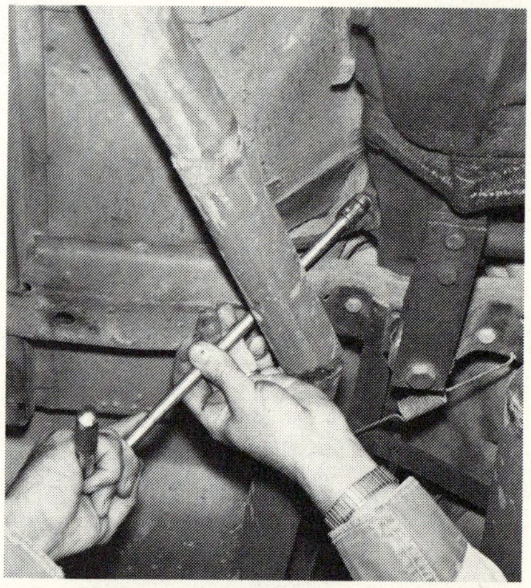

3. One bolt holds speedometer cable in place. Custom exhaust pipes are so close together that they must be removed before trans can be lowered.

4. You may have to remove exhaust pipes to gain clearance for lowering the trans or to permit lowering back of engine for access to bellhousing.

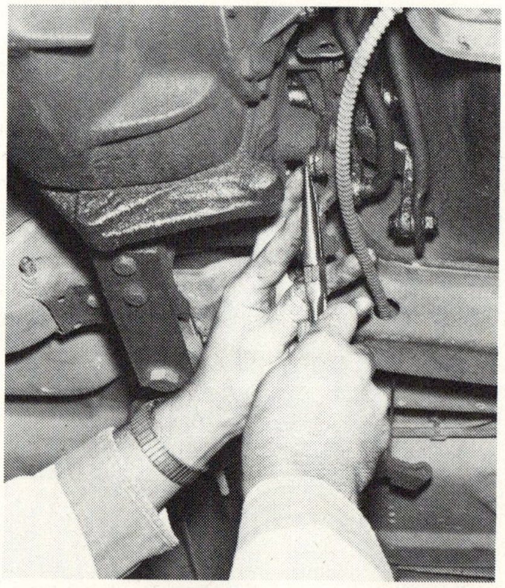

5. Disconnect shift linkage at transmission; what you disconnect will depend upon make and model of car. On floor shifts, stick may come out from top.

6. Disconnect shifter bracket where it bolts to the trans case. Even if bracket doesn't get in your way, this makes handling the trans easier.

The one troubleshooting area that is a little tricky is noisy gears. But if you clean up the gears and check the faces of the teeth for wear or chipping, you can usually spot the cause of the noise.

If you're not sure that your transmission problem will be obvious when you get the box apart, a good way to check is to remove the cover after you have the box on the bench and flush the inside of the case and gears with solvent. Once the grease is out and the transmission relatively clean, you'll be able to feel any play in the gears by moving them back and forth with your hands. If in doubt, replace the suspected part and check the play again. But don't be surprised if you find that your money and time were wasted; it's possible that the new part will have just as much play as the old one you replaced. As some transmission designs have excess play in certain areas without affecting operation of the box adversely, this is where experience in transmission work pays off.

Our standard transmission overhauls begin with a removal How-To on a Ford 4-speed in a Ranchero. This will give you some idea of what you're up against, although the exact amount of work (and difficulty) involved in removing a standard box depends a great deal upon the particular installation—some come out easily while others require that you unbolt half the underpinnings.

From that point, we show you how to overhaul the popular Pinto 4-speed, as well as the Ford 3-speed RAN top-loader. And if Vega is your thing, we've got a treat for you in the form of a '72 Vega box. Incidentally, Vega used the Buick/Opel "bottom-loader" trans during 1971-72 before switching to the standard Saginaw 3-speed in '73—and we've covered both for you.

Those who swear by the Warner T-10 will find an overhaul How-To for them, and we've even included the Muncie 4-speed to make the rest of you content. And to round things off, there's the same treatment given the Chrysler A-230. So, regardless of which make of standard box you call your own, you'll find it (or something very close) in the following pages—a treasurehouse of technique for the do-it-yourself transmission specialist.

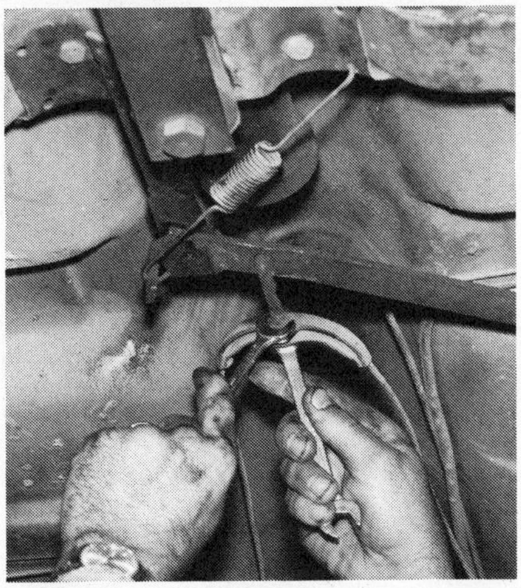

7. You may not have to disconnect all of emergency brake apparatus; adjustment is loosened to relieve some of the return spring pressure on crossmember.

8. Check engine support to see that it's adequate before you move on to the crossmember bolts; if this step is overlooked, it can result in injury.

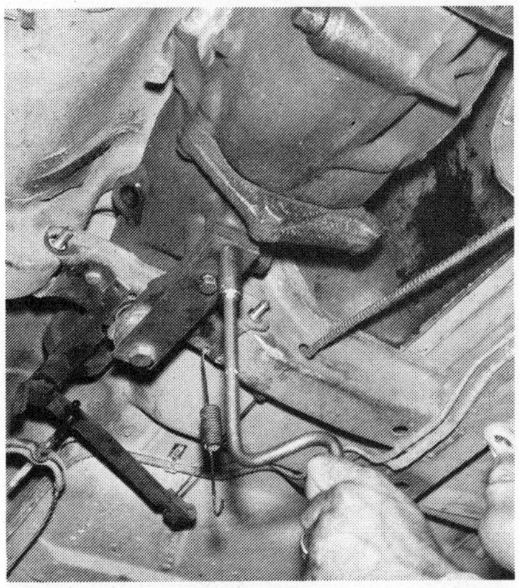

9. Remove crossmember nuts and two bolts that go into the trans case. Emergency brake spring is unhooked here but all else is on crossmember.

10. When those crossmember bolts come out, everything comes down as a unit. This means that if the engine isn't properly supported, trans could fall.

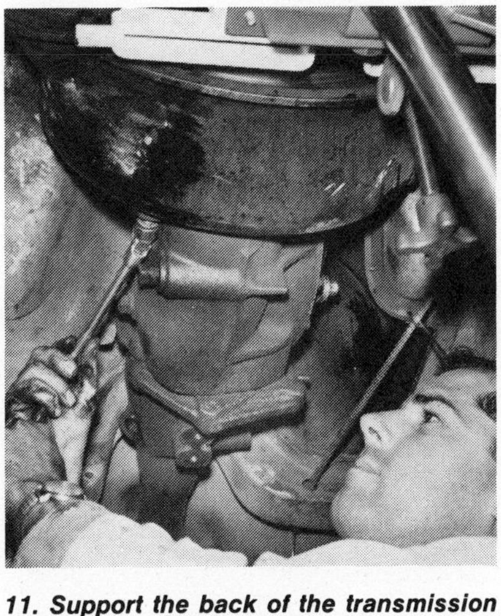

11. Support the back of the transmission so that its weight doesn't spring the clutch disc, then remove the bolts holding trans to bellhousing.

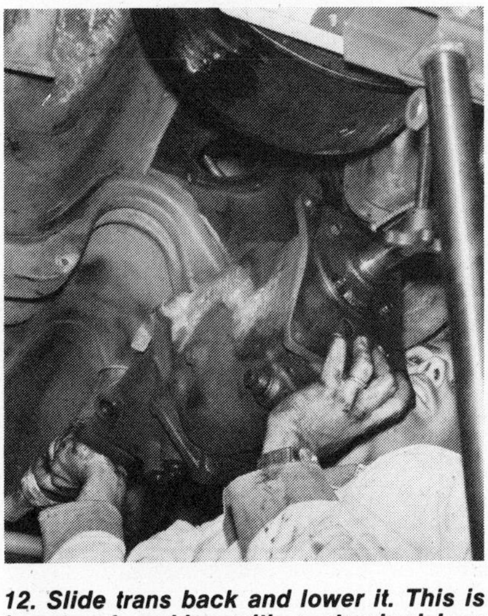

12. Slide trans back and lower it. This is beauty of working with a standard box; with an automatic, you need a trans stand to avoid a back sprain.

HOW-TO: Warner T-10 Overhaul

1. The T-10 is a side-cover design; remove 9 cap screws and cover lifts off bringing shift forks with it. Forks can be replaced without dropping trans.

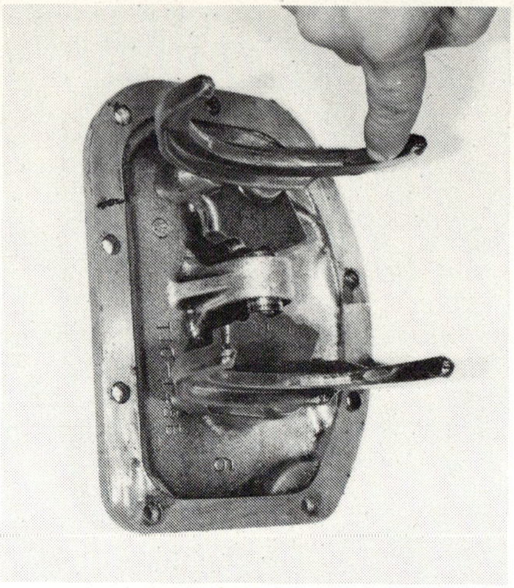

2. Inspect fork faces for wear; T-10 forks are identical and interchangeable in theory, but install them as they came out when replacing worn ones.

3. Unbolt the input shaft bearing retainer and slide it off the input shaft. Then remove the drift pin holding reverse lever shaft in place.

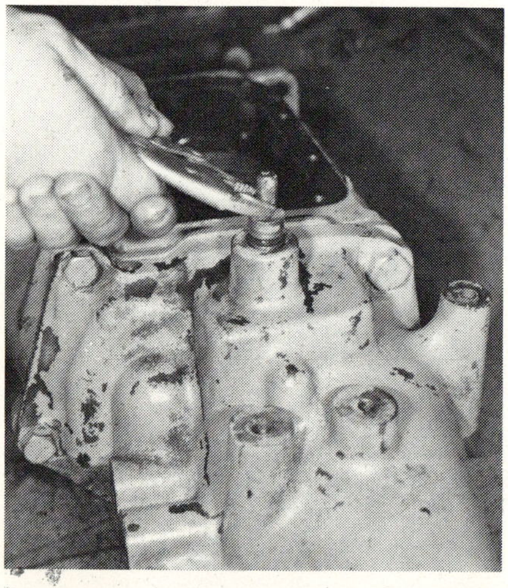

4. Pull the reverse shift shaft out about ⅛-inch to disengage the reverse fork lever from reverse sliding gear, then unbolt the tailshaft housing.

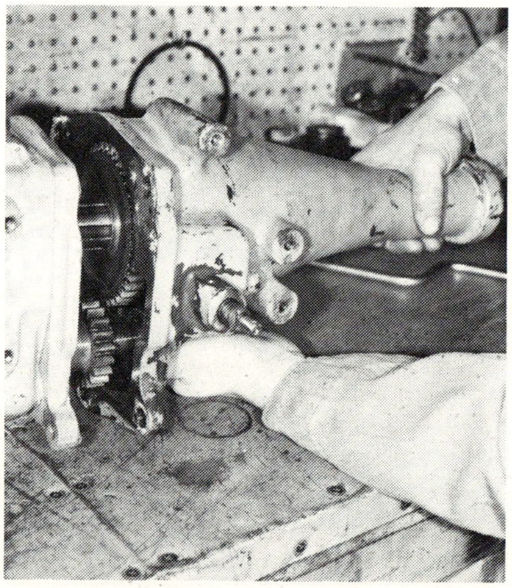

5. Tap tailshaft housing loose with soft mallet. Rotate housing and pull on reverse lever to let fork clear reverse gear as housing comes off.

6. Rotate and pull forward on reverse rear idler gear to remove from case. Then remove the self-locking bolt that holds the adapter plate to the case.

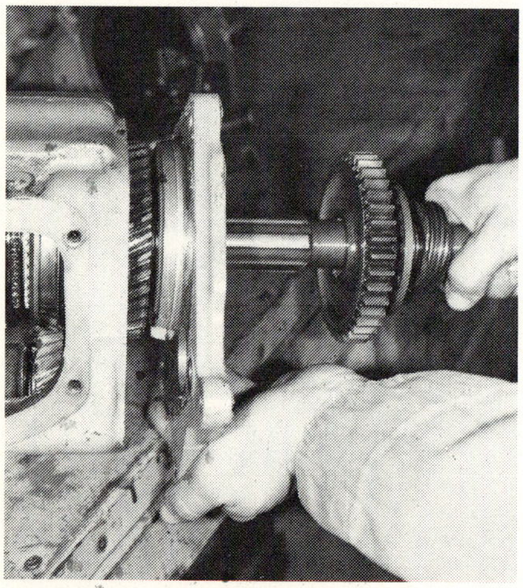

7. Entire mainshaft assembly can now be withdrawn from the case. The mainshaft needle bearings will fall into the trans case and should be retrieved.

8. Snap ring holding input shaft bearings must be removed, then flat steel washer comes off. You may have to tap shaft with a mallet to free it.

9. Once the bearing is loosened, the input shaft can be withdrawn from the case. Use care here, as parts can be damaged by careless handling.

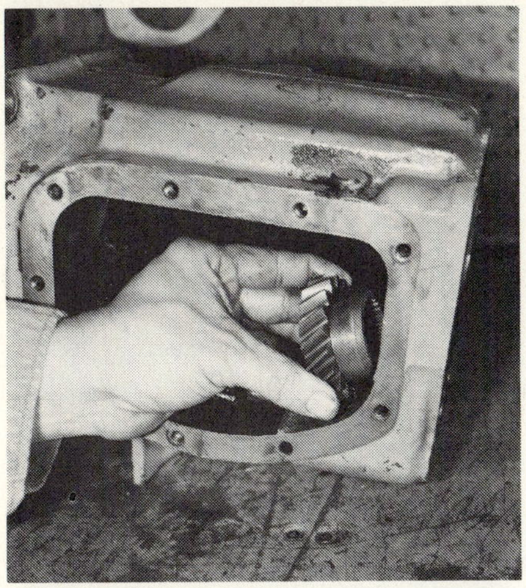

10. The reverse front idler gear and corresponding steel thrust washer is lifted out of position before the cluster gear shaft is driven out.

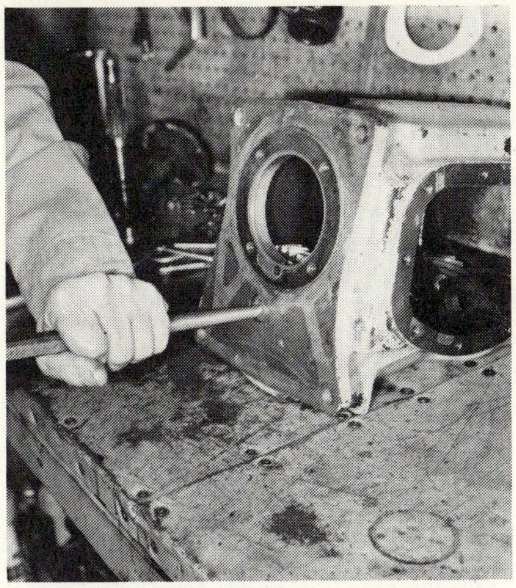

11. The countershaft (cluster gear shaft) is a press fit in the case and held in place by a Woodruff key. Tap shaft with drive pin to remove.

12. Remove the countershaft gears and the two bronze thrust washers (one at each end). Examine washers for signs of wear and replace them as necessary.

13. Before beginning reassembly of cluster gear, check for worn, cracked or chipped teeth. If any damaged areas are noted, the entire unit is replaced.

14. Some mechanics use a dummy shaft to hold the 80 needle bearings in place when cluster gear is reassembled, but who always has one at hand?

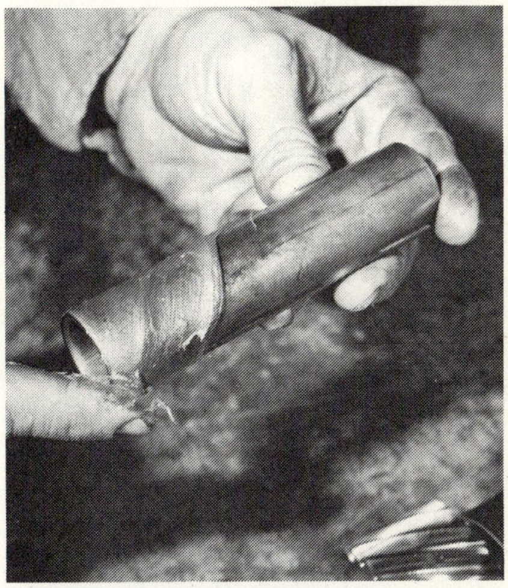

15. To work without one, wipe the outside of the spacer with grease and insert in cluster gear. This holds it in position to install bearings.

16. Insert first spacer and push up against the shaft end, then lay 20 bearings in place. Push them back and repeat process to install other 20.

17. Turn the cluster gear around and repeat the procedure at other end. Grease holds spacers and bearings neatly in place for shaft replacement.

18. Smear grease on both sides of bronze thrust washers and position at front and rear of case. Be sure to index washer tabs in recessed areas of case.

HOW-TO: Warner T-10 Overhaul

19. Carefully draw cluster gear through case from rear and set into position between thrust washers. You may have to rotate slightly to seat properly.

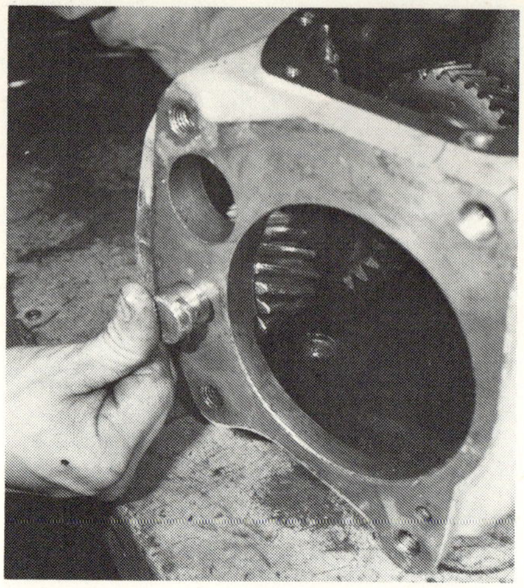

20. Now insert countershaft from rear of case and push through cluster gear. Set Woodruff key in place and push or tap the shaft the rest of the way.

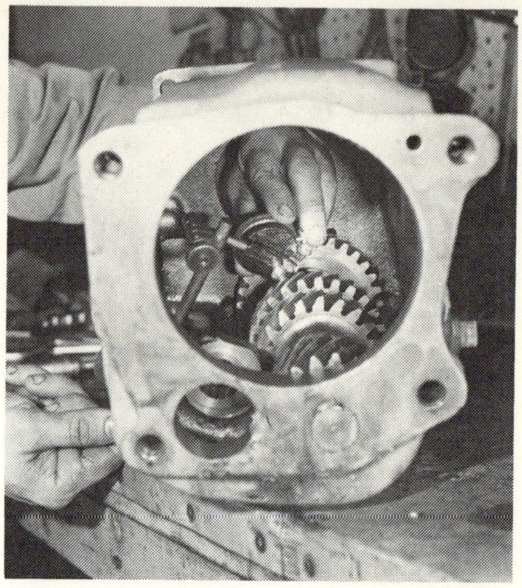

21. It's always best to check end-play at this point, although many no longer bother with this step. If end-play exceeds .025, change the thrust washers.

22. To begin gear tower disassembly, remove the synchro ring from front end of the mainshaft, and then remove the flat steel washer and snap ring.

23. The third and fourth speed synchronizer assembly and the third speed gear can now be removed by sliding off the front of the mainshaft.

24. Remove snap ring holding the other synchro-clutch in place and entire assembly comes off the mainshaft; note direction of synchro-clutch hub.

25. Check the synchro-clutch, friction ring and gear assemblies carefully for signs of wear while you have the gear tower apart. Clean each thoroughly.

26. In addition to inspecting for worn, cracked or chipped gear teeth, inspect for wear that might cause the gear to fit loosely on the mainshaft.

27. Check surfaces where shift forks ride (especially important if fork wear was noted). While forks are identical, each creates its own wear pattern.

28. Clean the bearings thoroughly—do not spin dry with air pressure—and examine them closely for signs of cracked races and/or worn balls.

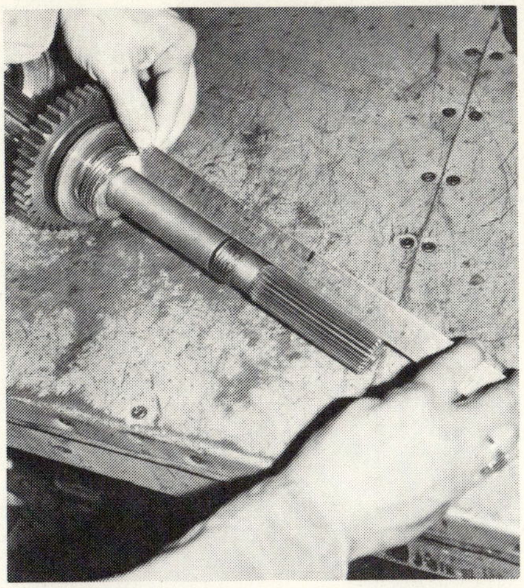

29. Before removing speedometer gear from shaft, measure so you can replace it exactly. Some T-10 shafts use a raised taper to indicate placement.

30. Press speedometer gear off or use a puller if a hydraulic press is not available, but never remove by hammering, as you'll ruin the gear teeth.

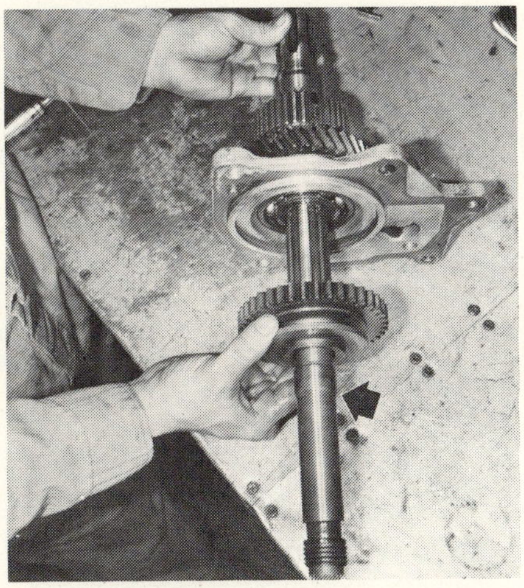

31. Once speedometer gear is removed, reverse sliding gear slips off the mainshaft from the rear. Note taper area for speedo gear placement on this shaft.

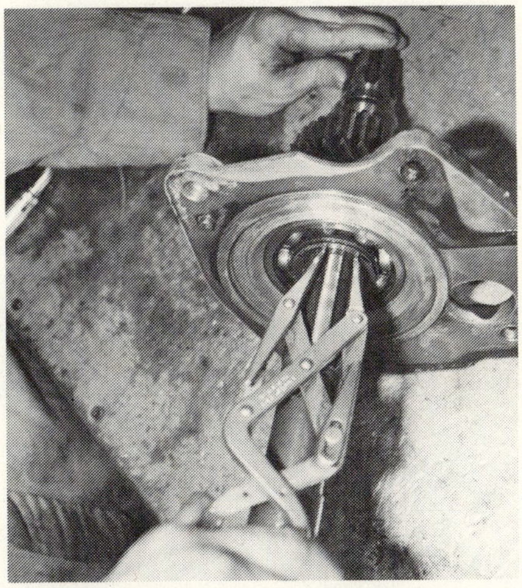

32. The adapter plate and rear bearing are held in place by another snap ring; remove and they slip off the shaft leaving just the first gear.

33. First gear and synchro rings can now be removed. Before replacing them on the shaft, clean and inspect them for wear as you did with the others.

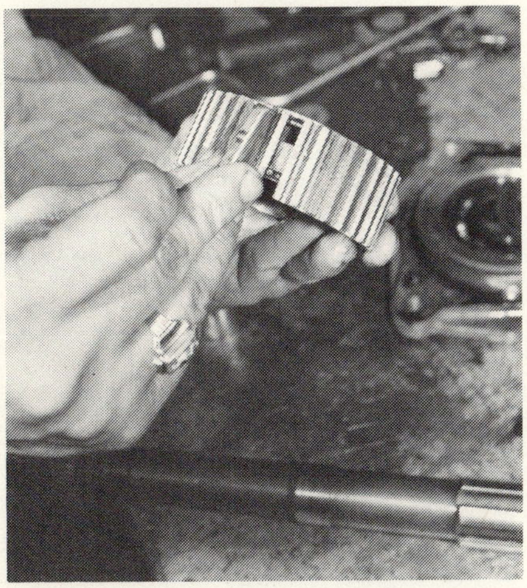

34. As synchro-clutch assemblies will come apart during disassembly, reinstall retainer rings on each side, and place three synchro-plates in position.

35. To reassemble the gear tower, begin by installing the first and second gear synchro-clutch assembly, then slide first gear into position.

36. Now slide adapter plate with bearing over the mainshaft. This is secured in place with a snap ring; make sure that it fits into place properly.

HOW-TO: Warner T-10 Overhaul

37. Reverse sliding gear is now replaced and secured with selective snap ring. It should fit in place with no play; if not, install one that does.

38. Speedometer gear is installed as a final step on the rear of the mainshaft. Use a length of galvanized pipe and hammer to position gear exactly.

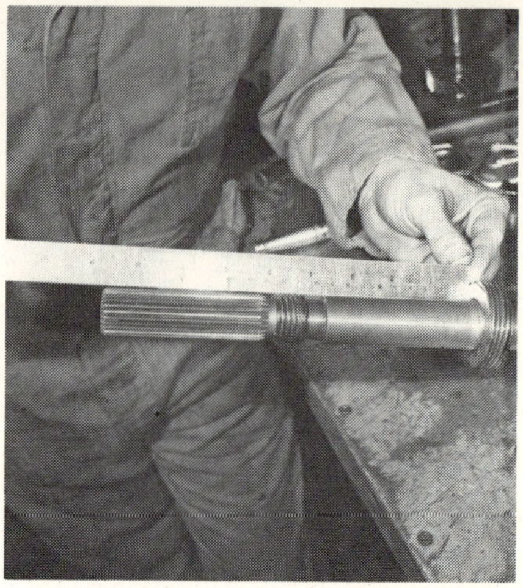

39. If you took our advice and premeasured distance from gear to end of shaft, all you have to do now is remeasure until positioning is exact.

40. Turn mainshaft around and slip synchro ring in place. Be sure that the tang on the ring fits into the corresponding cutout on the synchro-clutch.

41. Now slide second gear into position and check the mesh of the synchro-clutch assembly to make sure that it works properly before proceeding.

42. Install friction ring by coating with grease (as you did the thrust washers earlier) and slip it into position; grease holds ring in place.

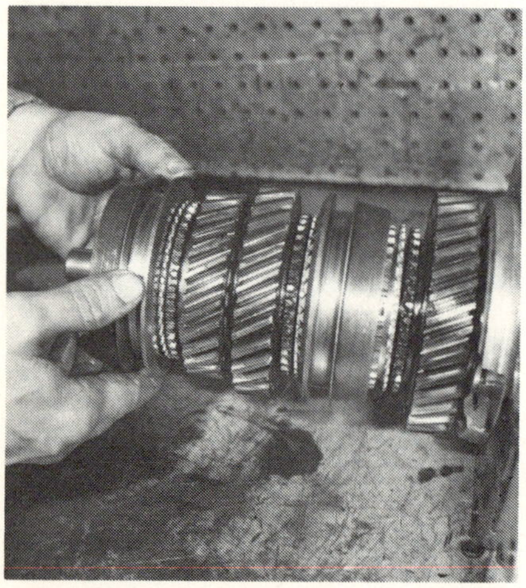

43. Third gear and the third/fourth gear synchronizer are now installed. Note direction of two synchronizers; beginners often replace them incorrectly.

44. Clipping the snap ring in place completes reassembly of mainshaft and gear tower. Now double-check the action of the synchro-clutch assemblies.

45. Replace the reverse front idler gear and its steel thrust washer in preparation for inserting the mainshaft assembly into the trans case.

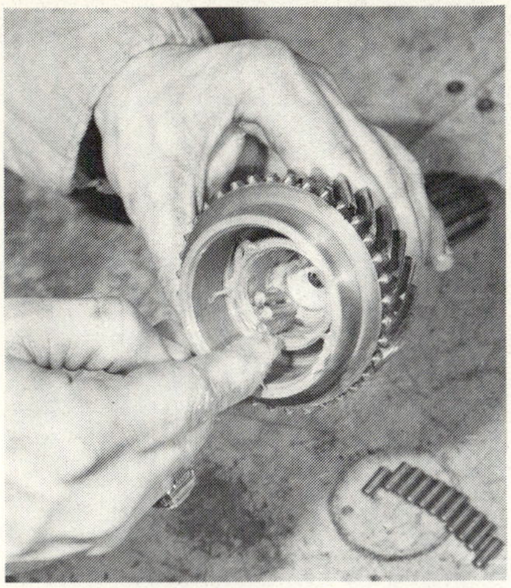

46. Reinstall the 16 needle bearings in bore of clutch shaft (input shaft) using grease to hold them in place as you did with the cluster gear.

47. The clutch shaft is now attached to the end of the mainshaft, and you are ready to reinstall the entire assembly in the transmission case.

48. This unit is darn heavy, so take it easy. Work the shaft and gears into the case and position by rotating slightly as you push forward.

49. Reinstall the self-locking bolt through the adapter plate to hold it to the case, then slide the reverse rear idler gear into place.

50. Fit a new gasket into place and rotate the tailhousing over the reverse sliding gear to the right to engage reverse shift fork and sliding gear.

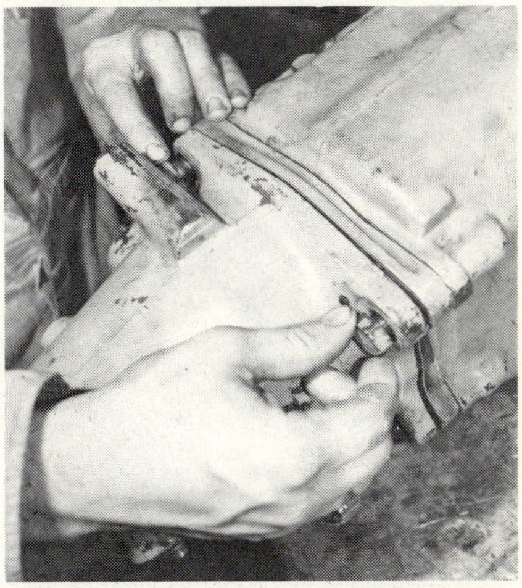

51. Bolt tailhousing in place. Don't forget to replace the identification tag on one of the bolts. It has information you'll need for future overhauls.

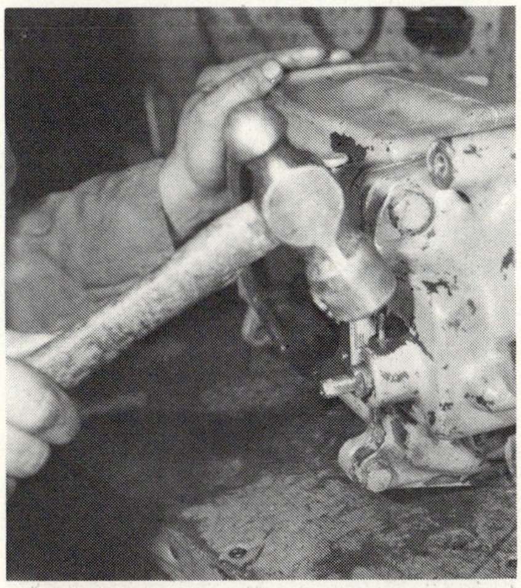

52. Replace the tapered reverse shift shaft or drift pin from the top of the housing and tap into place with a hammer. You should have no resistance.

53. Reinstall the front bearing and tap into place until it's properly seated; then secure the snap ring in place and you're almost home.

54. Fit the input shaft bearing retainer in place and bolt down. Replace the side cover and torque the cap screws to 20 ft.-lbs. Overhaul completed.

HOW-TO: Vega (Opel) 4-speed Overhaul

1. The '71-'72 Vega (Opel) 4-speed box is a "bottom-loader:" the case cover is at the bottom of the trans instead of side or top. Watch detent plug and spring when pulling cover.

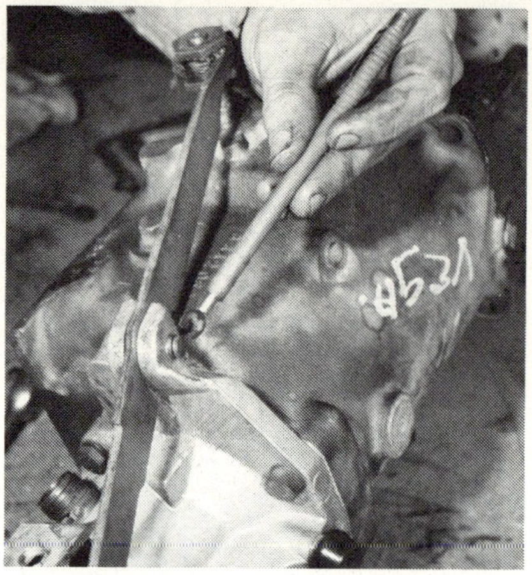

2. Lock nuts, cotter pins and C-clips hold selector lever and shifter shaft arrangement in place and must be removed as a first step before access to trans is possible.

3. Unbolt the extension housing from the trans case. Trans is a German design and requires a good deal of patience during disassembly as working space inside is limited.

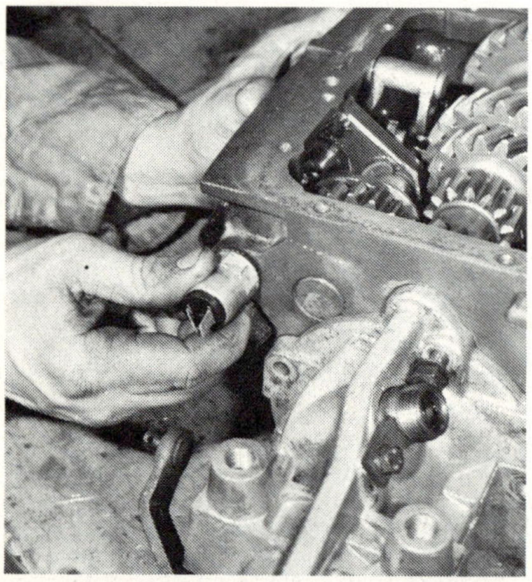

4. Loosen and unscrew the backup light switch from the trans case. "Bottom-loader" design makes countergear removal the next step in the disassembly procedure.

5. In order to provide easy access to the countergear shaft for removal, the extension housing must be rotated until its shaft end completely clears this cutout in the housing.

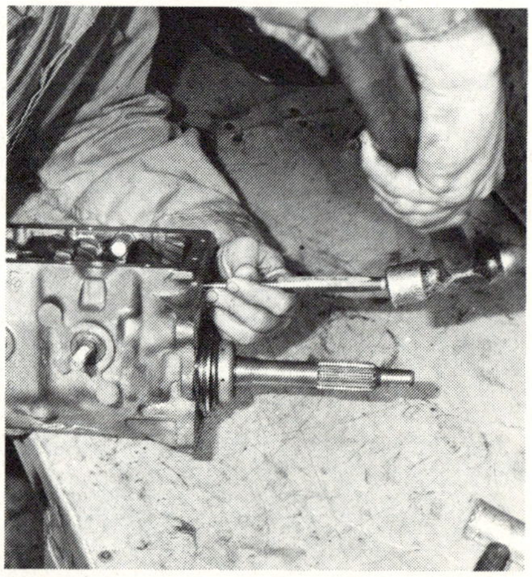

6. Countershaft is removed by driving to the rear with a punch and hammer from the front of the trans case. Watch for a lock ball in rear of countershaft and don't lose it.

7. Countergear can be lifted out easily with all needle bearings intact once the shaft is removed. Be sure and remember to clean and inspect all gear teeth for damage or wear.

8. The countergear thrust washers are removed next. Note the position of the cut edge, as well as the tangs which must fit into the recesses provided in the case when replaced.

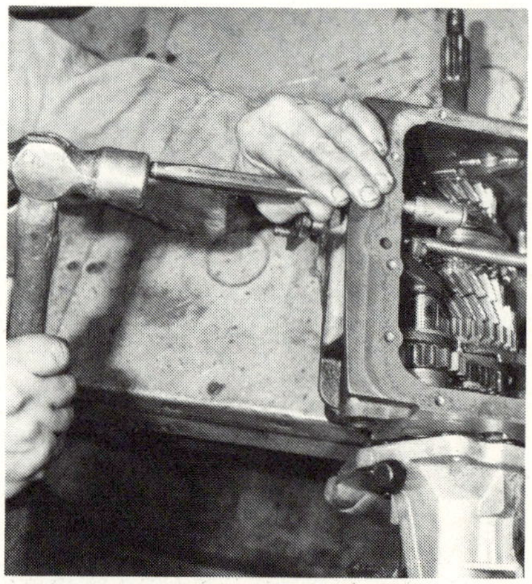

9. Now drive out reverse intermediate lever bedding bolt and remove the reverse speed shifter intermediate lever. The rest of Vega box is held together with drift pins.

10. Remove the input shaft and the main drive gear from the front of the trans by pulling straight off. It may be necessary to rock the input shaft back and forth as you pull it forward.

11. Two drift pins hold the selector shaft in place. This one is easy to see and work on, but access to the other pin is gained through hole (A) in case, and punch size used is important.

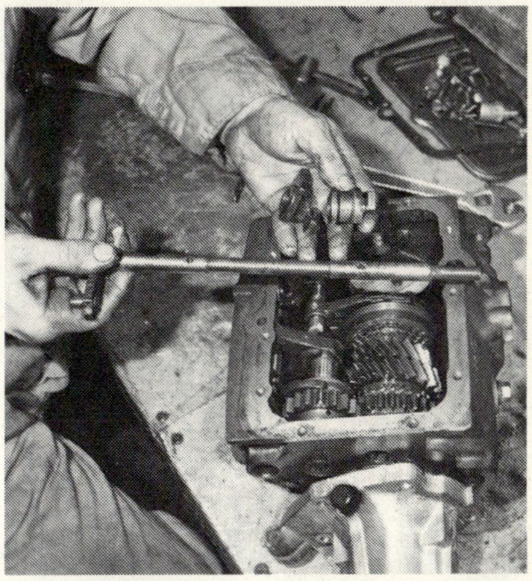

12. Once the drift pins are removed (use ⅛-in. pin punch), the selector shaft can be pushed out of the trans case and the intermediate levers removed with it. Check levers for wear.

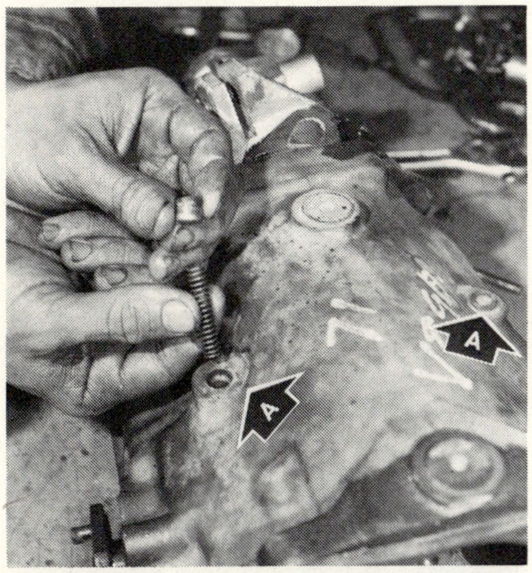

13. Remove the two aluminum gearshift interlock detents from the top of the case (A). Pry out plugs with care to prevent damage to them and remove springs and balls.

14. Knock out drift pin holding 3rd/4th shift fork to shifter shaft. Use care because the shift forks are rather brittle metal and can be easily broken by a misplaced blow.

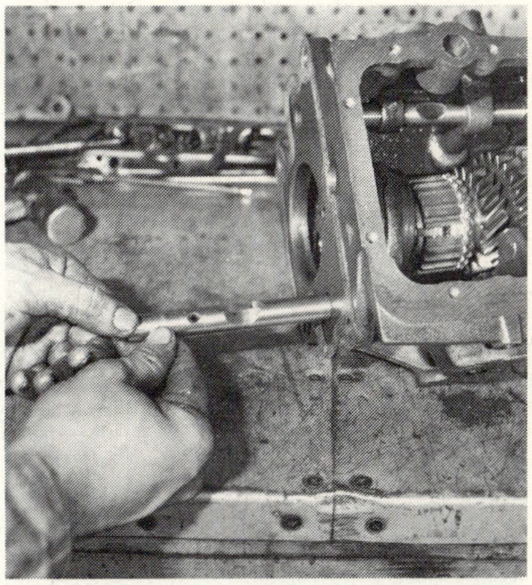

15. Turn the extension housing until the shifter shaft is exposed, and then tap shaft lightly from the rear of the case until it can be pulled out through the front and removed.

16. With shifter shaft removed, 3rd/4th fork is angled off clutch face and out of case. Rounded notch at shaft hole part of fork must face rearward when replacing fork in case.

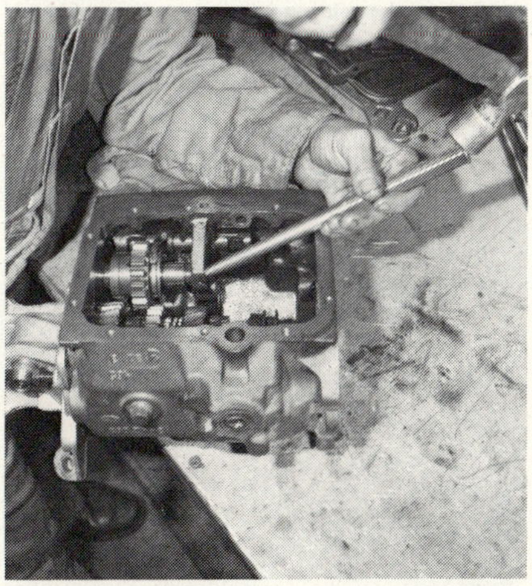

17. A special tool is made for driving reverse idler gear shaft rearward and out of the case, but if you don't have this tool, you can do the same with a punch and careful tapping.

18. Shaft uses a lock ball—watch for it when you remove the shaft and reverse idler gear from trans case. Inspect the shaft for burrs and the gear for worn or damaged teeth.

HOW-TO: Vega (Opel) 4-speed Overhaul

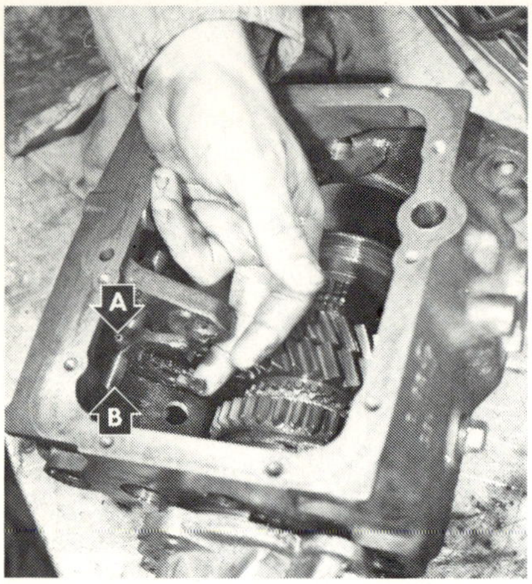

19. The reverse speed shift fork will remain in case unless drift pin (A) and shaft (B) are removed. If the fork is not damaged, there's no reason to remove this one from case.

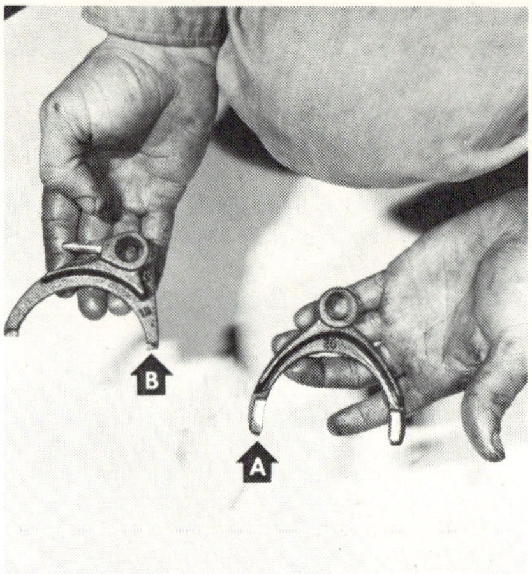

20. Forks are not interchangeable. The 3rd/4th fork (A) is worn, but legs on 1st/2nd fork (B) are broken. This is more likely to happen during disassembly than during operation.

21. Now withdraw the mainshaft assembly and extension housing from the rear of the trans case. The gear shifter sleeve will be pulled off as you remove the shaft.

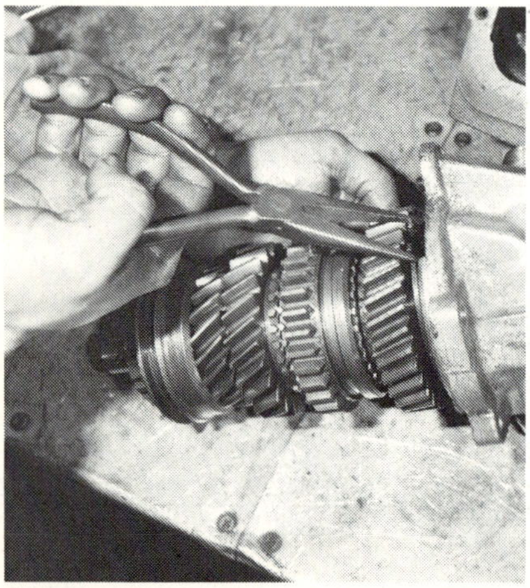

22. When you are ready to remove the mainshaft from the extension housing, you'll have to squeeze the mainshaft bearing snap ring together and then draw it forward.

23. After removing the snap ring at front of 3rd/4th clutch assembly, clutch and 3rd gear can be withdrawn from the mainshaft. Do everything you can to keep the clutch intact.

24. Turn mainshaft around and remove the snap ring and washer located behind the speedometer drive worm gear. Pressure clip with lock ball beneath is exposed as gear slides off.

25. Another snap ring is removed to continue disassembly of the gear tower from the rear of the mainshaft. Snap ring pliers are great help when you are doing manual trans work.

26. Washer and spacer are removed from rear of mainshaft inner bearing in that order. Bearing may have to be removed with puller. However, it usually slips right off.

27. The bearing, bearing snap ring and spacer can now be removed from the shaft. Note that the spacer cone faces toward the bearing instead of the 1st gear assembly.

28. When the 1st gear is pulled free from the 1st/2nd gear speed clutch, the clutch will come apart. Remove parts shown, then blocking ring and clutch key spring (A).

29. This 1st speed needle bearing sleeve comes off next and should be inspected carefully for loose, worn, missing or damaged bearings or deformed sleeve.

30. The remainder of synchro-clutch and 2nd gear are removed from the mainshaft, completing disassembly. Be sure and inspect all the gear faces and teeth for damage.

31. Reassembly of the mainshaft begins with the 2nd gear and the 1st/2nd speed synchro-clutch assembly. Set into position against the stop ring on the mainshaft.

32. Replace 1st speed needle bearing, blocking ring and 1st gear. Make sure that the blocking ring cutout engages the synchro-clutch properly and fits into place easily.

33. The mainshaft spacer (cone toward bearing), the bearing retainer snap ring and the inner bearing are replaced next. The bearing should seat easily without need for lubrication.

34. The spacer and washer are then fitted into position behind the bearing. Be sure to inspect the bearing for cracked races, worn balls and other defects before replacing it.

35. Metal spring clip slips under the speedometer drive worm gear. Lock ball is inserted in shaft and gear/clip positioned over it, then the rear snap ring is replaced and secured.

36. The next step is to turn mainshaft around and install 3rd gear and blocking ring, seating against the stop ring on shaft. Gear cone must face toward end of shaft for proper mesh.

HOW-TO: Vega (Opel) 4-speed Overhaul

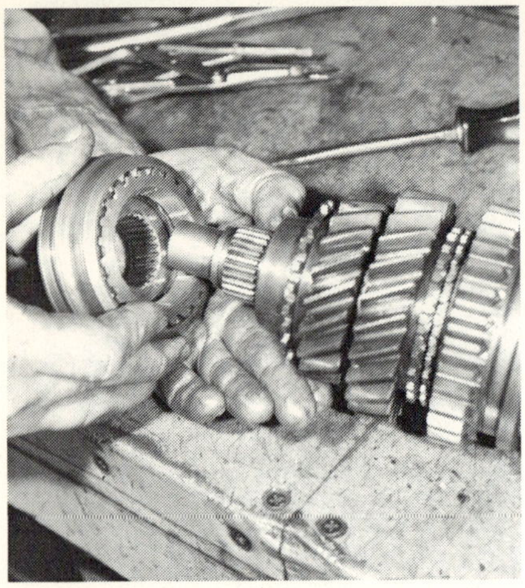

37. The 3rd/4th synchro-clutch assembly is fitted into place, seated and secured with the gear shifter sleeve snap ring, completing reassembly of gear tower on the mainshaft.

38. Place mainshaft assembly in the extension housing up to its stop, and then secure in place with the mainshaft bearing snap ring. Make sure that the ring engages completely.

39. Reinstall mainshaft in trans case, and then fit the main drive gear and input shaft through front of case to engage mainshaft. Use a twisting motion to fit together.

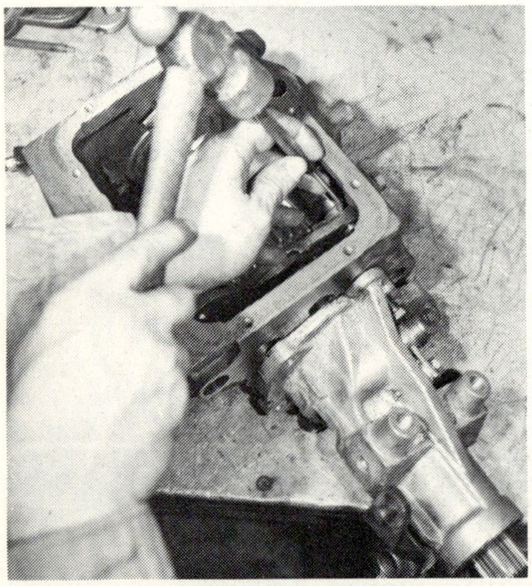

40. Fit the 1st/2nd gear shifter shaft into the transmission case and through the shift fork; then rotate until the holes line up properly. You are now ready to reinstall the drift pin.

41. Place lock ball into reverse idler gear shaft and push shaft through hole in rear of the transmission case to pass through reverse idler gear with groove end toward front.

42. Reinstall 3rd/4th speed shifter shaft through trans case and shift fork. Use a screwdriver to position the fork while turning the shaft until drift pin holes line up properly.

43. The drift pin holding 3rd/4th shifter fork to shaft is replaced. Be sure and use 3/16-inch or larger punch when reinstalling all drift pins in this particular transmission.

44. Insert new selector shaft oil seals in the appropriate holes on both sides of trans case, then slide the shaft into the case and through the intermediate lever. Install drift pin.

45. The reverse shifter intermediate lever is fitted into place to engage the arm on the 3rd/4th intermediate lever and the bedding bolt (A) is driven into transmission case to secure lever.

46. The countergear uses one set of roller bearings between two spacers at each end. Be sure and coat inside of gear with grease, and then set first washer in position.

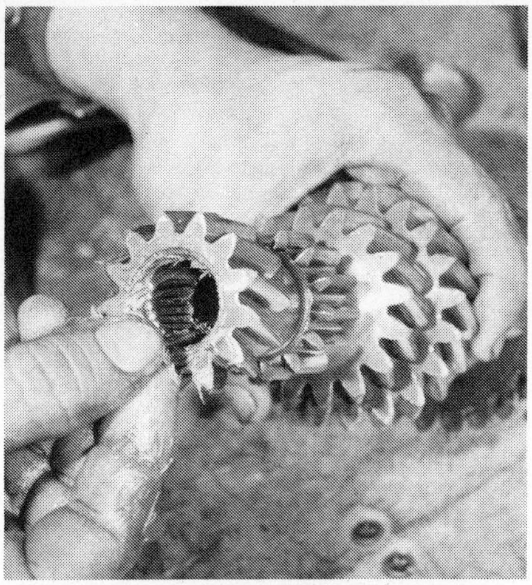

47. Place one row of roller bearings in the end of the countergear. Grease makes the use of a dummy shaft unnecessary in reassembling the gear for reinstallation.

48. After the bearings are positioned, you can place the second spacer coated with grease over them. Then repeat this procedure at the other end of the countergear in preparation for replacement.

49. Now you can coat the countergear thrust washers with grease and stick to transmission case. The thrust washer lugs must fit into the slots provided inside the case.

50. The next step is to place the countergear unit carefully in the trans case. This is the only real advantage of "bottom-loader" design as you don't have to struggle here.

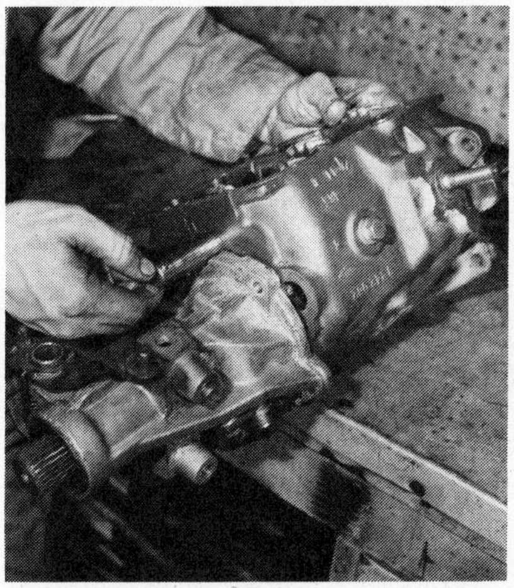

51. Place the lock ball in the countershaft recess and slide the shaft through the countergear from the rear of the case. Tap the shaft with a mallet until it's flush with the case.

52. Now rotate the extension housing until the bolt holes line up with the transmission case. You can then replace the bolts and torque each to 21 ft.-lbs. (Coat with a sealing compound).

53. Replace the gearshift interlock detent balls, springs and plugs at the top of the case; then turn transmission case over and insert reverse interlock detent ball, spring and plug.

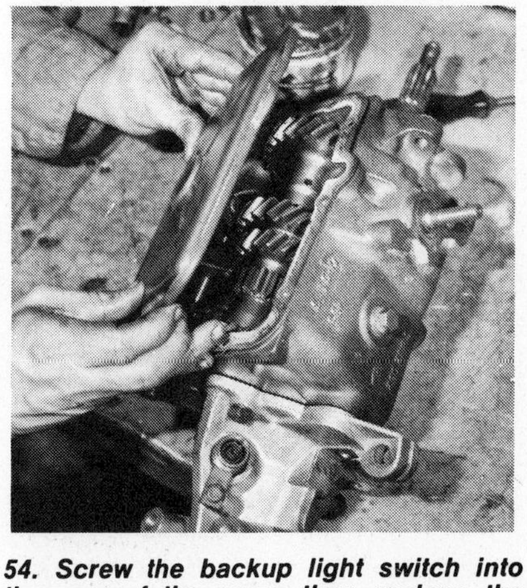

54. Screw the backup light switch into the rear of the case; then replace the cover and gasket. Tighten all 10 cover screws. You have just completed the Vega 4-speed overhaul.

HOW-TO: Pinto 4-speed Overhaul

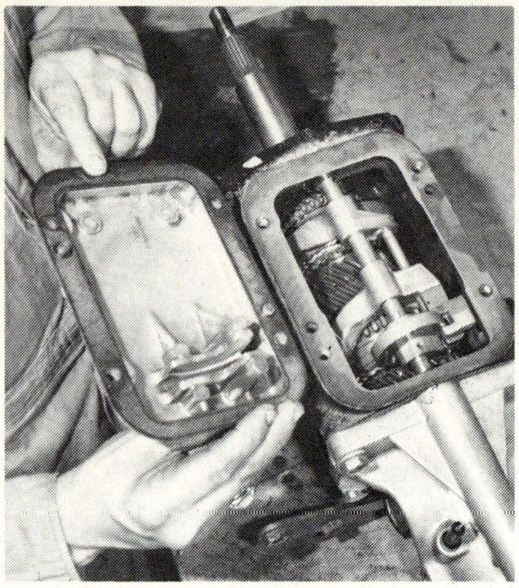

1. The 71WG 4-speed Pinto box is similar in design to the current 72WG and came in two versions—the AD made in Germany (steel top cover) and BC made in Britain (aluminum cover).

2. With gearshift lever in neutral, the backup lamp switch is unscrewed from extension housing. 1974 versions also use a seatbelt interlock switch—don't confuse the two.

3. Once the backup lamp switch has been removed, hold the shift lever lock-tab ring in place, unscrew the dome nut and lift the shift lever from the extension housing.

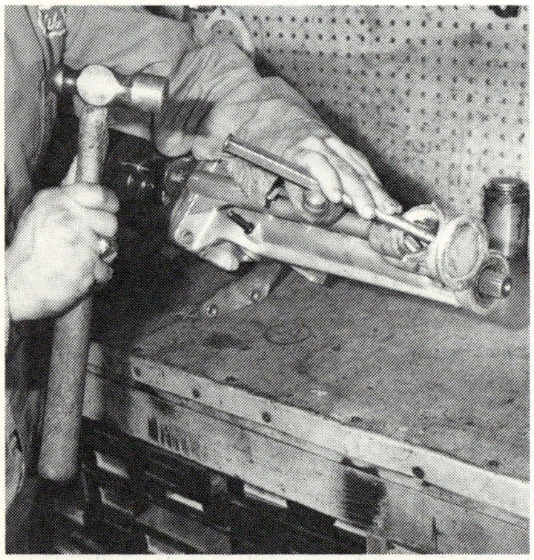

4. The extension housing plug can then be driven out by using a punch inserted carefully through the shift lever opening. You should tap it lightly to avoid distorting the plug.

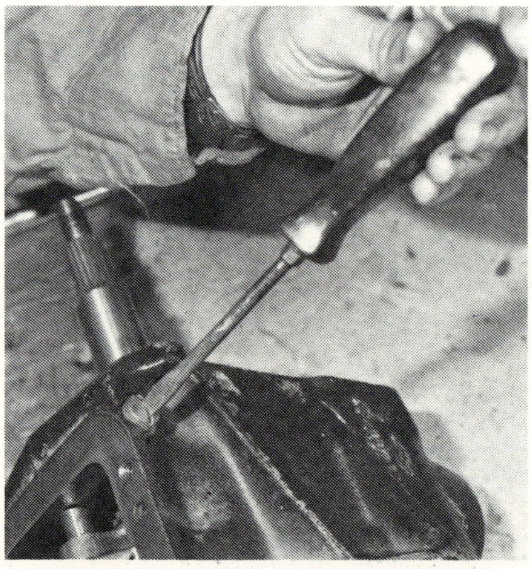

5. Under this threaded plug at the front of the trans case, you'll find a spring and the shift rail detent ball. A pencil-type magnet is extremely handy here for removing the ball.

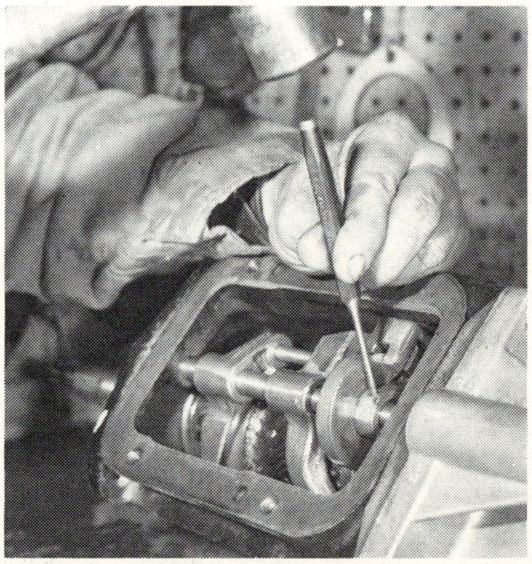

6. Now you can use a small punch and drive out the drift pin that holds the shift boss to the shift rail. Top loaders like this one use several drift pins to hold things in place.

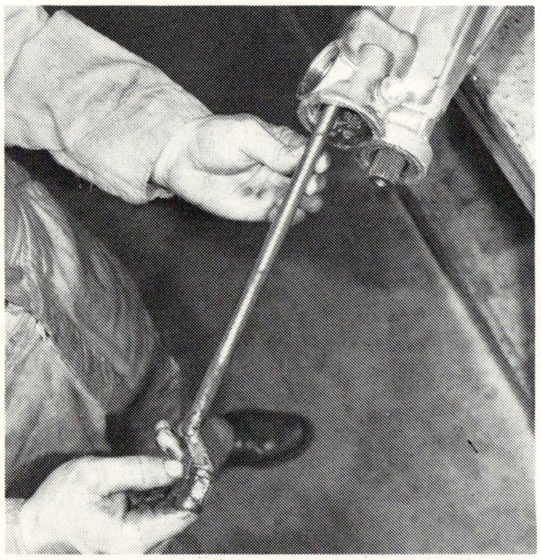

7. Withdraw the shift rail from the rear of the extension housing. If it doesn't want to come easily, grasp the rear of the rail and rotate as you pull. This should loosen it.

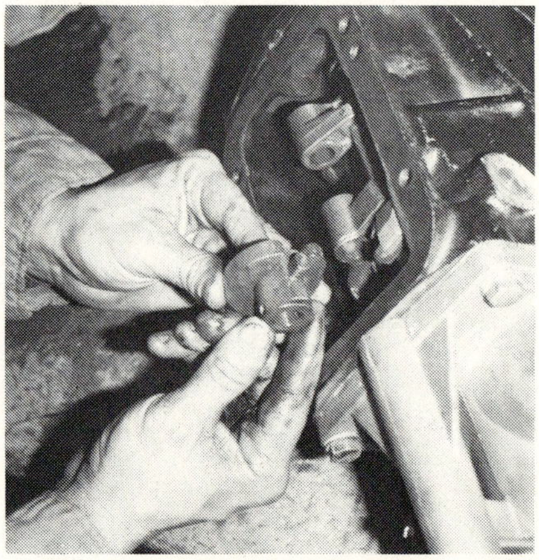

8. Once the shift rail is removed, the selector lever arm can be taken out by simply lifting. This is somewhat deceptive as it comes out so easily compared to the removal of other parts.

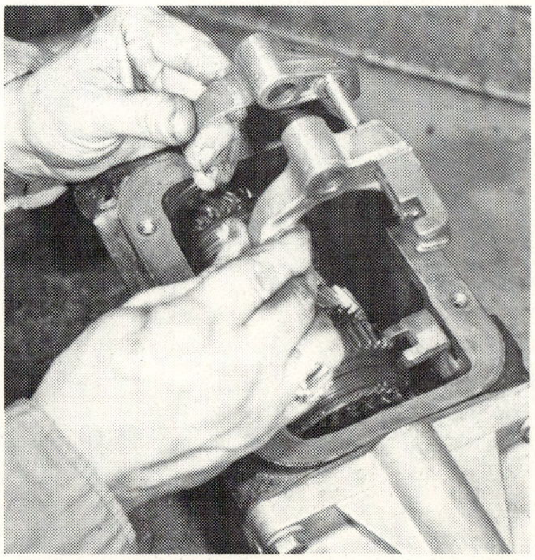

9. Removing the shift forks looks as easy, but you'll have to struggle a bit; rotate them back and forth and just when you're about to give up, you will find that they come out.

10. Remove the three 10mm cap screws and the bearing retainer slides off the input shaft. It's a good idea to throw the old gasket away and to use a new one for reassembly.

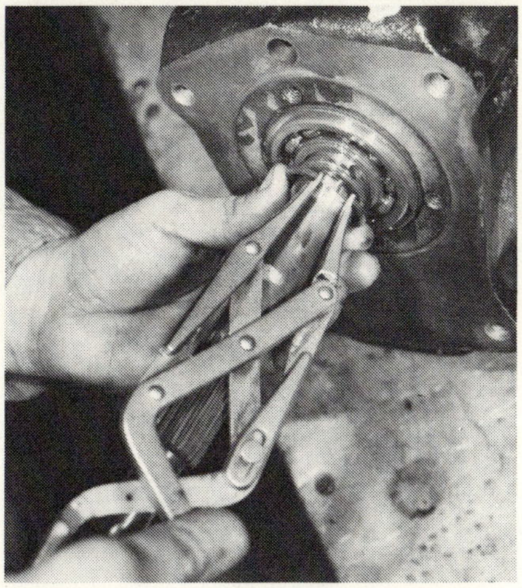

11. Be sure to remove the input bearing selective snap ring before attempting further disassembly. You'll find an inner and outer snap ring—remove the inner (small) one.

12. Once the snap ring is removed, the bearing is supposed to slip right off, but it doesn't always go by the book. Tap the input shaft with a screwdriver behind the outer snap ring.

13. This is how bearing and friction ring are supposed to come off. If you're really lucky, you'll avoid hammering on the input shaft. By the way—be sure to use a soft mallet.

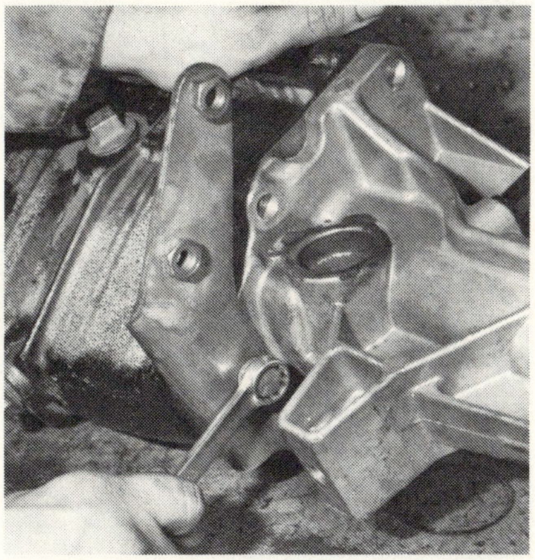

14. Remove the four extension housing bolts. If the trans uses this extra bracket on case, remove also. Then tap housing with mallet to loosen it from the transmission case.

15. Rotate the extension housing to align countershaft with cutaway in the housing. Hold the input shaft as you withdraw the mainshaft and the two will separate.

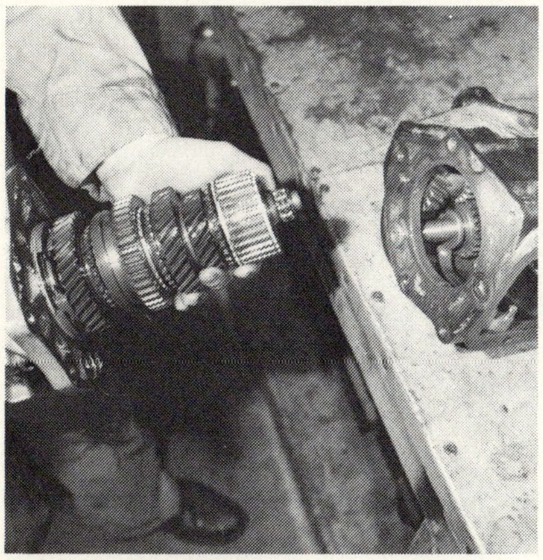

16. The mainshaft and input shaft separate at the input shaft pilot bearing. Unit is a bit awkward because of length, so use care in removing it so as to avoid damage to the gears.

17. Withdraw the input shaft and gear from the top of transmission case by angling and turning the unit as you remove it. This one's too easy, but other steps make up for it.

18. Drive the countershaft rearward until it clears the front of the case and remove. The countershaft gear can now be removed. Retrieve all bearings that fall out.

HOW-TO: Pinto 4-speed Overhaul

19. Tap reverse idler gear shaft once or twice from behind with a punch and then remove shaft and reverse idler gear from rear of the case. This is accomplished fairly easily.

20. The selective snap ring holds the output shaft bearing in the extension housing and must be released before the shaft and the gear tower can be removed from housing.

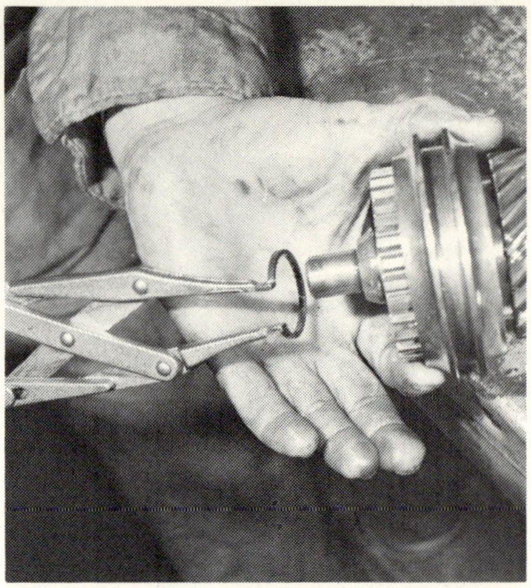

21. Remove the input shaft pilot bearing to gain access to the synchronizer hub snap ring. Then spread and remove this snap ring to begin disassembly of the gear tower.

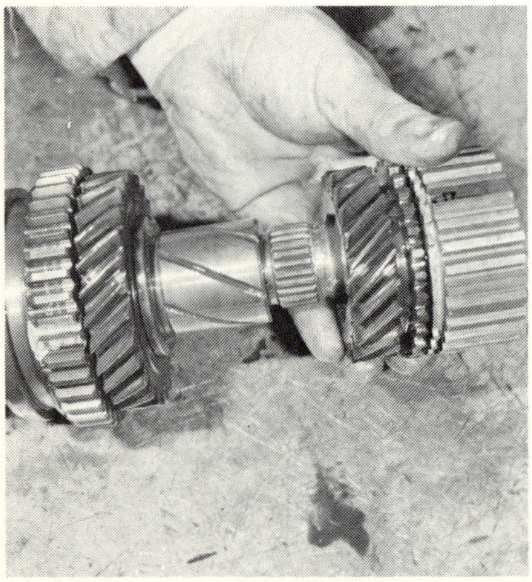

22. Turn shaft around for ease in handling and then remove the 3rd/4th gear synchronizer assembly, blocker ring and 3rd gear. If synchronizer comes apart, locate springs and inserts.

23. A synchronizer hub snap ring holds the speedometer drive gear in place. Remove this ring and carefully withdraw the gear to rear to expose the drive ball in shaft.

24. After speedometer drive gear is removed, release inner snap ring holding output shaft bearing and remove bearing, outer snap ring and sliding low/reverse gear.

25. Release another snap ring and 1st/2nd gear synchronizer assembly can be removed, as well as 2nd gear. All the parts should be cleaned and inspected carefully.

26. If the synchronizer assembly comes apart during disassembly, you can install the insert spring inside hub so that tanged end locates in the U-section of an insert.

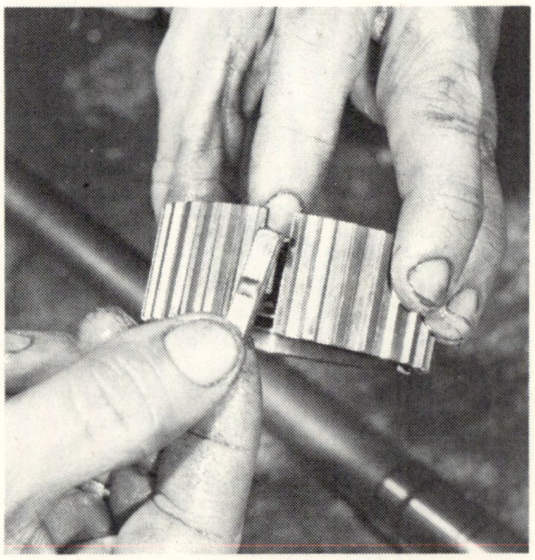

27. Install other spring on opposite side of hub and locate tag end in the same insert, then you can place the remaining two inserts in place and slide into sleeve.

28. In order to begin the reassembly procedure, you slide the 2nd gear and blocker ring onto the rear of mainshaft until they seat firmly against the stop ring on the shaft.

29. The 1st and 2nd gear synchronizer assembly is positioned on the shaft, then the sliding low/reverse gear and blocker ring is replaced with the ring cutouts matching.

30. The spacer, the outer snap ring and the output bearing are fitted in that order behind the low/reverse gear. Make sure that the tang on spacer fits into place properly.

31. To seat the output bearing, you can use a length of galvanized pipe to fit over the shaft and tap bearing into place with a mallet. Then you can secure the rear snap ring.

32. Place the drive ball in the output shaft detent and push the speedometer drive gear onto the shaft so that it clears snap ring groove. Then you can secure the ring.

33. Next, replace the third gear from front of the shaft with the cone and dog-teeth facing forward, then position against the output shaft stop ring in front of the 2nd gear.

34. The blocking ring fits over the third gear cone. At this point, the 3rd and 4th gear synchronizer assembly is fitted carefully in place along with the front blocking ring.

35. The synchronizer hub snap ring is secured in place and input shaft pilot bearing is then fitted in place. This is the final step to complete the assembly of the mainshaft.

36. Mainshaft is now ready for reinstallation in extension housing. Once it's properly seated, fit output bearing snap ring in place to hold shaft and housing together.

HOW-TO: Pinto 4-speed Overhaul

37. Input shaft and gear can now be fitted over the mainshaft and the unit is ready for reinstallation in transmission case. Set to one side and get the case ready.

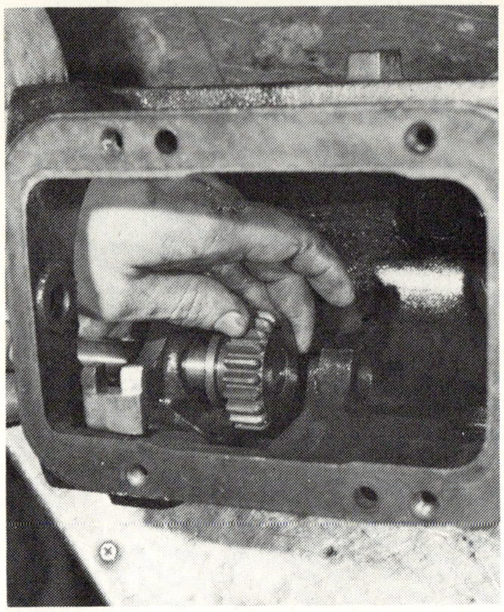

38. Place the reverse idler gear in case with the cone pointing to the rear and then insert the reverse idler gear shaft. Tap with a mallet carefully until shaft is flush with case.

39. It is not really necessary to have a dummy shaft to reinstall the 20 needle bearings and spacers in each end of the countergear; just run grease inside the shaft.

40. You can begin by placing a spacer inside the countergear and seating it all the way to the rear. This is done at both ends of the countergear before replacing bearings.

41. Insert the 20 needle bearings at each end and cap with a spacer. You're now ready to set the countergear back in the trans case—all without ever using a dummy shaft.

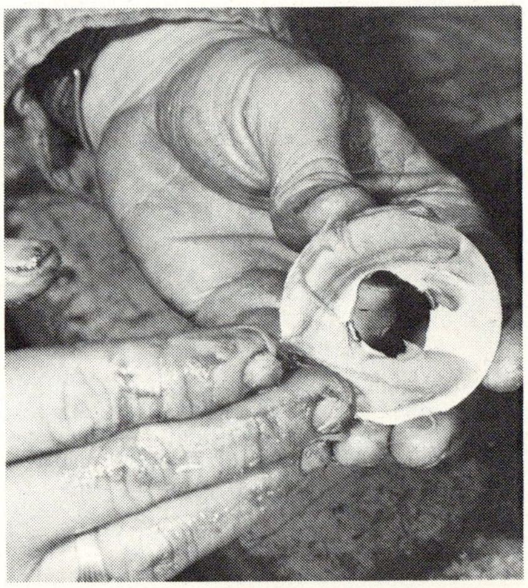

42. Next you can grease the front and rear countergear thrust washers and locate them inside the trans case so that the tab side seats into the recess that is provided.

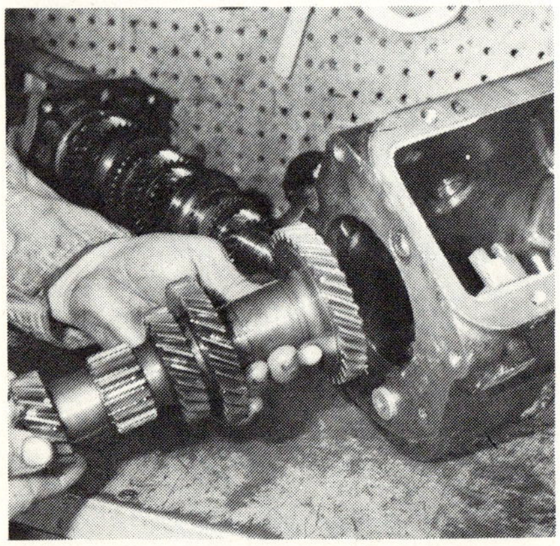

43. Insert the countergear through the rear of the transmission case and set between the thrust washers. Then you can replace the countershaft and tap it in place.

44. The mainshaft with input shaft attached can now be inserted in the case. The countershaft lug (A) must be horizontal to fit into the extension housing recess properly.

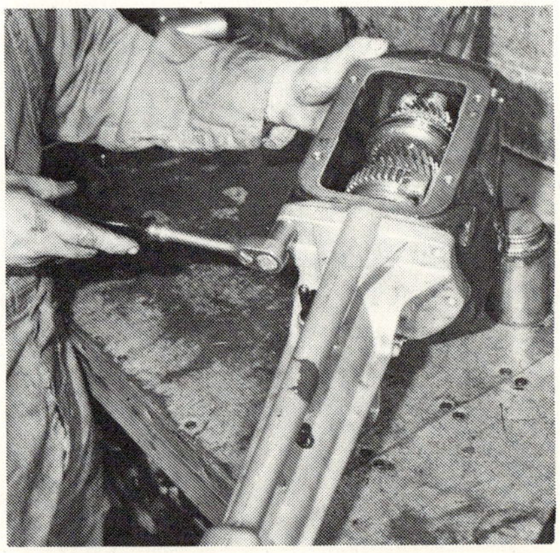

45. Rotate extension housing to align bolt holes, then push it fully home onto the trans case and torque all bolts 20-25 ft.-lbs. Replace bracket if the transmission used one.

46. The friction ring and the input shaft bearing are fitted in place. Because the bearing is an interference fit in the casing, it is extremely important that the outer race be tapped.

47. Use a length of galvanized water pipe and a hammer to seat bearing until the outer snap ring fits snugly against the face of the transmission case. Tap lightly to prevent damage.

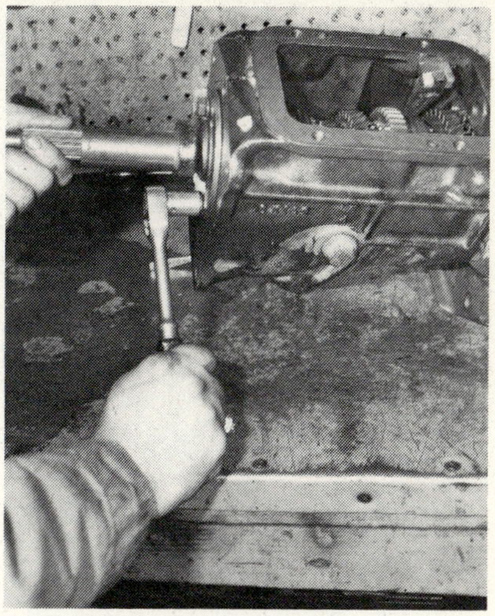

48. Replace the oil seal in the input shaft retainer, fit into place over a new gasket and then bolt the cap screws down. You should torque the cap screws to 12-15 ft.-lbs.

49. The shift forks with shift gate attached are now dropped in place. These will go back in place much easier than they came out, which may come as a surprise to you.

50. The shift selector arm with the interlock plate is set in position before the shift rail is inserted. You should be sure that the drift pin hole is toward rear of the case.

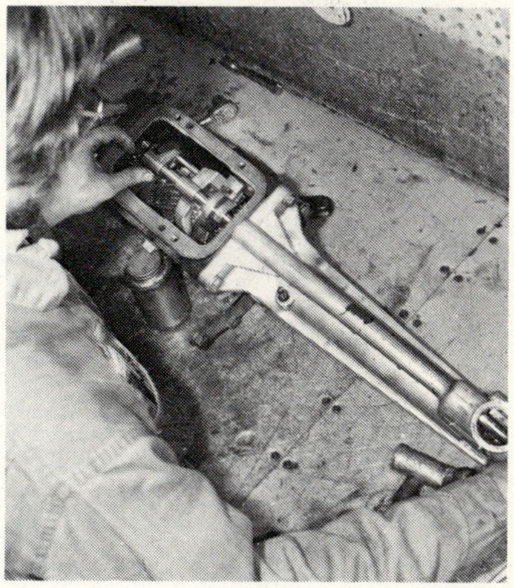

51. You can then fit the shift rail into the rear of the extension housing and push fairly slowly, fitting it through the selector arm, interlock plate and both of the shifter forks.

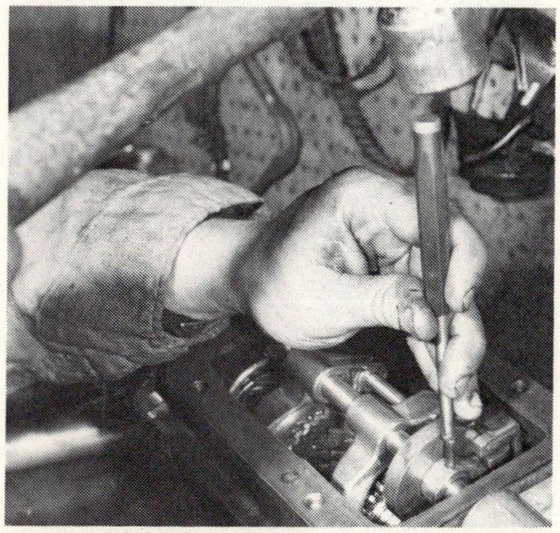

52. Reset shift selector arm drift pin using a punch slightly smaller than the pin. You should then tap it into place until the head of the pin is flush with the selector arm.

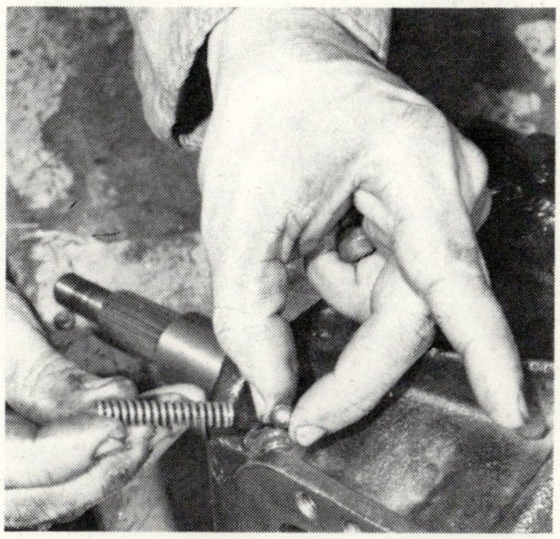

53. Drop the detent ball carefully into place, then fit the spring over it and screw the threaded plug back. Reinstall the cover and then torque bolts to 12-15 ft.-lbs.

54. Replace the extension housing plug by tapping carefully in place. Set shift lever in to engage shift rail, screw dome in tightly, and replace the backup lamp switch.

HOW-TO: Saginaw 3-speed Overhaul

1. Saginaw 3-speed is "side-loader" trans; shift forks are attached to cover and removed with it. Forks can be replaced without dropping trans.

2. After bearing retainer cap screws are removed, retainer and gasket are removed. Throw old gasket away and use a new one for the reassembly of trans.

3. After removing bearing inner snap ring, the slip-fit clutch gear bearing should slide right off clutch gear by pulling main drive gear forward.

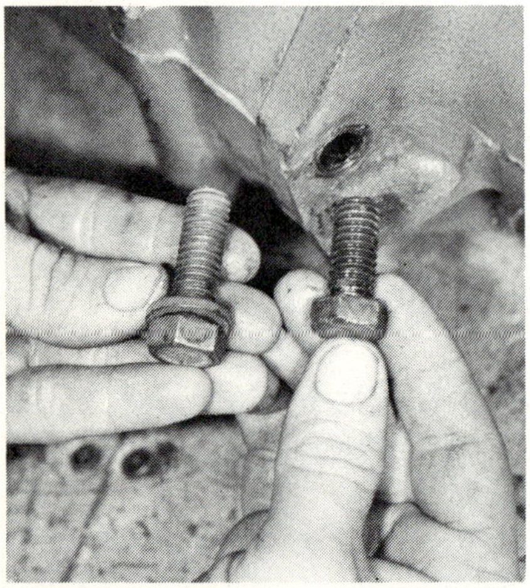

4. All but one of the extension housing-to-case bolts use a lock washer. Odd bolt fits here. Coat it with Permatex to prevent oil leaks.

5. Holding clutch gear with one hand, withdraw extension housing from case, separating clutch gear and mainshaft. Housing is heavy, so grip it carefully.

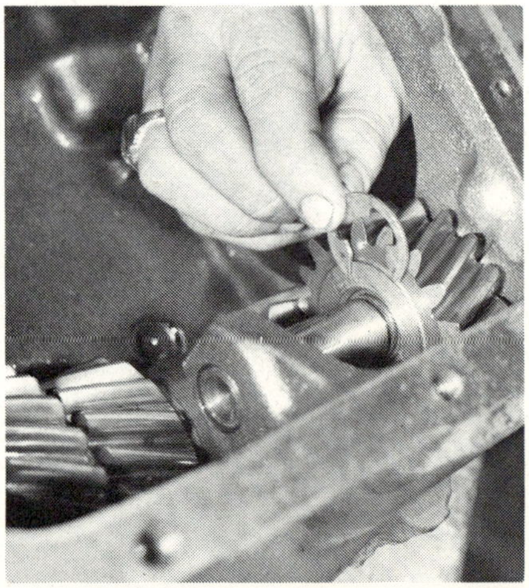

6. C-clip is pried off reverse idler gear shaft and removed. As this is the only C-clip used, you'll have no confusion during reassembly.

7. The reverse idler shaft is driven through the rear of trans case with a long punch. Shaft uses a Woodruff key where it mates with rear of case.

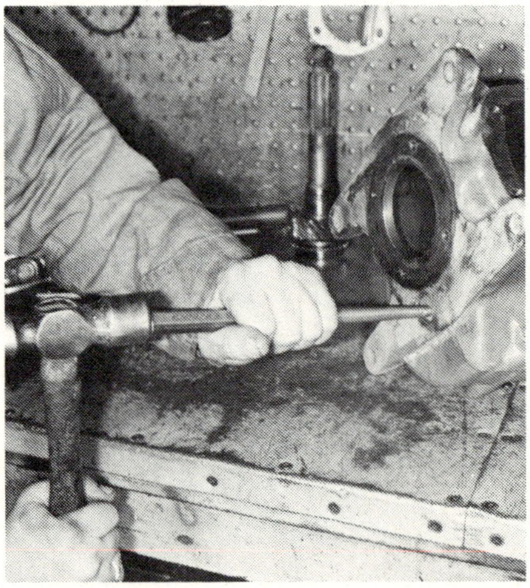

8. Countergear shaft is driven out from front of trans case. This shaft also uses a Woodruff key at the rear of the case, but it's a different size.

9. With countergear shaft removed, countergear is lifted out of the case. Needle bearings will fall out, but don't worry—just get them all.

10. Face of Saginaw countergear uses an anti-lash plate. Check the springs to make sure that none are broken or missing and make sure teeth are intact.

11. Expand snap ring in the extension housing that holds mainshaft rear bearing in place; then remove the housing by pulling to the rear.

12. Remove the hub-to-shaft snap ring and the 2nd/3rd sliding clutch assembly. Blocker ring and 2nd gear can all be withdrawn from the mainshaft.

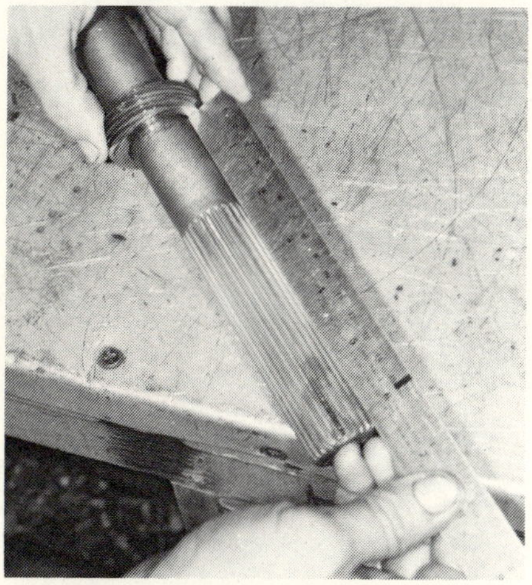

13. Measure distance from speedometer drive gear to end of shaft for correct reassembly; then depress drive gear retaining clip and press gear off shaft.

14. After removing rear bearing inner snap ring, the bearing, spring washer and reverse gear thrust washer are drawn off the back of the mainshaft.

15. Reverse gear can now be removed. Check gear teeth carefully for possible damage, as well as the face and cone of gear, after cleaning.

16. The 1st speed synchronizer and blocker ring are removed by spreading the snap ring that holds it in place; then the 1st gear comes off.

17. Countergear needle bearings are replaced at each end using a coating of grease inside the gear to hold them in place. Washer fits over bearings.

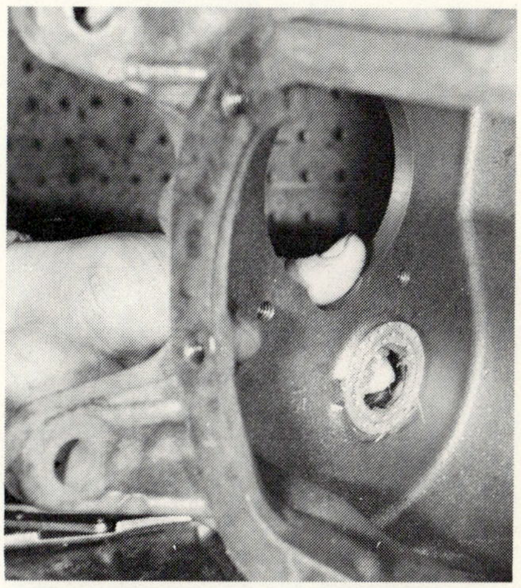

18. Countergear thrust washers are coated with grease and positioned in case. Tang on the washer must fit into case recesses as shown.

HOW-TO: Saginaw 3-speed Overhaul

19. Countergear is replaced through rear of case and seated in position between thrust washers. Greased bearings do not require use of dummy shaft.

20. Countergear shaft is installed through rear of case and countergear. Woodruff key is set in place and the shaft driven flush with the case.

21. Reverse idler gear thrust washers are coated with grease and fitted inside case. Like countergear washers, tangs must fit into case recesses.

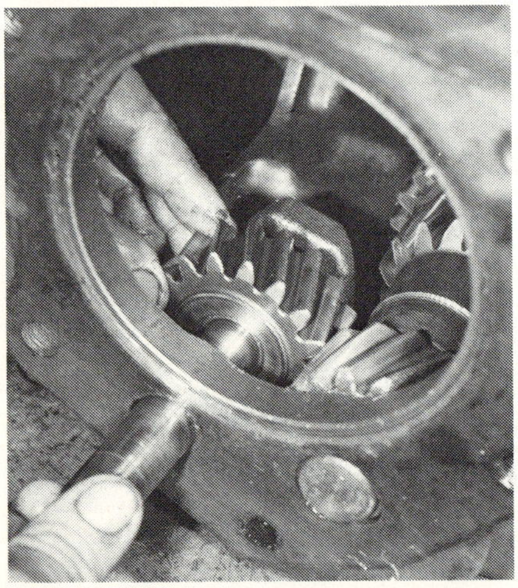

22. Reverse idler gear is inserted through side of case. Shaft is inserted with a rotating motion as you push it through the case and gear from rear.

23. Rotate shaft end until cutout matches that in case; then insert Woodruff key and tap shaft with mallet until it fits flush with case.

24. Gear reassembly on mainshaft begins with 1st gear. Install from rear of shaft and fit up against mainshaft shoulder—cone and teeth facing rear.

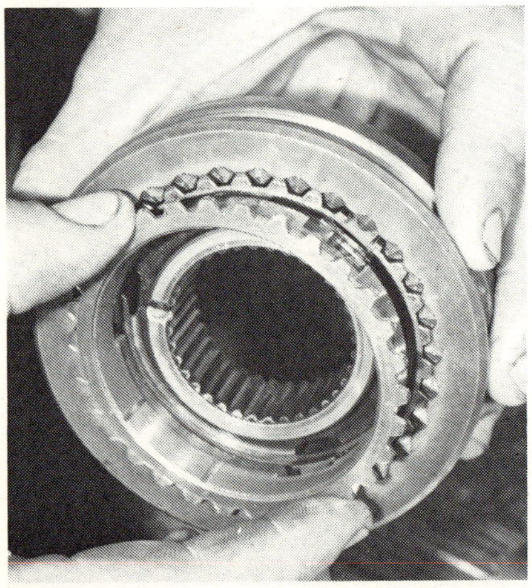

25. Check teeth of blocker rings and synchronizer, fitting them together to make sure that they mesh correctly and are not damaged in any way.

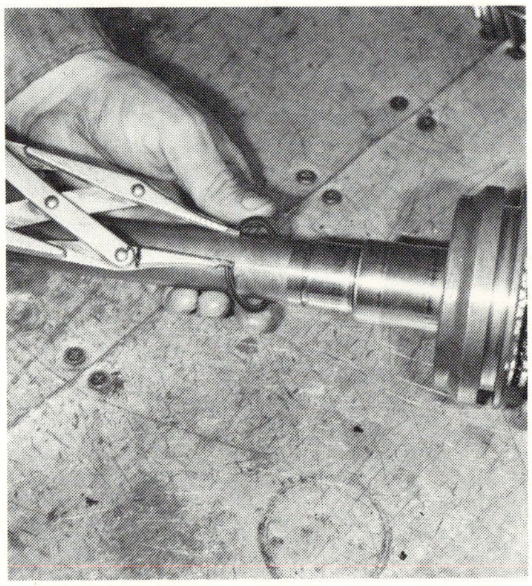

26. The 1st gear blocker ring and 1st/2nd synchronizer are positioned against 1st gear. Assembly is held in place by snap ring—don't forget it.

27. Reverse gear is replaced with its cone and teeth facing the synchronizer assembly. When fitted together, gear and synchronizer should mesh easily.

28. The reverse gear thrust washer, spring washer and rear bearing come next, all secured in place by a snap ring behind the bearing.

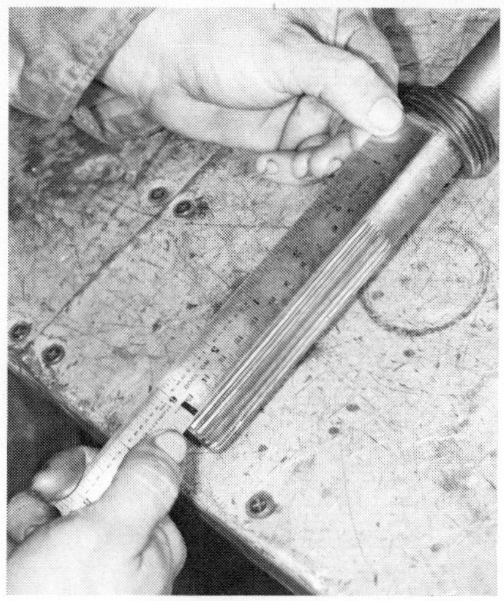

29. Distance from speedometer drive gear to end of shaft is measured and gear placement adjusted until you arrive at distance previously measured.

30. Expand the extension housing snap ring and fit mainshaft into the housing. Make sure snap ring is properly seated in the bearing groove.

31. Slide 2nd gear on front of shaft until it seats against the mainshaft shoulder. The 2nd gear blocker ring and synchronizer follow next.

32. Clutch gear bearings should be reinstalled before fitting gear to mainshaft. Wipe inside of gear with a thin coating of grease as shown.

33. Replace the pilot bearings inside the clutch gear. There are 14 of these and they are considerably larger than the countergear needle bearings.

34. After pilot bearings are in place, set clutch gear and 3rd speed blocker ring assembly over front of mainshaft, aligning blocker ring notches.

35. Position new gasket and insert clutch gear, mainshaft and extension housing into trans case, with clutch gear engaging anti-lash plate teeth.

36. With mainshaft in position, move reverse idler gear to rear of shaft and reinstall the C-clip in the slot in front of the gear.

HOW-TO: Saginaw 3-speed Overhaul

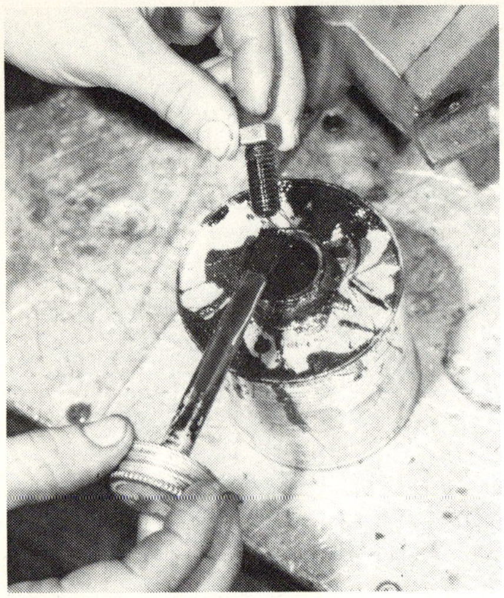

37. Bolt extension housing to case and torque to 45 ft.-lbs. Apply Permatex to smaller case bolt without lock washer and torque in place.

38. Install front bearing outer snap ring to bearing and position it over the clutch gear stem and into front case bore. Put inner snap ring in place.

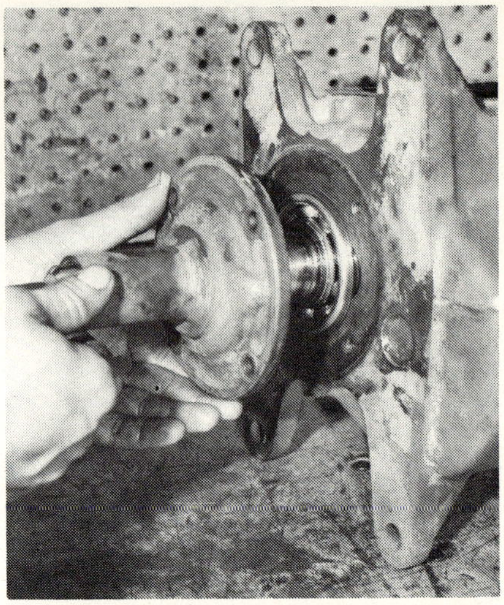

39. Bearing retainer and new gasket slide over clutch gear stem. Check oil seal inside retainer and replace if necessary. Torque bolts to 22 ft.-lbs.

40. Replacing the side cover is routine, but here's how to disassemble it if shift forks, detent cams or shifter shafts are bad. Forks pull out.

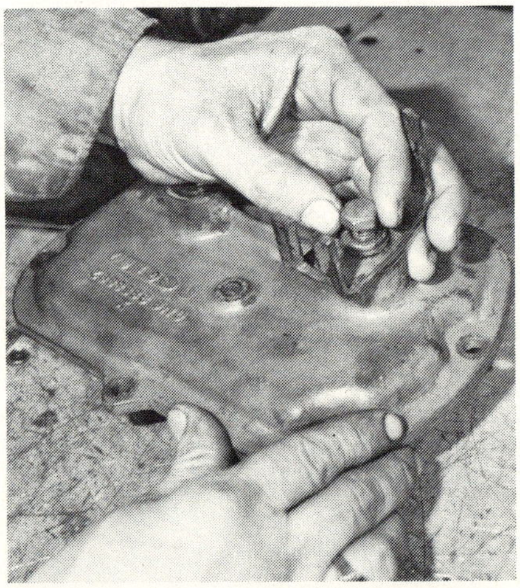

41. Shifter shaft assemblies are held in place by these cap screws. Loosen and remove screws and external shift lever connector.

42. Shifter shafts can now be pulled out of case cover, just as forks were removed from them. Saginaw design is simple and straightforward.

43. Spring holding detent cams in their operating position is removed by expanding it and lifting one end over the cam notches with pliers.

44. Remove C-clip from center stud and detent cams slip off. Reassembly of shifter cover is reverse—attach spring, secure shafts and insert forks.

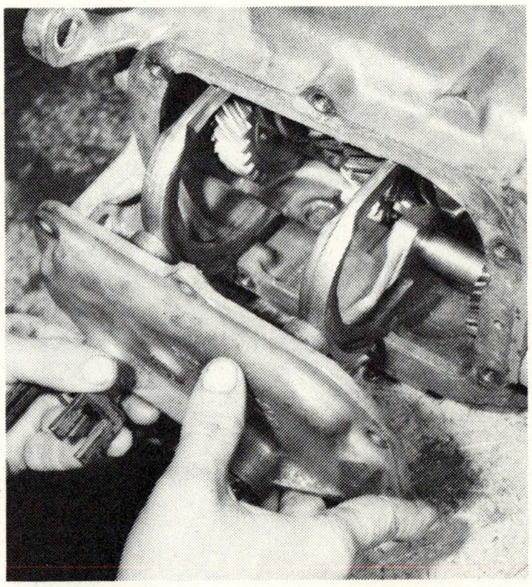

45. Fasten a new gasket to case or cover with a dab of grease. Replace cover. Torque all side cover bolts to 22 ft.-lbs. Saginaw overhaul is completed.

HOW-TO: Ford RAN 3-speed Overhaul

1. Begin overhaul of Ford RAN 3-speed box by draining lubricant. Then unbolt nine cap screws that hold the cover to the case. Lift cover up and off.

2. As cover comes off, long spring pops up. Underneath is detent plug. Removed with pencil-type magnet. Older models use set screw and short spring.

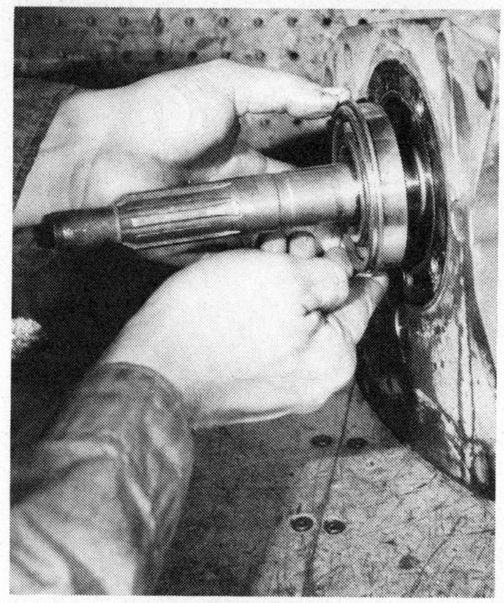

3. Remove the bearing retainer and unclip inner snap ring. If this is difficult to remove, place screwdriver behind outer snap ring and tap input shaft.

4. Remove five cap screws and lock washers holding extension housing to trans case, then withdraw housing from the case by pulling straight back.

5. Old gasket should be removed and discarded. It is a good idea to always use new gaskets when overhauling any trans; the cost is slight to prevent leaks.

6. You'll need a ⅛-inch Allen wrench to remove the set screws holding the shift forks to shift rail. Vise grip pliers are handy to break initial set.

7. Input shaft and gear come off by pulling straight out. A blocking ring and handful of roller bearings come with it. Set to one side—continue disassembly.

8. You should have no difficulty in removing the 1st/reverse shift rail as it slides out from rear of the case. If you find it necessary, tap lightly from behind.

9. Use a pencil-type magnet to remove the interlock plug. Another detent plug and spring are found below the interlock plug, and these must also come out.

HOW-TO: Ford RAN 3-speed Overhaul

10. Rotate the 2nd/3rd shift rail 90°, insert a punch through 1st/reverse rail opening in rear of case and tap until 2nd/3rd rail pops expansion plug at the front.

11. Draw the 2nd/3rd shift rail from the front of the trans case and lift out the two shift forks. Examine fork ends for signs of wear that might signal the need for a replacement.

12. Remove the snap ring at the rear of the speedometer drive gear and pull gear back to expose the output shaft lock ball. You should then remove gear and lock ball.

13. After output shaft bearing inner snap ring is removed, bearing should come off easily. If not, use a puller or insert screwdriver behind outer snap ring and tap shaft.

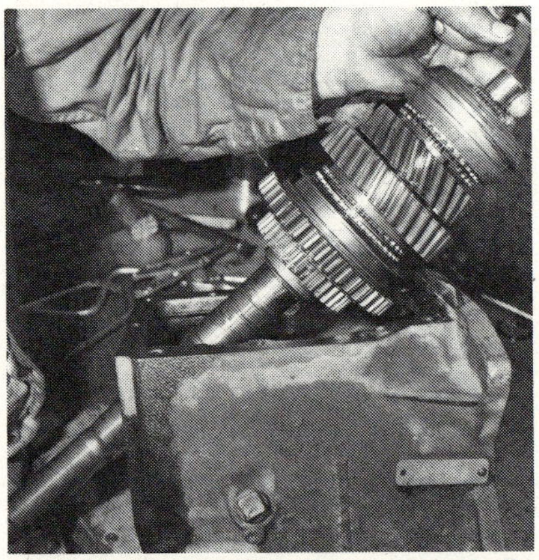

14. With the rear bearing removed, the output shaft and gear tower can be lifted out through top of case. Be sure to handle carefully to prevent damage to the assembly.

15. With the lubricant filler plug removed from the case, you can insert a ¼-in. punch through the plug hole. This will enable you to drive the countershaft roll pin out of the case.

16. Tap the countershaft from the rear of the case and withdraw shaft and countergear. Drive out reverse idler gear shaft and remove reverse idler gear and thrust washers.

17. The countershaft gear uses 25 needle bearings and a retainer at each end. Greasing the inside of the gear holds the bearings in place without using a dummy shaft.

18. Examine the countershaft gear thrust washer for wear and replace in case. Coat washer with grease and install at each end, making sure tangs fit into case recess.

19. Lay the countershaft gear in position between the thrust washers and then insert the countershaft from the rear of the case, then replace the roll pin in the shaft.

20. Check the countershaft gear end-play. If you find that it exceeds .018-in., thrust washers should be replaced to bring end-play within limits of .004-in.—.018-in. specifications.

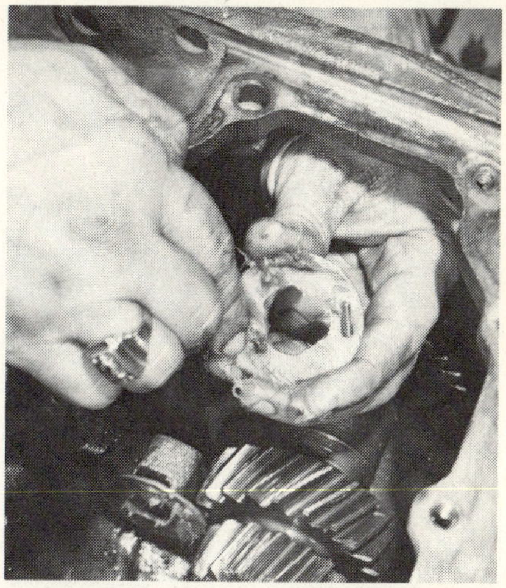

21. Coat the reverse idler gear thrust washers with grease and then position both inside the case. The tangs must also fit into the case recesses for proper installation.

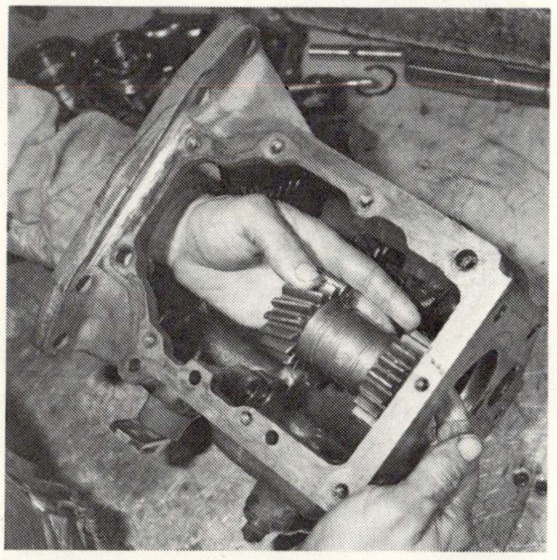

22. The reverse idler gear is replaced, and the shaft is installed. End-play should also be checked. The specified range is the same as for the countershaft gear.

23. Remove the snap ring from front of the output shaft, then slide the synchronizer, blocking rings and 2nd gear off the shaft. Examine all parts for wear or signs of damage.

24. Another snap ring and a tabbed thrust washer come off next, then the 1st gear and the blocking ring can be drawn from the output shaft without too much difficulty.

25. A third snap ring must be removed before 1st/reverse synchronizer and reverse gear can be removed. Synchronizer is a press fit and should not be hammered or pried off.

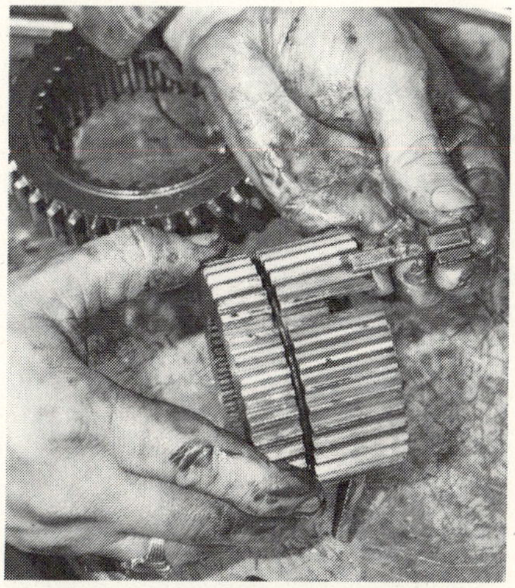

26. Although the synchronizer design is essentially the same, the 1st/reverse unit uses T-shaped inserts rather than the usual cap-type that are found on 2nd/3rd unit.

27. After examining all gears for worn, chipped or damaged teeth, reassembly of gear tower begins with reverse gear and 1st/reverse synchronizer. Secure the snap ring.

HOW-TO: Ford RAN 3-speed Overhaul

28. The blocking ring is fitted to the synchronizer. Then the 1st gear is reinstalled with cone to rear. This fits into the blocking ring and combines to form an integral unit.

29. The tabbed thrust washer fits in position in front of the 1st gear. You should make sure that tab engages slot as shown; then you can replace the snap ring.

30. Now slide the 2nd gear on the output shaft with its cone facing forward to engage the 2nd/3rd gear synchronizer assembly properly. This often confuses beginners.

31. The 2nd/3rd synchronizer with blocking rings at each end fits on output shaft. As the last step, you can replace the snap ring. This will serve to secure the assembly.

32. Reassembly of the transmission continues with replacement of output shaft. Insert shaft through the top of case and lower gears into position. Use a delicate touch.

33. Slide the rear bearing into position and seat until the outer snap ring fits snugly against the rear of the transmission case. Then you're ready to replace inner snap ring.

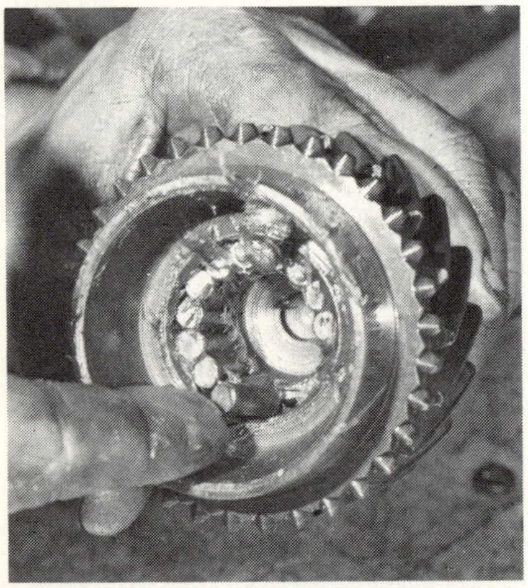

34. Use grease inside the input shaft to hold the roller bearings in place, then you can fit the input shaft through the front of the case and attach to end of output shaft.

35. Input shaft bearing is replaced just as rear bearing—outer snap ring must fit snugly against the front of the transmission case. After this has been done, reattach the inner snap ring.

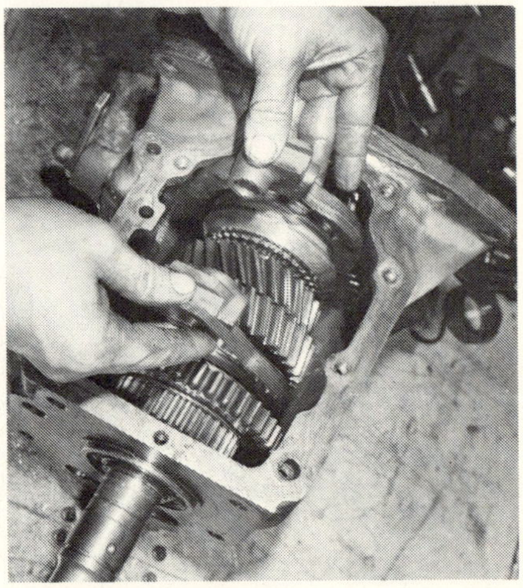

36. Reposition both shift forks over their respective synchronizer assemblies. You will find that the really tricky part comes next when you tackle the problem of the detent plug installation.

37. The next step is to move the synchronizer to 2nd gear position. Then fit the 2nd/3rd shift rail through the shift fork and drop a detent plug spring and detent plug in case.

38. Depress detent plug/spring and move shift rail until plug engages forward notch. Drop in interlock plug and reinstall 1st/2nd shift rail with synchronizer in 1st gear.

39. Replace the set screws in each shift fork, then you can work the synchronizer units by hand to make certain that the detent plug engages the shaft in the proper manner.

40. Replace the expansion plug at the rear of the case to cover the end of 2nd/3rd shift rail. Tap lightly in place with a mallet until the plug is flush with the case.

41. The front bearing retainer is now replaced using a new gasket. All cap screws should be torqued to 30-36 ft.-lbs., another thing many beginners forget or ignore.

42. Insert the lock ball in the output shaft recess and slide the speedometer drive gear in position on shaft just past snap ring groove so that it covers the lock ball.

43. The final snap ring is replaced at the rear of the speedometer drive gear. Snap ring pliers are a real must for anyone planning to overhaul this particular transmission.

44. After replacing its oil seal, the extension housing can now be fitted to transmission case. Be sure to use a new gasket, also be sure to torque all bolts to 42-50 ft.-lbs.

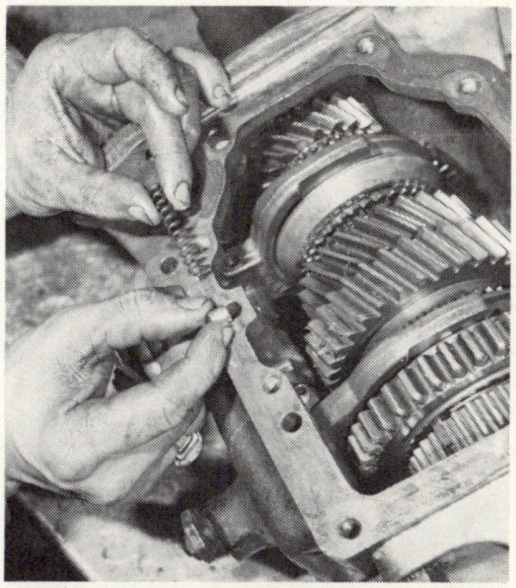

45. Replace the detent plug and spring (plug, spring and set screw on older models), then fit the cover into place and torque cap screws to 14-19 ft.-lbs. That's it.

HOW-TO: Muncie 4-speed Overhaul

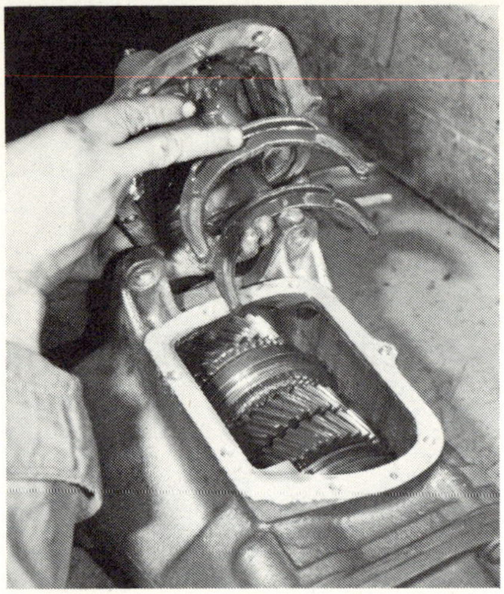

1. Muncie 4-speed design is very similiar to that of Saginaw 4-speed. Remove the side cover with the shift forks intact to begin disassembly.

2. Shift forks are removed from shifter shafts by simply pulling them out. Check forks for wear, and then replace in the shaft by pushing in.

3. Remove the four bolts and two bolt lock strips from the front bearing retainer. Slide retainer and gasket off the main drive and discard gasket.

4. Lock up trans by shifting it into two gears at once. This lets you loosen the main drive retaining nut; if the gears are not locked, shaft will turn.

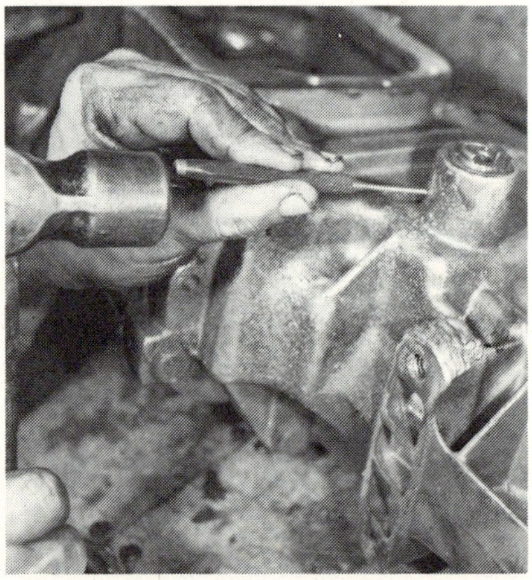

5. Return the gears to neutral position, and then you can drive out the reverse shifter lever boss retaining pin with an appropriate size punch.

6. Now you can withdraw the shifter shaft about ⅛-in. You will find that this disengages the reverse shift fork from the reverse gear.

7. Remove main drive bearing. If it does not come easily, use a screwdriver to pry on the snap ring or pull the shaft forward against bearing a few times.

8. Unbolt the extension housing and tap with a soft hammer. Move housing to the left until the reverse fork clears reverse gear and then remove.

9. Removing the extension housing will pop the rear reverse idler gear out where it can be easily grasped. Pull out of case and inspect teeth.

10. Speedometer drive gear is held in place by spring clip that fits into hole in main shaft. Depress the clip and slide the gear off.

11. The reverse gear can be removed by simply sliding it off the mainshaft. Gear should be cleaned and checked for worn or damaged teeth.

12. Before attempting to remove the mainshaft assembly from the case, move the 3-4 synchronizer clutch sleeve to 4th gear and tap bearing retainer.

13. Now reach inside case and lift out the front reverse idler gear and tanged thrust washer. This clears the way for countergear removal.

14. Drive the countergear shaft rearward until it clears the front of the case, then pull shaft out from rear of case with a revolving motion.

15. Lift countergear up and out of case. Pick out the two tanged washers and any roller bearings or spacers that might have dropped from gear.

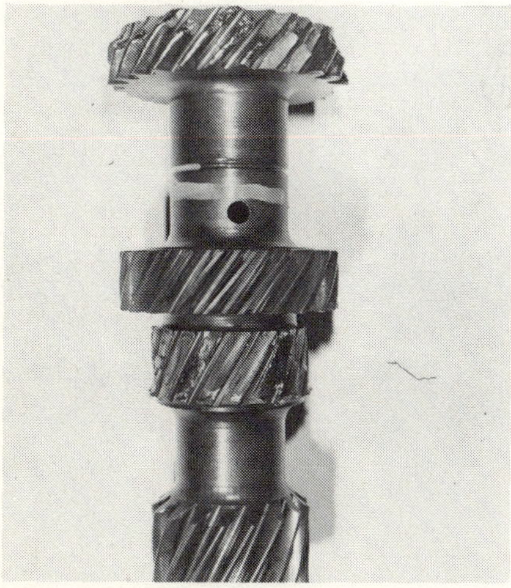

16. Inspect countergear. It takes a lot of doing to ruin one this badly, but as with anything else, if you really set your mind to it, well . . .

17. Main drive gear separated from the mainshaft when it was removed from trans case. Slide off 3-4 sliding sleeve, synchro ring and 3rd gear.

18. Check the clutch hub, the synchro ring and the sliding sleeve to make certain that the teeth are not worn and the unit meshes properly.

HOW-TO: Muncie 4-speed Overhaul

19. Remove rear bearing snap ring and tap retainer with soft hammer until it's free on the mainshaft, then slide the retainer off.

20. Pull the 1st gear and the synchro ring off the rear of the mainshaft. Then check the gear teeth and the synchro ring for wear or damage.

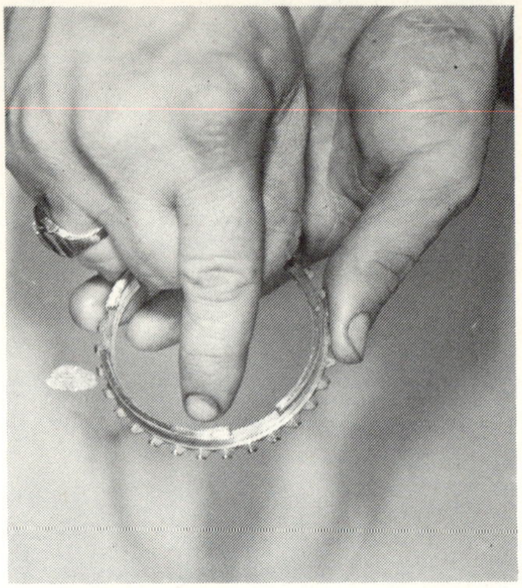

21. Synchro rings should be inspected carefully for defects. If they are worn or otherwise damaged, they can cause shifting problems.

22. Remove 2nd gear and 1-2 clutch assembly with a hydraulic press—don't try to beat the shaft out of the gear with a hammer.

23. The mainshaft/gear tower reassembly begins with 2nd gear, 1-2 clutch assembly and 1st gear sleeve. Seat against shaft stop.

24. The synchro ring and the 1st gear are replaced with the gear hub facing the clutch assembly. Seat in position against clutch hub.

25. Clean rear bearing with solvent and blow dry while holding race. Lubricate bearing with light engine oil and replace retainer on shaft.

26. Slide the selective snap ring onto mainshaft and lock in place. Up to .005-in. end-play is acceptable here—change snap ring if it's greater.

27. Slide reverse gear onto mainshaft. As it's a sliding gear, there's no snap ring used to secure it, so breathe easy—you didn't lose it.

28. Fit the tang end of the speedometer drive gear spring clip into the mainshaft hole, and then slide the gear over the clip until it stops.

29. Turn the mainshaft around and install the 3rd gear, synchro ring and 3-4 clutch hub. Slide entire assembly up against the mainshaft stop.

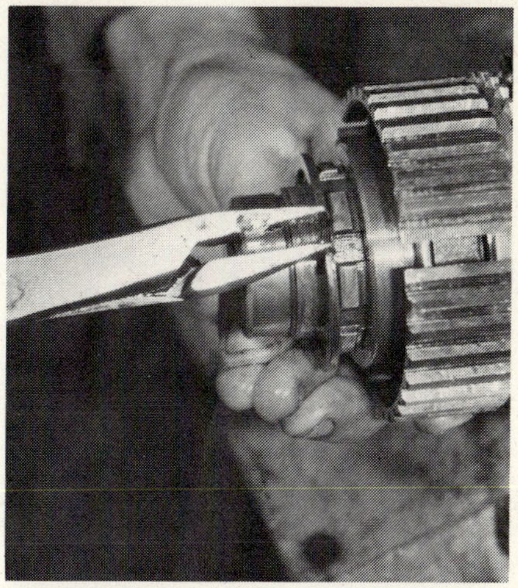

30. Replace the snap ring to hold the clutch hub and gear in position. Make sure to check that it fits completely into the groove.

31. With the two springs in position (one on each side of the hub), set all three keys into place so that both springs engage them.

32. With the keys in position, slip the 3-4 clutch sliding sleeve over the hub. Holding all three keys in place to start sleeve can be tricky.

33. Main drive gear bearing is composed of 17 rollers and a holding cage. Place rollers in the cage with grease and install bearing in shaft.

34. Now slip main drive and synchro ring over front end of mainshaft to complete gear tower reassembly. Set to one side while you work on countergear.

35. The countergear is equipped with anti-rattle plate. Check the teeth of the plate for wear or damage before reassembling the gear bearings.

36. Countergear uses 2 rows of 28 roller bearings at each end, separated by spacers and a roller spacer. Use grease to stick bearings in shaft.

HOW-TO: Muncie 4-speed Overhaul

37. Insert the roller spacer, a flat spacer, 28 bearings, another flat spacer and 28 more bearings. Cap each end with a flat spacer.

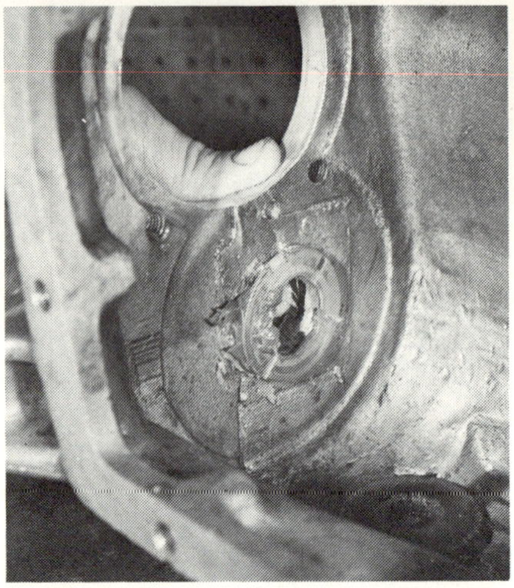

38. Grease is also used to hang the thrust washers in place for countergear replacement. Make sure tangs fit into case recesses.

39. Insert countergear and set in place between the thrust washers. Some Muncie 4-speeds are a tight fit here, and it's well to check end-play.

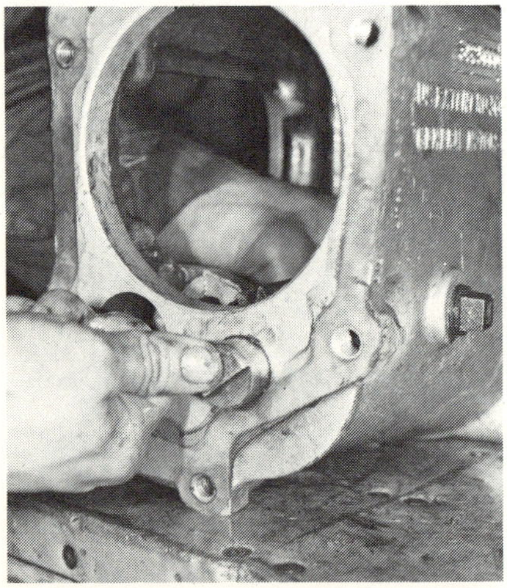

40. Install the countershaft in position shown and tap in place flush with case. Use a dial indicator to check end-play; it should be .025-in. or less.

41. Position the tanged thrust washer in trans case, and then replace the front reverse idler gear with its hub facing toward the rear of the case.

42. Slide the 3-4 synchronizing clutch sleeve forward into 4th gear, and then insert the mainshaft assembly into the transmission case.

43. Line up guide pin in rear bearing retainer with hole in the rear of the case and tap retainer into place with a soft hammer. Fit new gasket.

44. Replace the rear reverse idler gear in trans case and then seat into position. Some versions use a flat thrust washer here.

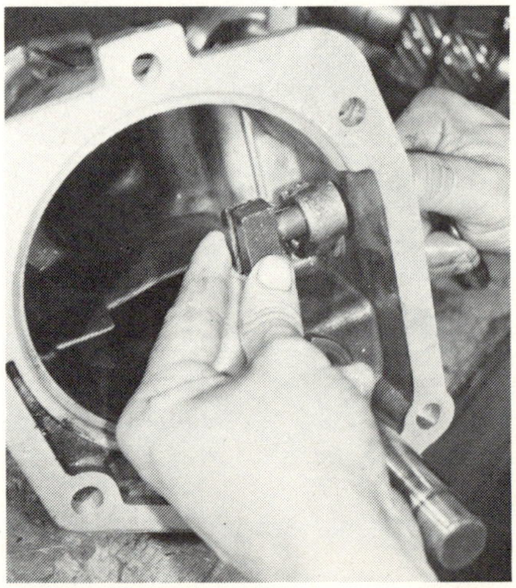

45. Pull reverse shifter shaft to the left side of the extension and rotate shaft until the reverse shift fork is forward in the housing.

46. Slide the extension housing onto the mainshaft while pushing in slowly on shifter shaft to engage fork with reverse gear shift collar.

47. Install the six extension housing bolts. Torque the upper three to 20 ft.-lbs. and the lower three to 30 ft.-lbs. Be sure new gasket is used.

48. Push/pull the reverse shifter shaft to line up the groove in the shaft with the holes in the boss and then drive the lock pin in place.

49. With the snap ring groove facing to the front, press the main drive bearing into trans case until several threads for the retaining nut can be seen.

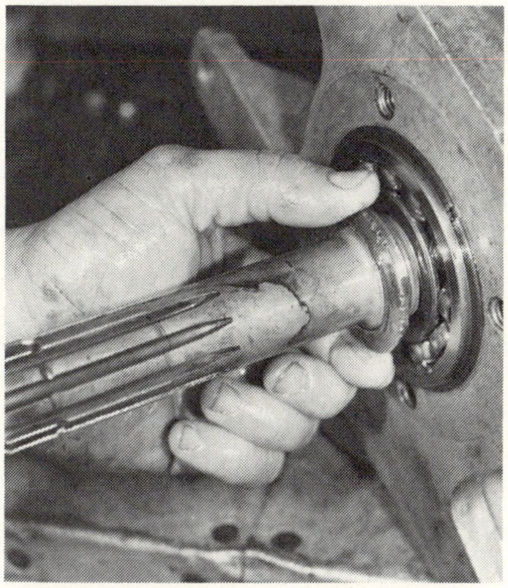

50. Thread the retaining nut in place on the main drive shaft and tighten the nut securely. Note: there is no torque requirement here.

51. Fit bearing retainer in place over the main drive shaft. Be sure that a new gasket is used here, and a little sealing compound won't hurt either.

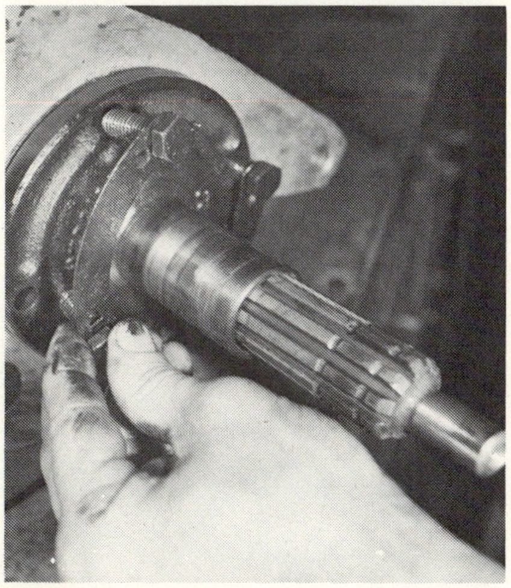

52. Thread retainer bolts through the lock strips and tighten in place by hand. Tighten evenly to prevent distortion and torque bolts to 25 ft.-lbs.

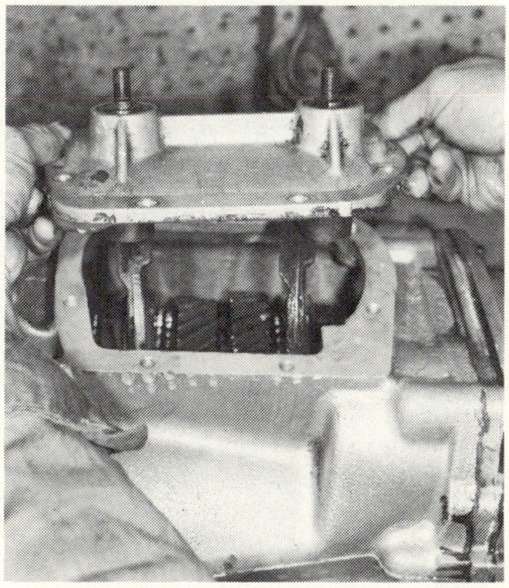

53. Shift the 3-4 clutch into neutral and the 1-2 clutch into 2nd gear. Match up side cover shifter shaft detents to same positions and slip cover in place.

54. The dowel pin in the side cover must engage the trans case. Replace bolts and tighten carefully to prevent cover distortion; torque to 20 ft.-lbs.

HOW-TO: Chrysler A-903/A-250 Overhaul

1. The A-903/A-250 is used on Chrysler products with 198 or 225 CID engines and is essentially the same except for gear ratio and countergear support.

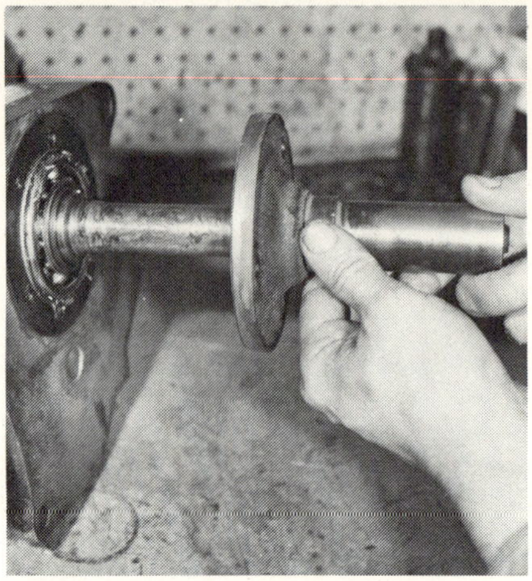

2. Remove the top cover and unbolt the front bearing retainer. Slide retainers off drive pinion shaft and discard the old gasket.

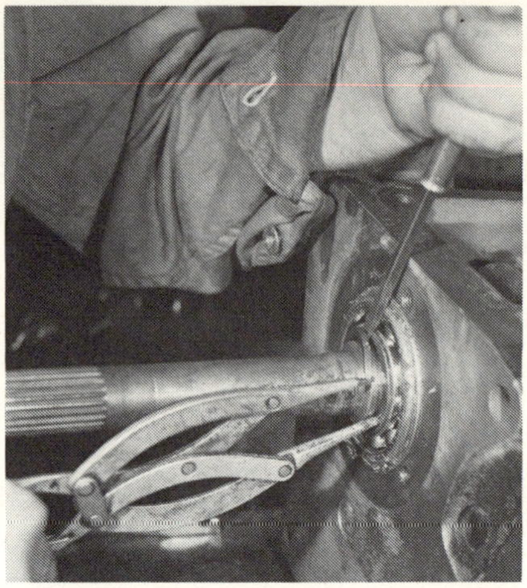

3. Small selective snap ring gives access to bearing. As snap ring has a similar size washer behind it, you may need help of a screwdriver.

4. With small snap ring removed, bearing should come out of case easily. If not, use screwdriver to pry against the bearing's large snap ring.

5. Drive pinion has omitted tooth area to clear countergear for removal. Rotate pinion shaft and pull—front inner synchro ring comes out too.

6. Unbolt and remove U-joint connector from mainshaft. Some applications use this solid connector while others use a sliding yoke.

7. Unbolt and carefully remove speedometer pinion adapter from extension housing, then unbolt and slide housing from the mainshaft.

8. Remove clutch gear retaining ring from the front of the mainshaft and the mainshaft bearing retaining ring from the rear of shaft.

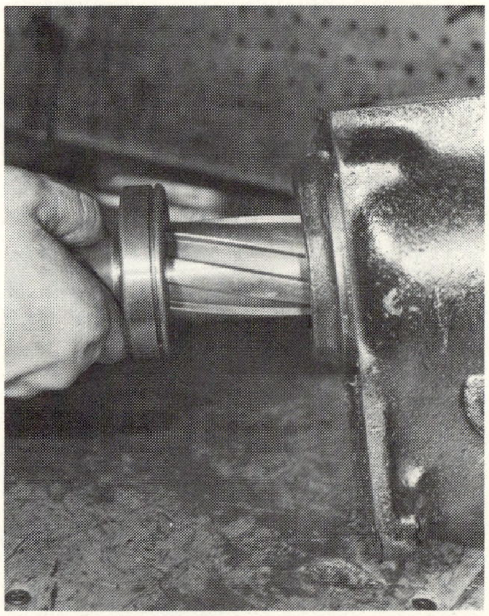

9. Pull the mainshaft and bearing from the rear of trans case. Steady the gears as they'll drop free inside the case—no disassembly required!

10. Remove and inspect the clutch gear, gear sleeve and two outer stop rings. This is the only standard box still using pin-type synchronizer.

11. Lift out and inspect 1st/reverse and 2nd gear. You'll also find another inner stop ring and a spreader spring that fits against 2nd gear face.

12. The shift forks pull out of case and are completely different. As this trans has only one synchro unit, a 1st-to-2nd shift is not synchronized.

13. Measure end-play of countergear with feeler gauge (.005-.022-in.), and then drive the countershaft toward rear of the case and remove.

14. Lift countergear and thrust washers from case. The A-250 uses a needle roller thrust bearing at the front for additional support.

15. Drive reverse idler gear shaft to rear and out of trans case. Lift gear and thrust washers from case. Remove roller bearings, clean and inspect.

16. Countergear uses one set of 22 roller bearings at each end separated by a roller spacer. Lubricate with grease and then reinstall.

17. Cap the bearings at each end with a bearing spacer ring and the countergear is ready for replacement in the transmission case.

18. Reverse idler gear also uses 22 roller bearings. Install with grease as you did countergear bearings and cap each end with thrust washers.

HOW-TO: Chrysler A-903/A-250 Overhaul

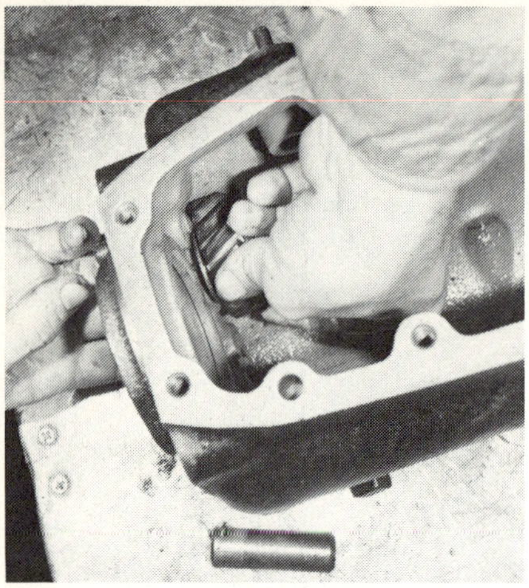

19. Position reverse idler gear in trans case with chamfered end of gear teeth toward front of case and insert reverse idler shaft.

20. Push shaft toward front of case and position the Woodruff key. Align key with case recess and drive shaft in flush with the case.

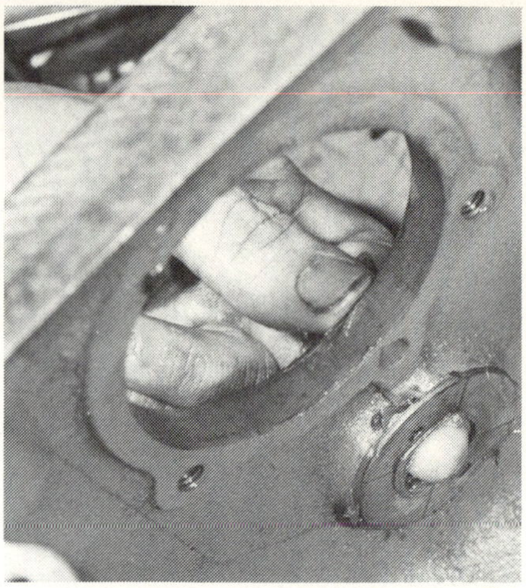

21. Coat the countergear thrust washers with grease and insert in trans case. Make sure that washer tabs slide into case grooves provided.

22. Lower the countergear into position and install countershaft. Insert Woodruff key and drive shaft in until key is bottomed in case recess.

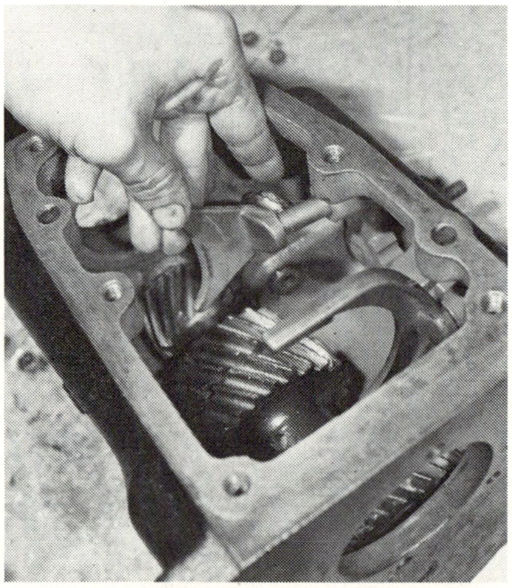

23. Shift forks simply push back into place. If shift lever shafts were removed to replace leaking seals, replace the detent spring and balls.

24. Forks are offset toward rear of trans when correctly positioned. Install 1st/reverse gear so that the gear hub engages fork.

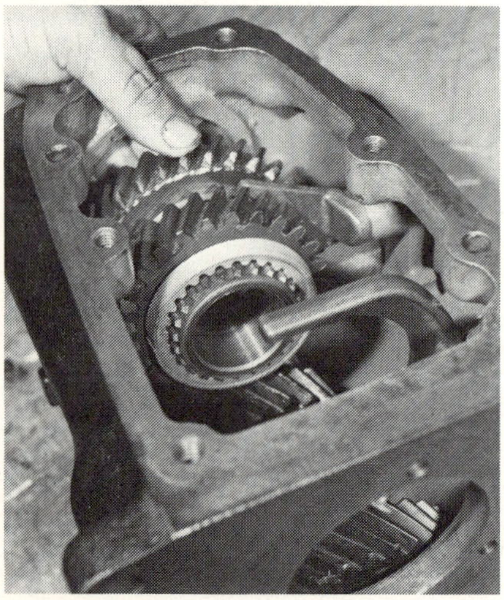

25. Now install 2nd gear with the spreader ring facing front of trans. Inner stop ring of the synchro unit fits onto gear hub.

26. Arrange synchronizer parts (minus stop ring already installed) into a single unit and position in trans case to engage 2nd/3rd shift fork.

27. Carefully work mainshaft through the gears and synchronizer assembly from the rear of trans case until the bearing bottoms into place.

28. Reaching through the front of the case, install the synchronizer clutch gear snap ring on the mainshaft and seat completely.

29. Replacing mainshaft bearing snap ring secures mainshaft at each end. Not having to disassemble the gears from mainshaft makes it easier.

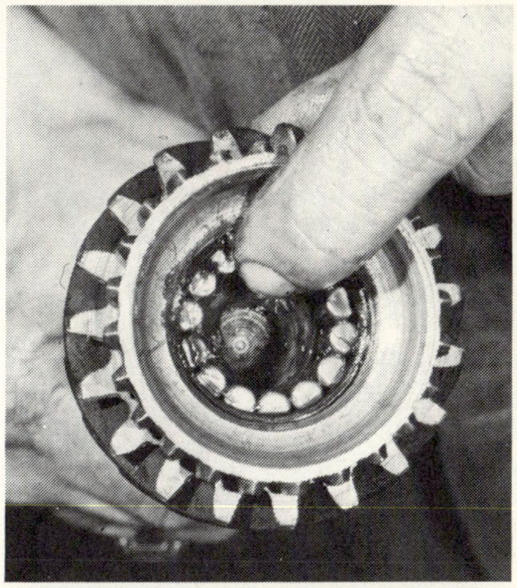

30. Mainshaft pilot bearing in end of drive pinion uses 14 roller bearings. Install with grease to hold in place and then insert bearing retainer ring.

31. Rotate drive pinion until omitted clutch tooth area is next to countergear. Guide into case to engage inner stop ring with clutch teeth and seat.

32. Be sure to replace the oil slinger before seating the bearing. Bearing is fully seated when its snap ring is in full contact with case.

33. Washer has tab that must engage slot in pinion shaft. Seating the small snap ring in place properly completes the reassembly.

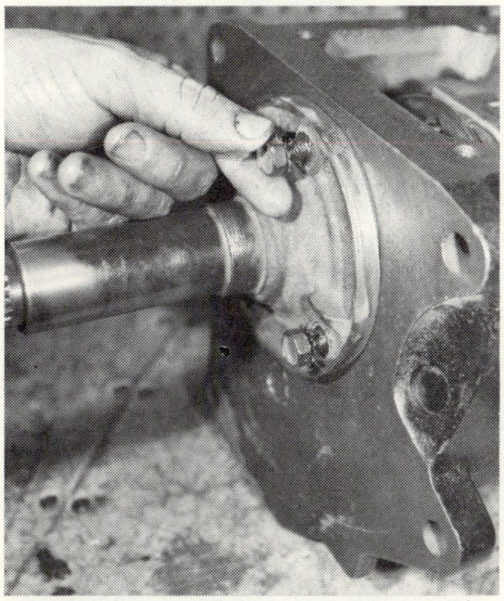

34. Position the bearing retainer and a new gasket on the trans case. Install bolts with thread sealing compound and torque to 30 ft.-lbs.

35. Use a thin coating of grease to hold the new gasket in position before replacing extension housing. Then slide housing over the mainshaft.

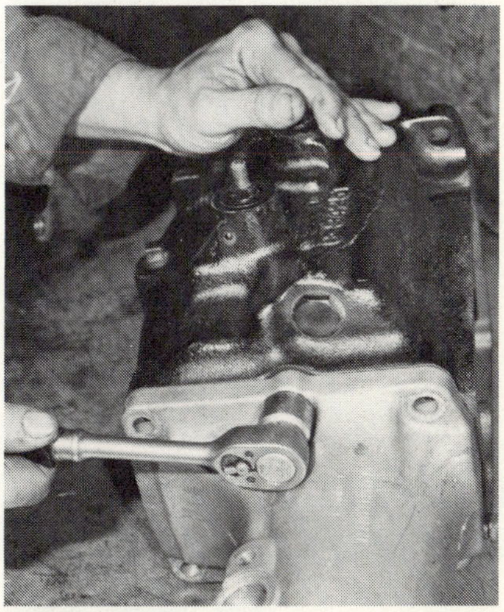

36. Reinstall the bolts and torque to 50 ft.-lbs. Then replace cover and torque bolts to 12 ft.-lbs. Install U-joint yoke and speedometer adapter.

Hot Rodding Standard Transmissions

Does your car have 4-on-the-floor? If it doesn't, you probably wish it did. The 4-speed transmission, with its wobble stick jetting out of the carpet or perched prominently on top of a fancy console, is the way to go. There isn't any better method of getting the most power out of your engine.

In order to give you the best acceleration and speed, an engine has to be running at an rpm where it produces its maximum horsepower. The more gears you have to select in a transmission, the easier it is to find a gear that will keep the engine rpm up where there are great gobs of power. In accelerating from a standing start, it would be nice if you could use a Granny gear that would put the engine at its maximum power rpm the second you put your foot in the carburetor. With such a gear you would probably get no wheel-spin and be out in front at the end of your first 50 feet. Unfortunately, races are not won in the first 50 feet, but over the whole quarter-mile.

In order to keep your engine from being all wound out and completely out of breath just after it gets off the line, you have to get your gear ratio out of the basement and up to a compromise figure. With a little bit higher gear you may not get off the line with as much snap, but once you are moving in the first few feet then you can really grab hold of the track and have much more wind left in your engine.

If this business of engine rpm and gear ratio is new to you, then you owe it to yourself to take a drive in a manual transmission car and listen to what's happening to the engine in the different gears. You can only go so fast in low gear no matter how much you open the throttle. Once the engine has reached the rpm where it produces maximum horsepower, it doesn't do any good to give it more throttle, and in fact it is possible to literally destroy the engine by overspeeding it. Shift into 2nd gear, and you get the rpm back down to where the engine can pull at the speed the car is moving. Then the same thing happens again in 2nd—you run out of gear and have to shift to high.

You can see that for any speed the car is traveling, there is an ideal gear ratio that will put the engine rpm at the point where it can push the car

1

2

3

4

the best. The more speeds we have in the transmission, the better we can do this.

There are various theories on the best way to space the gear ratios in a 4-speed box. A wide-ratio box gives a low gear that will get the car off the line like a striped ape, but when you shift to 2nd, there is too much of a jump, and the engine can't pull as well because the car didn't get going fast enough in the low gear. With the wide-ratio transmission, the gap between the gears can be ideal for some shifts, and horrible for others. You may have a perfect low gear, but when you drop it in 2nd—nothing. You may have real good gaps between 1st and 2nd, and 2nd and 3rd, but then the jump from 3rd to 4th is so great that the engine can't pull the car when you drop it into high.

Most road racing applications use what is called a close-ratio transmission. The theory there is that standing starts are not important, but the ability to select the right gear in all of the many corners on a road race course is vital. A good driver with a close-ratio 4-speed in a Corvette can really play a tune on the gears, always keeping his engine at the rpm where it can really propel his car. Close-ratio boxes also work pretty well in a drag race, if you can break the rear tires loose or slip the clutch enough to get the car moving from a dead stop. Once you get everything going, then the engine is in an rpm range where it can really put out, and it's away we go from then on.

If a 4-speed box is so great, why not 5 or even 6? They would be great. The more speeds the better. But you reach a point where you don't need any more speeds because your engine is flexible enough to carry you with the gears that you have. An engine with great gobs of power over a wide rpm range does not need as many gears to do its job as a small European engine that has to be buzzed to put out any horsepower. Dragsters put out over 1000 hp; they only use one gear and do a pretty good job of it. In most cases, they would lose more time making a shift than they would by leaving the engine in a single gear and allowing the engine to over-rev. The new Lenco overdrive is going to solve the problem of the time to make a shift, because it changes speeds without any demeshing of gears.

Some hot rodders may feel that they can throw a shift without losing any time. No matter how fast they are, it still takes a certain amount of time to throw the shift and, when you are working in the 6-second range, that additional shifting time may cost you the race. Another reason that a 5-speed box is not necessary in a lot of cars is the weight of the vehicle. In a light sports car the engine can pull pretty well, even when it is outside its best rpm band, because it isn't pulling much weight. This is one reason that the Corvette can get away with the close-ratio box, with a 2.20:1 low gear. The car is light so the engine doesn't have much load to pull around.

When the Corvette first came out, there was a close-ratio 3-speed available with a 2.20 low gear, the same as the later 4-speed. Low in the standard 3-speed was 2.94. The close-ratio 3-speed box was also available in other Chevrolet passenger cars in 1957. It was great if you wanted to go racing, but in a heavy street job that had to stop and start a lot, that 2.20:1 low gear could really be a problem. If you don't believe it, just try starting up from a dead stop on a hill with a big heavy passenger

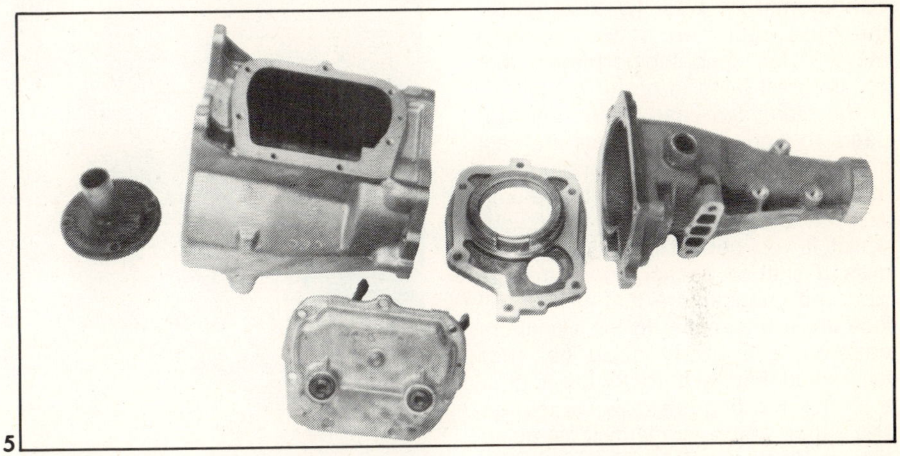
5

6

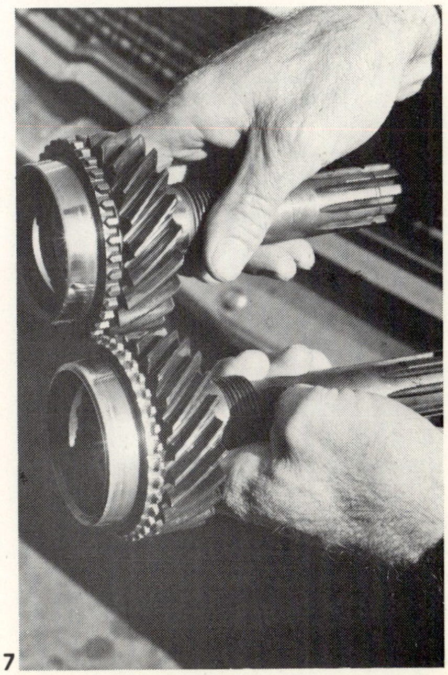
7

1. A recognized authority on stock as well as modified transmissions (and rear ends) Pepe Estrada took time to consult with us on this chapter.

2. Even the stout Warner T-10 can't withstand all that can be built into an engine. Obviously, this mainshaft "went" when someone really got on it.

3. Interesting cutaway display model of a Ford 4-speed. Better than a drawing, this one can be shifted while you watch the internal goings-on.

4. A Muncie with the shift cover removed. You won't see synchro sleeves like this in a regular Muncie. These are special sleeves adapted by Pepe Estrada.

5. The parts that hold a Muncie together. The plate between the tailshaft and the case is a support for the rear bearing.

6. Here's a slick way to remove a pressed-on speedometer gear on any transmission. Just use the reverse gear as a slide hammer. It doesn't take many blows before the speedometer gear comes right off.

7. Look closely and you'll see what happens when you miss a shift. The teeth on the lower gear are half gone.

car equipped with a close-ratio box. You'll be lucky if you can get going without burning up the clutch. On Corvette there was no problem even in street traffic on the steepest hills, because it was light enough so that the engine could easily pull it.

THE T-10 AND THE MUNCIE

The first 4-speed was the Warner T-10 which appeared on Corvettes in the middle of the model year 1957. In 1959, and maybe on a few '58's, the T-10 appeared on Chevrolet passenger cars with the 283 fuel injection or with the 348. It was a good transmission, but there wasn't any great rush to buy 4-speeds. That would come later. Actually the T-10 was a redesign made up from an earlier Borg-Warner 3-speed.

1

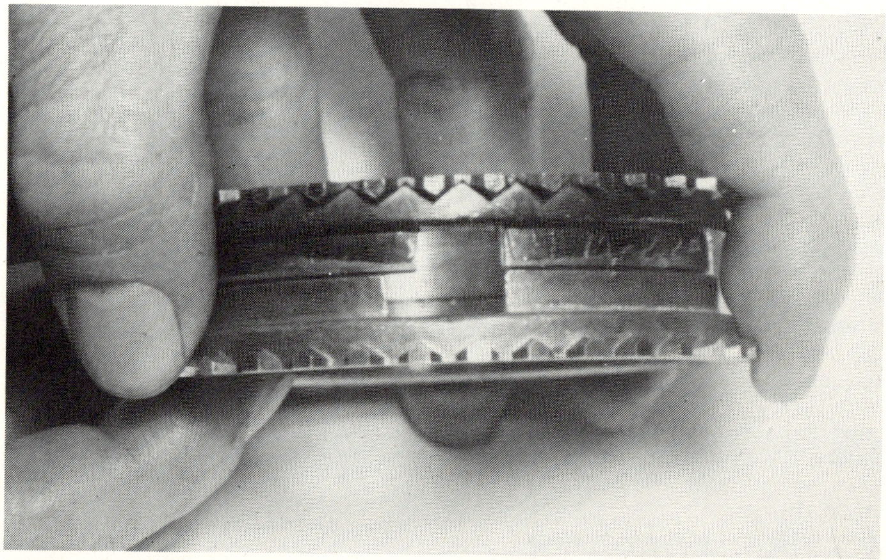

2

In 1961, if you ordered a 4-speed on your Corvette, you got a T-10 with an aluminum case, and the following year the aluminum case T-10 appeared on Chevy passenger cars. The aluminum was 15 lbs. lighter than the cast iron. It was a good move as far as lightening the car, but not the best thing to do for reliability.

The aluminum case is fine and appears to work just as well as the cast iron when everything is operating correctly. The problem with the aluminum case is that it really hurts your wallet if you cobble a shift. You can miss a shift in a cast iron transmission and chances are that all you will lose are a few gears. In the aluminum case, it is a pretty good bet that when you knock a tooth off a gear and the tooth goes down between two other gears—something is going to give. It is very unfortunate, but the aluminum is the weakest part of the transmission.

In 1963, the famous Chevrolet Muncie 4-speed came along. It was also in an aluminum case, but it was a much tougher transmission than the old T-10, which saw its last year on Chevrolet in 1963. If you ever have a chance to compare a Muncie with a T-10 side by side, you will notice there is quite a bit of similarity. The main difference internally is that the T-10 has straight-cut gears on reverse, and the Muncie has helical for quieter running.

With the two transmissions fully assembled, the easiest way to tell the Muncie from the Borg-Warner T-10 is to look at the shift linkage cover. Both transmissions have two levers on the main cover, with the reverse lever on the extension housing. But the Muncie has one bolt in the center at the top of the cover. At the bolt hole, the cover has a slight projection like an arrow pointing straight up. Look for that projection, and you'll know you are looking at a Muncie. Other 4-speed boxes may have a similar projection, but you will know that they are not the Muncie because they will have more than two shift levers on the shift cover.

The Muncie transmission has been a successful high-performance box for quite a few years. Since its inception in 1963 up until the end of 1965 model year, it was available in either a wide-ratio or a close-ratio version with a 2.56 low or a 2.20 low. Generally speaking, low-horsepower engines, compact cars or intermediate cars used the 2.56 low gear transmission; and the higher horsepower engines used the 2.20. In fact, the 2.20:1 low gear was standard on the upper horsepower Corvette versions.

In 1966, the wide-ratio Muncies were improved with coarser teeth on the main drive and cluster gear. They also had a riveted plate on the big

1. A layout of the Muncie gears with the synchronizers removed from the mainshaft.

2. All synchro rings look alike from a distance, but notice the difference in the width of the slot on these two rings. If you put the wrong ring in, you are going to have shifting problems.

3. Here's the T-10 gear train; note the straight-cut gears on reverse.

4. The ring on the left has a thicker shoulder and it is much stronger than the one on the right.

5. These two rings are interchangeable, but notice the difference in thickness of the upper portion. The one on the right is considerably stronger.

6. A synchro ring must sit on the gear cone solidly without rocking. Dropping a ring on the floor will make it egg shaped, and then you might as well throw it away.

7. This synchro ring is no good, because it goes too far onto the cone of the gear. Notice the considerable lack of space between the two rows of teeth.

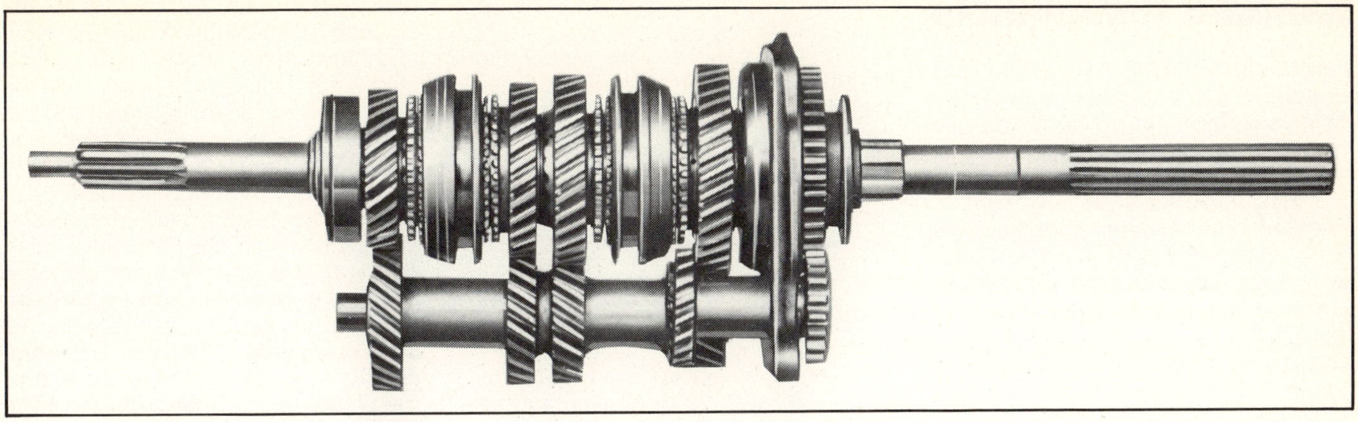

3

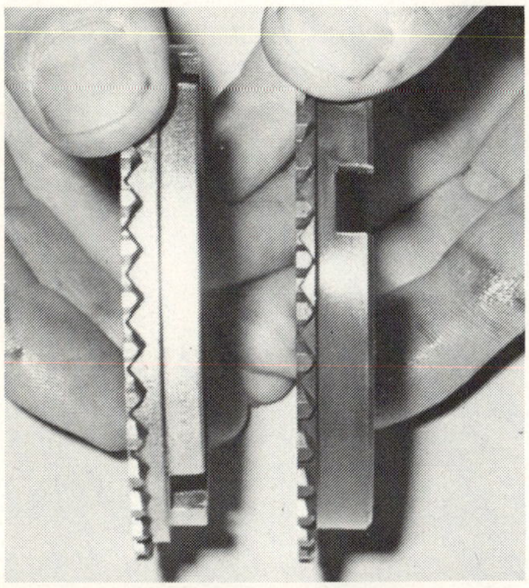

4

5

6

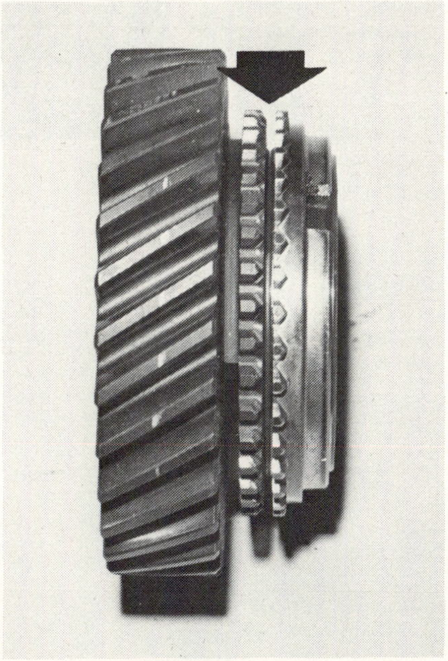

7

end of the cluster that was spring-loaded and prevented cluster rattle when the power flow was straight through the transmission, as in high gear. Several other small changes were made in both the wide- and close-ratio Muncies, such as reinforced synchronizer blocker rings and a larger diameter cluster gear shaft. The changes in the wide-ratio model resulted in a slight difference in low gear ratio from a 2.56 low to a 2.52 low. From 1966 on, the Muncie 4-speed has been available in the 2.20 and the 2.52 low gears.

Complicating the situation in 1966 was a new 4-speed, medium-duty transmission called the Saginaw. It was actually the same old 3-speed trans with the 4-speed added to the box. It is easily recognizable because it has all of the shift levers on the cover, whereas the Muncie and the Borg-Warner 4-speeds have the reverse lever on the extension housing.

The Saginaw 4-speed is available with two different sets of ratios. Low-horsepower engines have 1st, 2nd and 3rd gears of 3.11, 2.20 and 1.47. Bigger engines use the same transmission but with different gears that give ratios in 1st, 2nd and 3rd of 2.54, 1.80 and 1.32. The year 1966 was the first time you could get a Muncie close-ratio 4-speed with the 2.20 low gear in a Chevy II.

In both 1966 and 1967, you may find a listing of a 4-speed transmission with the only identification being the gear ratio of the low gear. It could be 3.11, 2.54, 2.52 or 2.20. The 3.11 and 2.54 are the medium-duty Saginaw 4-speed, and the 2.52 and 2.20 are the Muncie 4-speed. In 1968, the Saginaw with a 3.11 low gear was replaced with a new gearset which resulted in a 2.85 low, a 2.20 2nd and a 1.35 3rd.

Anybody who wants to do a little juggling of transmissions and engines can achieve just about the ideal ratio for any application. But you must remember that the Saginaw transmission is for medium-duty use only. If you want a high-performance transmission that will stand up under a lot of heavy use, you must use the Muncie unit.

Standard Transmissions

Through the years there were slight changes in the Muncie that really don't amount to much as far as use of the transmission goes. In 1968 they added an oil slinger to the bearing that holds the main drive gear in the front of the case. Synchronizers were changed slightly in 1968 also, and on all passenger cars except Corvette the manual lockout on the linkage for reverse was changed to a spring that made it easier and faster to get into reverse gear.

In 1969 came the infamous transmission locking device connected with the key. The ignition cannot be turned to the lock position unless the transmission is in reverse. After years of preaching that a manual transmission car should never be left in gear because a bump from another vehicle might break the transmission, it seems kind of silly to force the driver to leave his car parked in reverse gear. We can't blame the Chevrolet engineers for the transmission locking device, however. They had to do it under pressure from the United States government. Perhaps in later years, some better sense will prevail and the locking device will be removed from manual transmission cars but kept on automatics where it has some value.

In 1970 the Muncie went from a separate magnet attached inside the case to a magnetic drain plug. In both cases the idea is to pick up any particles of metal that come off the gears. Also included in all 1970 transmissions is transmission controlled spark. It's a little switch on the 3-4 shift lever that eliminates the vacuum advance in the distributor unless the transmission is in high gear. It's strictly an emission control gadget and has nothing to do with performance. When the engine is wide open you don't have any vacuum advance anyway, so the switch does no harm and there is no point in attempting to remove or disconnect it.

In the last few years a special heavy-duty Muncie 4-speed has been available on various Chevrolet Division cars as option M-22. It has low angle gears and heavier synchros. The reduced angle of the gears makes them much more noisy than those in the regular Muncie, so the trans has been dubbed "The Rockcrusher."

The Rockcrusher comes with the close-ratio gearset, which has a 2.20 low gear. It can be swapped with a regular Muncie, but not easily, because the front and rear splines are different, making it necessary to change the clutch disc and the U-joint yoke.

In conversations with owners of M-22 transmissions, we discovered that a switch to an M-22 (or taking it as an option with a new car) may not be worth the trouble. It doesn't appear to last any longer than the regular Muncie, and in some cases has an even shorter life.

GEAR RATIOS

With all the different ratios available in the various transmissions, you'd think there would be some way you could look at the transmission and tell what ratio its gears were. There is a way—you take the shift cover off and count the teeth. It's messy but it is positive. On the Borg-Warner T-10 transmission with 2.20 1st gear, the big end of the cluster has 28 teeth and the main drive gear which meshes with it has 27 teeth. On the T-10 with the 2.54 1st gear, the big end of the cluster has 30 teeth and the main drive gear has 25. The 1963-65 Muncie with the 2.56 1st gear has 29 teeth on the front cluster and 24 teeth on the main drive gear. The '66 and later Muncie with the 2.52 1st gear comes with 25 teeth on the front cluster and 21 teeth on the main drive gear. All of the 2.20 Muncies have 27 teeth on the big end of the cluster and 26 teeth on the main drive gear.

All of this information on numbers of gear teeth is available at any parts house by looking in their gear cata-

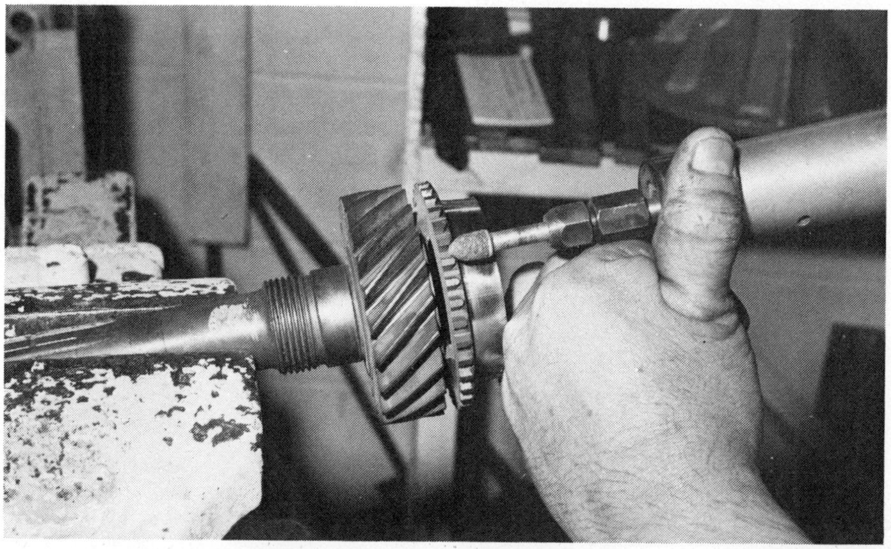

1

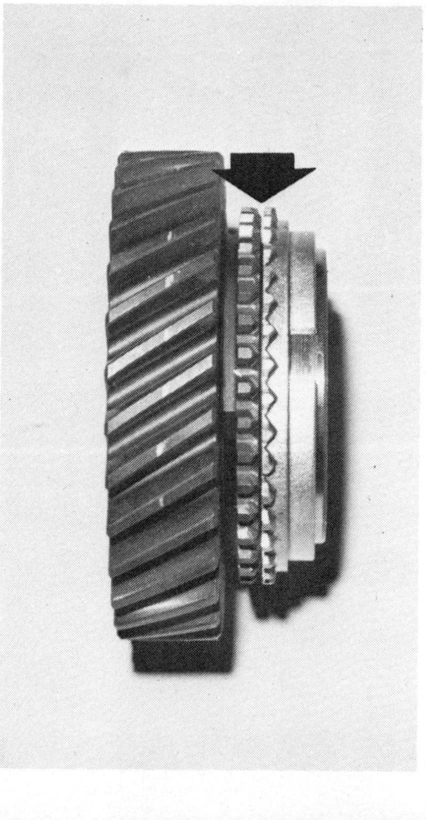

2

3

log. The parts book at your local Chevrolet dealer will also tell you the number of teeth on the various gears, and in most cases the ratio that went with those gears.

You can figure out the ratio yourself just by counting the teeth and doing a little arithmetic. In 1st, 2nd or 3rd gear, the power flow is through four different gears. Simply figure out which is the driving gear and which gear is driven; then divide the number of teeth on the driving gear into the driven gear. You will come up with one figure for each pair of gears. Multiply those two figures together and you'll end up with the overall ratio of the combination.

In some cases the ratio you get by figuring will not agree precisely with the listed ratio. Your calculations might come out to 2.24, but the listed ratio might be 2.20. The reason for the discrepancy is some slight design change in the transmission. Rather than spend the rest of their lives explaining why they made a ratio change, the engineers will keep calling the ratio by the old figure, because the change is so slight that it won't affect the performance of the transmission noticeably.

Almost any transmission that is strong enough to do the job will work fine as long as it is in gear. But when you shift it, you find out if it is a good transmission or not. Sometimes the trouble is in the linkage, or in the basic design of the shifter, but—whatever it is—the combination of transmission and shifter has to be right, or you will miss shifts.

We have spoken with drag racers who feel that the stock 4-speed shift linkage is inadequate. It seems that rather than try to modify the stock shifter, everybody we talked to would prefer to go to a different brand. There are almost as many brands of shifters as there are custom wheels, and everybody seems to think that any of them is better than the stock Chevy shifter. We doubt that very much. If you buy a major brand of shifter and install it carefully, you won't have any problems, but buying any shifter just so you can get rid of the stock Chevy one would well be asking for trouble.

SYNCHROS

Inside the box it's a different story. You can't throw away the transmission if it doesn't shift properly. Modification is the name of the game, and it all takes place in the area of the synchros. But before you can modify them, or understand how the experts do it, you have to know what makes them work.

In a modern transmission, all forward gears are in constant mesh. Power flow comes from the clutch to the main drive gear and directly to the cluster gear. The 1st, 2nd and 3rd gears are all meshed with the cluster, and mounted on the mainshaft that goes out the back of the transmission and connects with the driveshaft.

None of the gears on the mainshaft is splined to it. They all spin freely on the shaft. Each gear has a ring of teeth on its side which are separate from the teeth that mesh with the cluster. A synchro that is splined to the mainshaft sits alongside each

4

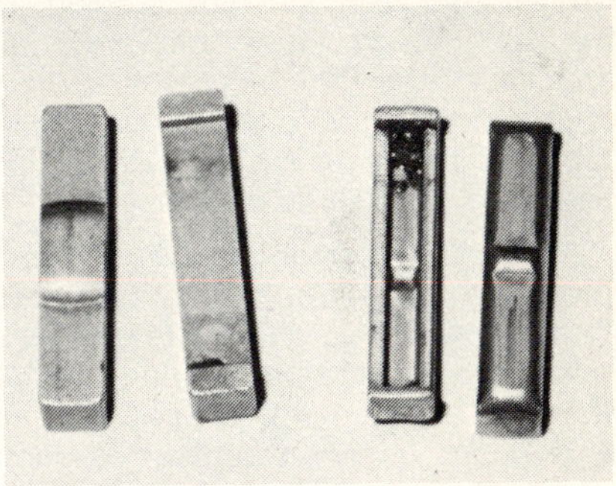

5

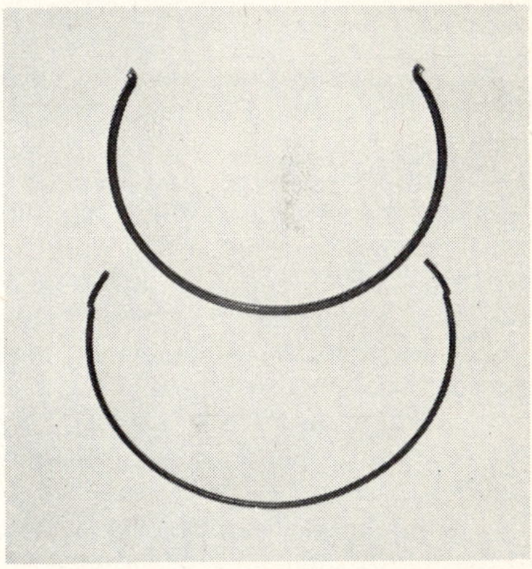

6

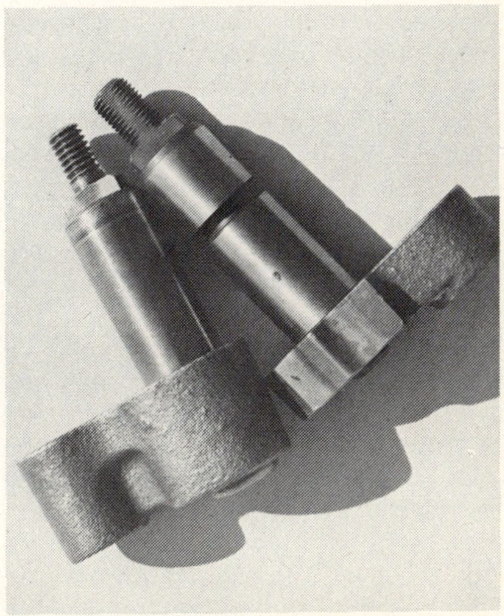

7

1. Renewing the teeth on the gear is recommended for experts only. Pepe can do it neatly with a hand grinder, but only once.

2. This synchro ring is a good one. It stands high on the cone, leaving space between the two rows of teeth.

3. To maintain proper distance between the mainshaft and main drive gear, Pepe installs a flat washer. Result is easier shifting from 3rd to 4th.

4. Synchro sleeves come with two different widths of detent notches. On the left is wide notch which takes key to match. On the right, key with narrow projection fits in the narrow notch.

5. Selection of right key may make the difference between good and bad shifting. Here are top and bottom of two kinds of keys. The one on the left is solid. At high rpm it weighs several pounds, and makes the shift difficult. The one on the right, being hollow, will shift easier at high rpm.

6. These are the springs that go underneath the keys. The upper one is too heavy and should not be used. The lower, lighter one, is much better for hot-rodding, but sometimes Pepe leaves it out, depending upon use of transmission.

7. Muncie transmissions have used two different types of shifter shafts with threaded studs. Early models used a 5/16 coarse thread which has a tendency to break. The later ones have a ⅜-in. fine thread which is the one to use.

gear. When the driver makes a shift, he moves the synchro outer sleeve so that it engages with the ring of teeth on the side of the gear. With the gear locked to the synchro, and the synchro splined to the mainshaft, power is transmitted to the driveshaft.

The problem is that unless the synchro sleeve and the gear are turning at the same speed, the teeth don't want to engage. It's like trying to shove a stick through a picket fence when you are moving past the fence at high speed. All you get is a lot of noise, which doesn't do either the stick or the fence any good.

It is the synchro's job to slow down the stick or speed up the picket fence, whichever is necessary to get the synchro and the gear turning at the same speed. Once they are both functioning at the same speed, the teeth on the synchro sleeve will easily slip over the teeth on the side of the gear, and the shift is completed.

When the sleeve of the synchro moves toward the gear, it moves three keys along with it. These keys push on slots in a bronze ring so that the ring goes against a cone on the side of the gear. Of course, as soon as the ring hits the cone, it tries to spin with the gear, but the keys prevent this from happening.

As the sleeve continues to move toward the gear, it slides over the keys and starts to mesh with the teeth on the bronze ring. But the teeth on the ring do not line up with the spaces between the teeth on the synchro sleeve, and that is the reason that this type of synchronizer works. Each tooth is pointed, so that the teeth can be off mesh by several thousandths of an inch. When the synchro sleeve starts to engage with the bronze ring, its teeth slide along the teeth on the ring, forcing the ring to back up, against the direction of rotation. In pushing against the ring, the sleeve forces the ring against the cone on the gear, which makes the gear rotate at the same speed as the synchro. After the sleeve has forced the bronze ring up against the cone, it continues to move toward the gear and engages the teeth easily.

Modifications to the synchros are about as numerous as you can imagine. There are bronze rings on the market that have no teeth at all. Using the toothless rings eliminates any synchronizing, which will result in an easier shift if the driver waits for the gear to slow down before he makes his shift. If he doesn't wait, the teeth will clash, but they will mesh, providing the driver has a good strong arm. The trouble with the toothless setup

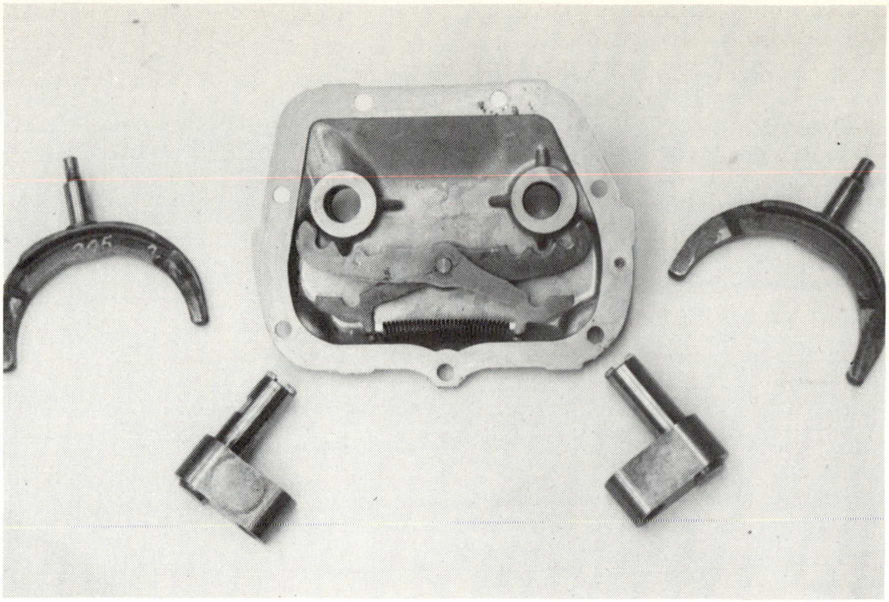

1

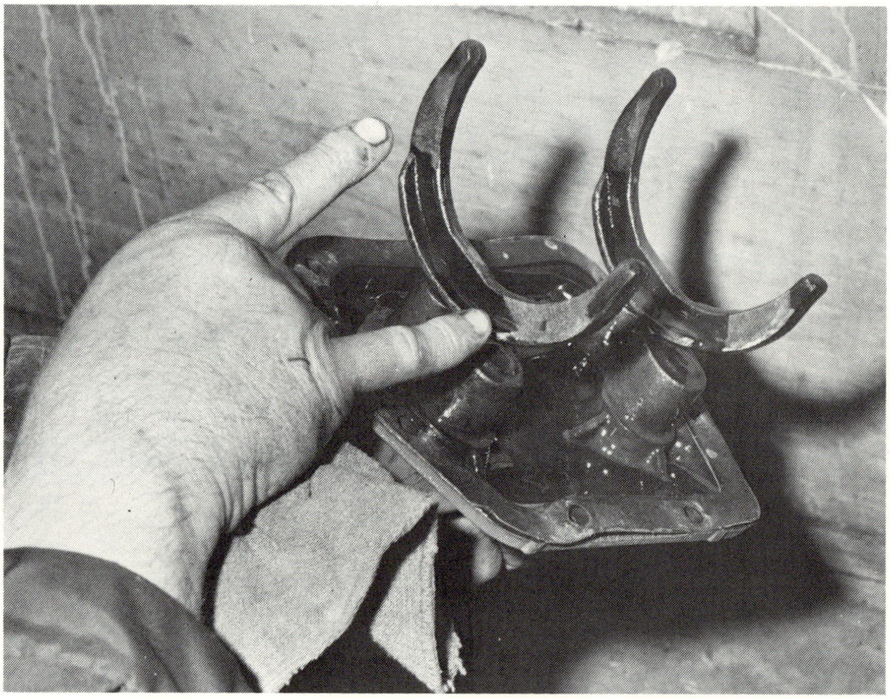

2

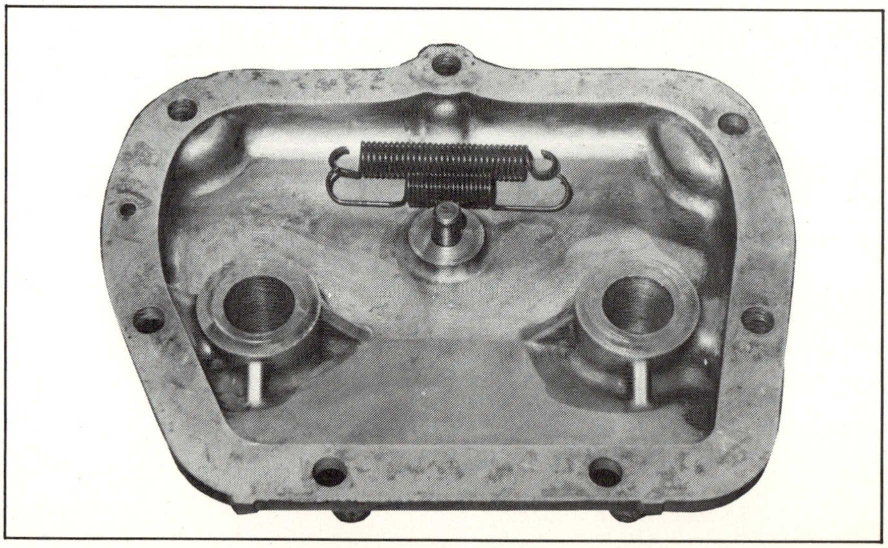

3

is that the clashing of the teeth chews them up something terrible. After a few full-power shifts, the teeth can get so bad that they won't even mesh, and then you get missed shifts and lost races.

To find out the right way to modify synchros, we went to see Pepe Estrada, a former transmission instructor who has recently expanded into a new facility; Pepe's Gear Shop, 14130 Orange Ave., Paramount, Calif. 90723. Pepe builds, among other things, 4-speed boxes for drag racing. He says that a good transmission should be able to withstand hundreds of full-power shifts, the kind you make by popping the clutch and yanking the stick through the gears without taking your foot off the gas.

A stock transmission won't do it, because it isn't made to handle power shifting. But Pepe's transmissions will. He calls them his "slickshifts," and they are so slick and work so well that he actually guarantees them for racing.

Pepe changes clearances throughout the transmission. The end clearance, or the amount of slop between the synchros and the gears, is set precisely where Pepe wants it by using different thickness washers. He changes the amount of offset between the teeth on the synchro sleeve and the bronze ring. The synchro still works, but much easier, so that missed shifts just don't happen.

Pepe builds Muncie transmissions in different ways for different applications. There are circular springs behind the synchro keys in the stock boxes. Pepe takes the springs out, usually, but not always. He polishes the cones on the gears so that the bronze rings will not seize on the cones. Pepe has found that a little slippage between ring and cone is better than perfect synchronization. The shift is easier that way.

Different types of oil are used for different applications, depending on the amount of slipperiness needed to make the transmission work right. In fact, everything that Pepe Estrada does to a Muncie varies according to the way the transmission is going to be used. You tell him what you want to do with the transmission, and he will build one that will do the job. Because the technology is constantly changing, we have deliberately not given any specific recommendations for what you should do with your trans. In a recent issue of a magazine, Pepe recommended using a motorcycle chain lubricant as an assembly fluid on 4-speed boxes. Since then, he has found out that the chain lube can combine with the grease used to pack the roller bearings into position, and form a compound that destroys the lubrication qualities of the transmission oil.

Our pictures will point out how the transmission fits together, and some of the things that Pepe does. If you want to develop your own slickshift, these hints will get you started in the right direction, and perhaps will save you a blown box or two.

4

5

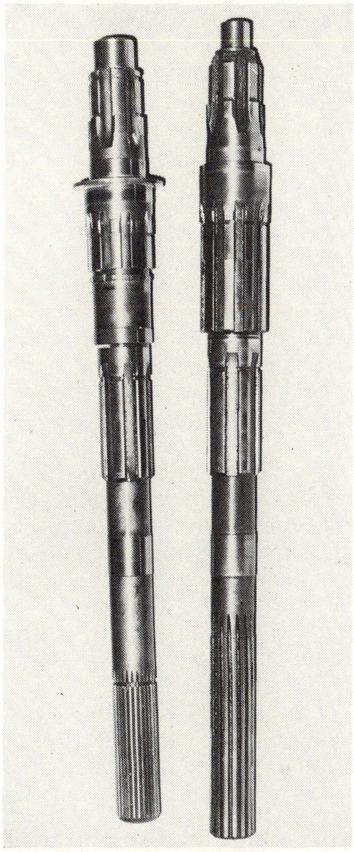

6

7

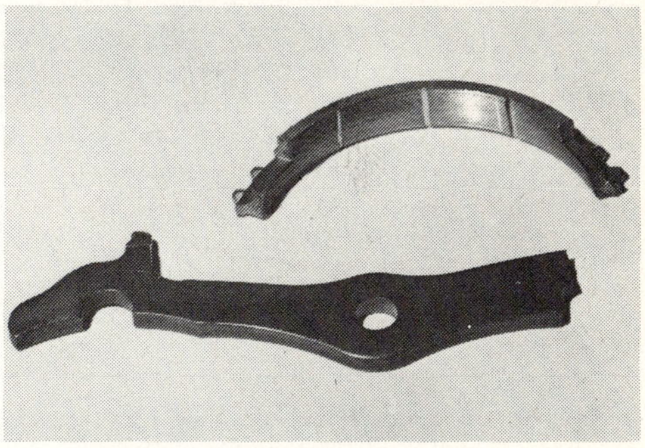

1. The later shifter shafts have a threaded hole instead of a stud.

2. Shift forks should push on the shift collar at opposite points, not on one edge. Pepe grinds the forks for clearance to be sure that they will push evenly on the shift collar.

3. The heavy spring is stock in a Muncie cover. Pepe changes to a lighter spring; it makes shift easier.

4. The Muncie second gear is on the left, and the T-10 on the right. Because it has a smaller and lighter gear, the T-10 gear train has less inertia and, therefore, it will shift easier.

5. The left mainshaft is from a Super T-10, the right one from a regular T-10. The finer splines at the end of the Super T-10 shaft give it greater strength. Other design refinements were made in the Super T-10 in the hope of surpassing all other hot rod transmissions.

6. The T-10 with its smaller gear cone is on the left. The smaller cone makes the T-10 an easier shifting transmission, but the T-10 is not as strong as the Muncie. The new Super T-10 may change that.

7. A broken lock-out lever and synchro ring. This won't happen even under extremely hard usage if the transmission is set up correctly.

Examining the Automatic Transmission

Despite its complexity, you can understand an automatic transmission and how it works. Sure, there are a lot of pieces in it and if you had to know what each little part does you would have to study automatic transmissions for the rest of your life. But we don't want to turn you into an automatic transmission expert; all we want to do is to give you a good basic understanding of how automatic transmissions work so that you can talk intelligently about them and understand what a transmission man means when he says the friggit is stuck on the glammis and it's going to cost $200 to fix it. Of course modern automatic transmissions don't have any friggits or glammises. They are composed of three main sections: a torque converter, a gear system and a hydraulic control system. Let's take them one at a time and you'll see that it's really easy to understand what goes on in that big box under your floorboards.

TORQUE CONVERTER

A torque converter is really nothing more than a fancy fluid coupling, and a fluid coupling can best be described as a couple of propellers fighting each other inside a housing. Suppose you had a propeller mounted on a shaft not hooked up to an engine or anything but just sitting there under water. If you put your hands down close to the blade of the propeller and moved your hands so that the water was forced against the blades, the propeller would turn. What you had, actually, was a fluid coupling between your hands and the propeller.

To make it a little more efficient, you could use two propellers spaced fairly close together so that one prop would force water against the other and make it turn also. Enclose the two propellers in a housing, hook the car's engine to one, the rear wheels to the other, fill the housing with a light oil and you have a fluid coupling that will move the car.

The propeller that is driven by the engine forces oil against the other propeller and that causes the rear wheels of the car to turn. There is no actual mechanical connection between the engine and the rear wheels. You can put the brakes on, stop the car dead, and the engine still keeps turning its own propeller, but the one that's connected to the rear wheels, of course, is stopped. That is the advantage of a fluid coupling, whether it's hooked to an automatic transmission or to a standard transmission with the usual three or four speeds. When the car is stopped, the engine does not have to be disconnected by shifting the transmission into neutral. The engine idles with its driving propeller forcing oil against the driven propeller that's connected to the rear.

The propeller that's connected to the engine may be called the driving member, the impeller, the pump, the driving torus or the driver. The propeller connected to the rear wheels is called the turbine, driven member, driven torus or the runner.

If you ever drove one of the Dodges made in the '40's, you know that it was possible to start off in high gear from a dead stop. As torque was applied to the fluid coupling by the engine, the driven member would slowly get going and eventually the car would get up to speed. Acceleration was not the best in the world but it was a good lazy man's car because the driver didn't have to shift the transmission.

You can see that if the fluid coupling were really efficient, it wouldn't allow you to stop the car with the engine still engaged or to start off in high gear. Slippage in the fluid coupling was its one big disadvantage. It resulted not only in poor acceleration but in poor gas mileage too because the engine was doing a lot of work churning up that fluid, and the car wasn't getting anywhere.

Chrysler and Dodge later went to a manual transmission with an automatic shift still incorporating a fluid coupling. It was quite an improvement over the earlier models but still that slippage in the coupling was its big disadvantage. Of course if the fluid coupling slips on acceleration it'll slip on deceleration too, and that was another bad point of the fluid coupling transmissions—their inability to hold the car back when going down a hill. It caused excessive wear on

1

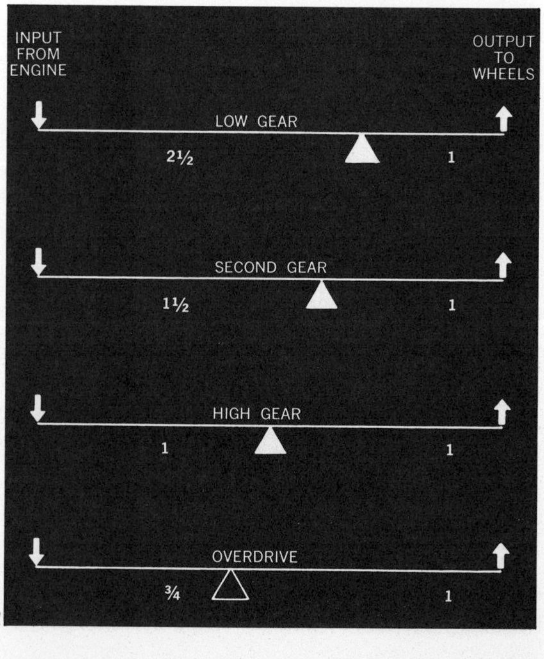

2

the brakes because the driver had to use them more.

Actually the impeller and the turbine do not look like propellers at all but they perform the same function. Instead of having two or three or four blades as a propeller does, the impeller and the turbine have many blades set very close together. The angle of the blades is carefully set so that the fluid coupling will operate with the greatest efficiency. The difficulty is that when the angle of the blades is set for maximum efficiency, the fluid—as it leaves the turbine—shoots back at the impeller at an angle that has a tendency to slow the impeller down. In other words the engine is trying hard to turn the impeller and the oil is squirting at the impeller to slow the engine.

Oil inside the fluid coupling is not just churning away as it would if we had an egg beater in a bowl. It is actually being directed by the blades on the impeller to strike the turbine with a maximum amount of push, and is constantly circulating in a definite well-planned pattern.

STATORS

After the engineers reached the limit of their design abilities on the fluid coupling, they decided to put in some extra stationary blades so that the fluid leaving the turbine would not come back against the impeller with retarding force. The extra blades are mounted in a full circle within the fluid coupling and are called a stator. With the stator we don't call it a fluid coupling anymore; when it has the extra stator blades it is called a torque converter.

By putting a stator into our fluid coupling we not only got rid of the retarding effect of the oil leaving the turbine but we gained an unexpected bonus in that it ceased to be a plain fluid coupling and was now a torque multiplier. More torque actually came out of the torque converter than was put into it.

You know that a manual transmission multiplies torque because it gives the engine leverage to work. It's the old story of fulcrums and levers. Try to pull a nail out with your fingers and you can't do it, but use the leverage supplied by a hammer and the nail will come right out. However, your hand moves through a much greater distance in pulling the nail than the nail does.

Torque multiplication in a transmission accomplishes the same thing. If you are in low gear, your engine may turn over 10 times for each revolution of the rear wheels. But in doing this the torque is multiplied so that we get much more push at the rear wheels than we do at the back of the engine.

Leverage is something we use every day without even thinking about it, whether it's flicking a light switch, opening the rings on a notebook binder, or opening a door. Your muscles are connected to your bones on the principle of leverage.

With leverage we can multiply torque or we can multiply distance. The ordinary playground see-saw multiplies neither because the pivot point is equidistant from both ends. Move the pivot point toward one end and the child on the long end has the advantage. He doesn't have to push as hard but of course he moves a greater distance, which is just like being in low gear.

The child on the other end of the see-saw is in high gear. He has to push harder, but he accomplishes more because the other end of the board moves further than he does. However, when he meets any resistance from the other end of the board he can be easily slowed down.

What leverage amounts to is that you have the engine on one end of the see-saw doing the work and the weight of the car on the other end opposing it. The transmission merely shifts the fulcrum back and forth between the engine and the car so that the engine can either put out more torque for pushing and starting or more distance for moving over the highway.

All of this multiplication-of-levers business is easy to see if you have gears. A small gear hooked to the

1. The automatic transmission is one of the more important automotive developments. Though it's been with us now for over 30 years, its potential is just being reached. Here's one of the best, the TorqueFlite, shown with varying tailshaft housing lengths.

2. Transmission gearing represented by a see-saw. In low gear, it's easier for the engine to push down on its end of the see-saw. The job becomes progressively more difficult as the gears are higher. In overdrive the engine has its hardest job; the wheels travel further for each engine revolution than in lower gears.

3

3. Some torque converters can be disassembled by unbolting them. Others are welded units, such as this Turbo Hydra-Matic converter. It was cut apart with a torch so you could see what goes on inside.

4

4. This is the turbine half of converter, and below it is the stator with the blades that make it function.

5

5. An actual stator, removed from a Turbo Hydra-Matic torque converter that was cut apart. The blades are in the high angle. The splines go on the stator support, which is the roller clutch, which allows the stator to turn in one direction only. Most transmission makers have dropped variable pitch stators in favor of fixed blades. Variable pitch is a good feature, but the general public doesn't even know it's there, so the manufacturers are quietly dropping it, and saving some cash in the process.

6

6. Stator blades in low angle. As the Turbo Hydra-Matic converter is a welded unit, the stock mechanic never sees the stator or any of the converters inside.

engine, with a large gear connected to the wheels, allows the engine to turn faster and the rear wheels to turn slower. But the rear wheels in sacrificing distance have gained torque, and they are pushing harder against the road than the engine can push if there is no gear reduction.

We figure gear reduction from the size of the gears. Because the teeth mesh with each other on any pair of gears, the teeth must be all of the same size, so it's a simple matter of counting the teeth on the gears to get the gear reduction. If the gear on the engine, or driving gear, has 15 teeth and the driven gear has 30 teeth, we divide the driving gear into the driven gear and we get the numerical value two, which means we have a 2-to-1 reduction.

If the driving gear has five teeth and the driven gear 15 teeth, we have a 3-to-1 reduction. It's simply a matter of dividing the number of teeth on the driving gear into the driven gear on any pair of gears that are meshed. The torque converter accomplishes the same type of reduction only it does it through fluid mechanics and the angle of the stator blades, instead of through gear size.

To get a better idea of how fluid can give a torque multiplication, suppose you had a garden hose and were squirting water through the nozzle into an ordinary tea cup out on the lawn. Let's suppose that the cup is firmly fastened to the top of a roller skate. By directing the flow of water at the best angle you can move the roller skate and the cup along the ground at a pretty good rate of speed. While doing this you will notice that some of the water is flying back and getting into the stream from the nozzle. This slows down the stream from the nozzle and impedes the progress of the cup. Water flying backwards from the cup is going in exactly the opposite direction to the water coming out of the nozzle the same way that oil leaving the turbine shoots back at the impeller blades.

If you mount some more tea cups so that they catch the water coming out of the cup, turn it and force it to go back in the same direction as the water coming out of the nozzle. Then you will use the same water twice to move the cup and roller skate over the ground. The force of this water joining with the stream from the nozzle will of course hit the cup and roller skate that much harder. The result is that you have gotten a torque multiplication out of redirecting the water, the same principle a stator uses. You will find that not only can you push the roller skate over the ground but now with your added force you can actually push it in any direction and even uphill.

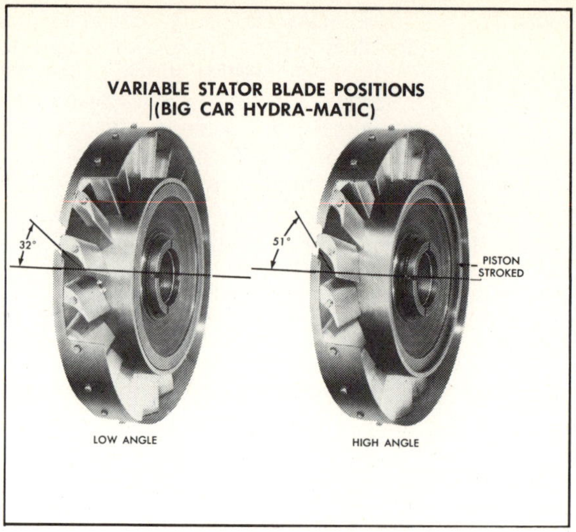

1

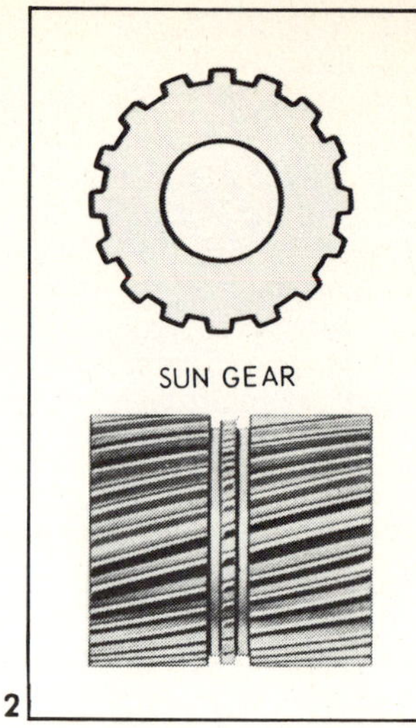

2

1. Here you see how we arrive at the terms high and low angle. Note how the piston strokes to change the angle. It's done with oil pressure, which can act on either side of the piston.

2. Gear train buildup on the C-4 Ford transmission. This is not the way a mechanic would assemble the transmission. We do it this way for illustration purposes, and we start with the sun gear.

3. Add the planet pinions, meshing with the front part of the sun gear.

4. The ring gear goes around the planet pinions and meshes with them. Note the projections on the ring gear. They will come in handy later when we connect the ring gear to some other pieces.

5. Here we have added a planet pinion carrier. The carrier supports the pinions, but does not touch either the ring gear or the sun gear. The output shaft goes through the sun gear, and is splined to the carrier.

That, in effect, is what the stator does with its non-rotating blades. It redirects the oil back into the flow from the impeller so that more force goes against the turbine and we end up with a torque multiplication, the same as you find in a geared transmission unit.

This torque multiplication comes into play whenever the impeller is turning faster than the turbine. Suppose we back off the throttle so that the engine is just pushing enough to keep the rear wheels going, then the impeller and turbine are turning at about the same speed. All of the fluid in the housing is turning at the same speed as the impeller and turbine but there is no flow through the blades at this particular time.

Another problem rears its ugly head under that condition. While the impeller and the turbine and the fluid are all turning at the same speed, the stator blades are not turning so they are creating a drag that tends to slow everything down. To get around this problem the hub upon which the stator blades are mounted is a one-way roller clutch. When the stator blades are redirecting the flow of fluid back into the impeller, the roller clutch does not permit rotation of the stator. But when the turbine catches up with the impeller, then the fluid rotating inside the housing strikes the back side of the stator blades and the stator goes along for a free ride, with the roller clutch permitting it to turn in the same direction the impeller, the turbine, and the fluid are rotating.

Multiplication through a torque converter can run as high as 2-to-1 or a little bit more. It all depends on how much coal you're pouring to the engine. The faster you rotate that impeller the harder the fluid is going to press against the turbine, and the faster it will come back through the stator and be redirected against the turbine again.

Because of the torque multiplication, the torque converter does not slip nearly as much as the old fluid coupling. But it does slip a little at all times, even at highway speeds. Anytime the slippage exceeds 10% the torque multiplication starts because the flow of oil through the turbine blades strikes the front side of the stator, the stator locks up on its roller clutch, and the oil is redirected into the impeller to help it along.

So while slippage in a fluid coupling is a pain in the neck, slippage in a torque converter is really something we look forward to because it results in this torque multiplication and gives us greater acceleration. Maximum torque multiplication comes at full stall when the car is stationary and the throttle is wide open.

In all modern transmissions thi torque multiplication is used along with gear changes in the automatic transmission that multiply the torque

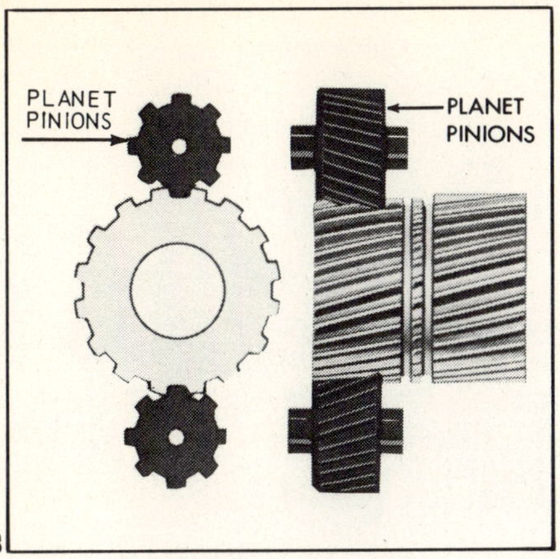

3

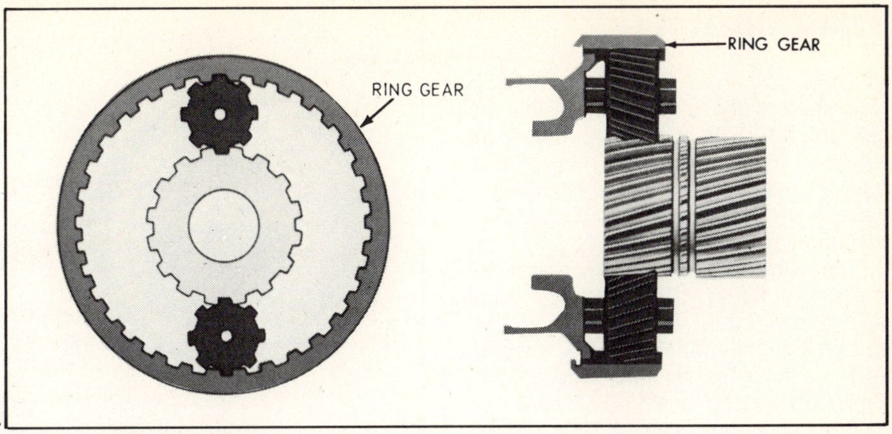

4

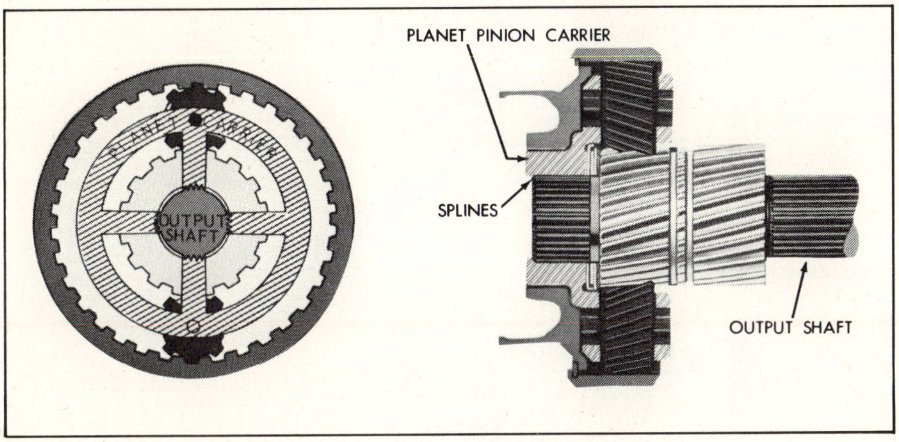

5

even further. The gear changes are made by the hydraulic control system at the proper speed and load so that we have what we know as a completely automatic transmission.

Many modern torque converters have the stator blades set at a fixed angle. But some have a hydraulically controlled piston within the hub of the stator assembly which changes the angle or pitch of the blades depending upon the load and speed of the engine. For example: in some models of the Hydra-Matic there are two positions on the stator blades: low angle or high angle. Low angle is the normal position of the blades and they are in that position from above idle until 50% throttle is reached.

At idle the stator goes to high angle which would normally result in greater torque multiplication but because the engine is turning so slowly it is a less efficient position and idle creep is reduced. Above approximately 50% throttle opening, the stator also goes to the high angle. This results in a greater redirection of oil to the impeller for more torque multiplication. When you floor the throttle with a variable stator you get much better acceleration from your car.

The variable stator, as we said, is controlled by a hydraulic piston within its hub, but the valve which admits oil to this piston is electrically controlled and may receive an electric signal either from a switch on the throttle linkage or from a pair of points in the speedometer. When controlled by the speedometer, throttle opening has no effect on the angle of the stator blades, and they will go into the high angle at whatever speed the speedometer is set. That setting is fixed and cannot be adjusted.

Converter design has not always been as simple as it is on modern transmissions. In an attempt to provide the best acceleration with the least amount of slippage, designers have gone to great lengths with converters that have as many as three turbines. The Buick Flight Pitch transmission used in 1958 and '59 was such a design. Chevrolet's Turboglide also had three turbines.

Those torque converters are commonly referred to as 5-element or 5-member converters, because they have three turbines, one impeller and one stator. In those designs, which are now obsolete, there is no actual gear changing within the transmission as there is in the modern automatics. The first turbine drives the output shaft of the transmission through gears that give a reduction similar to what you would get in low gear of a standard transmission. The second turbine corresponds with second gear, and the third turbine is the high gear which is connected directly to the output shaft of the transmission for a 1-to-1 ratio.

When starting from a dead stop the first turbine does almost all of the work because of the difference in speeds between the impeller and the first turbine. The stator has variable pitch controlled by the throttle linkage. As the first turbine begins to catch up with the impeller, the oil flows in a different manner so that most of its force is directed against the second turbine. Eventually the second turbine also catches up with the impeller and is turning at about the same speed, and then the oil applies most of its force against the third turbine which is the same as being in high gear.

Only the third turbine is connected directly to the output shaft of the transmission. The others are connected through an overrunning or roller-type clutch, with the hub attached to the turbine and the housing attached to the output shaft. When the second turbine starts taking over from the first, the output shaft—because of the second turbine's higher gear—rotates faster. This increased speed of the output shaft causes the overrunning clutch housing to disconnect from its hub so that the first turbine which is connected to the hub is allowed to lag behind. A second overrunning clutch connects the second turbine to the output shaft and the same thing happens to it when the third turbine takes over.

At highway speeds, when the converter is acting as a fluid coupling, all of the torque is being applied from the third turbine to the output shaft, and the first and second turbines are actually idling along at a slower rate of speed. Anytime the engine puts out more torque this causes increased slippage within the torque converter. Increased slippage causes the oil flow to react on either the second or first turbine depending on how much torque the engine is producing. Thus we have infinitely variable gear ratios without any actual gear changing in this type transmission.

Change-over points between first,

second and third turbines are very smooth and it's almost impossible to tell which turbine is doing the work at any one time. You would think that it was a great system but its complexity gave a lot of trouble and the design was later discontinued.

The first Chevrolet Powerglide transmissions in 1950 and '52 also had five-element torque converters but of a slightly different type. Instead of three turbines, a stator and an impeller, they used one turbine, two impellers and two stators. The effect was similar to the three-turbine type of torque converter. A low gear ratio was also provided in the transmission, but the driver didn't have to use it if he was not in a hurry. Chevrolet had 6-cylinder engines in those days, and coupled with the inefficient early Powerglide they just didn't get the job done. Later Powerglides went to a normal 3-element converter with one impeller, one turbine and one stator.

In any fluid coupling or torque converter you can understand that there must be quite a bit of churning of the fluid, like a propeller in a bucket of oil. All of this churning creates a tremendous amount of heat, and from the beginning of fluid coupling and torque converter design, eliminating this heat has been a big problem. Some of the early designs had fins on the torque converter housing and a big air scoop that would pick up air from underneath the car, rotate it through the housing, picking up heat off the fins, and discharging it out the other side.

In order to keep a torque converter full at all times an oil pump in the transmission keeps the oil in the converter under pressure and circulates it back and forth between the body of the transmission and the converter. As long as that oil was under pressure and circulating, somebody got the bright idea, why not run it up to the radiator and let the radiator water cool the transmission oil? That is now the accepted method on almost all modern automatic transmissions.

Two lines come out of the transmission. One of them is direct from the converter and goes up to a small radiator inside one tank of the engine's radiator. The other line returns the fluid to the body of the transmission. It works fine as long as the radiator coolant is cool enough to absorb heat from the transmission oil. When the engine is pulling hard, it's creating more heat in the transmission, but it's also creating more heat in the engine's cooling system. Therefore, there isn't as much differential between the temperatures of

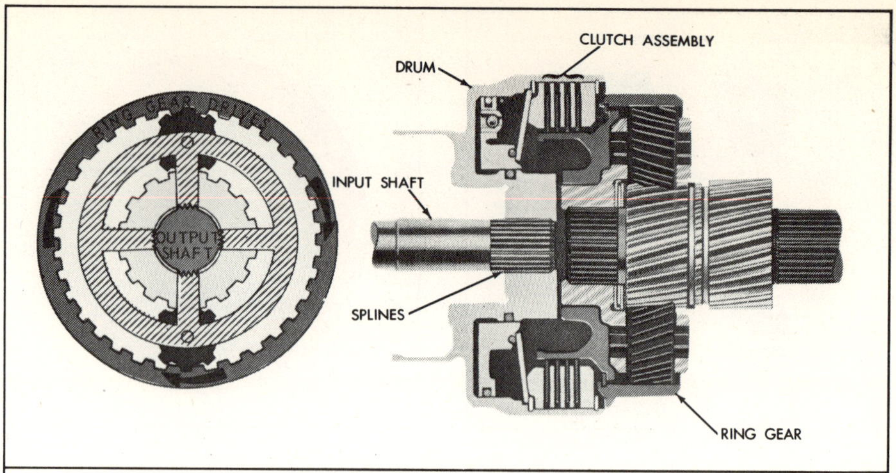

1

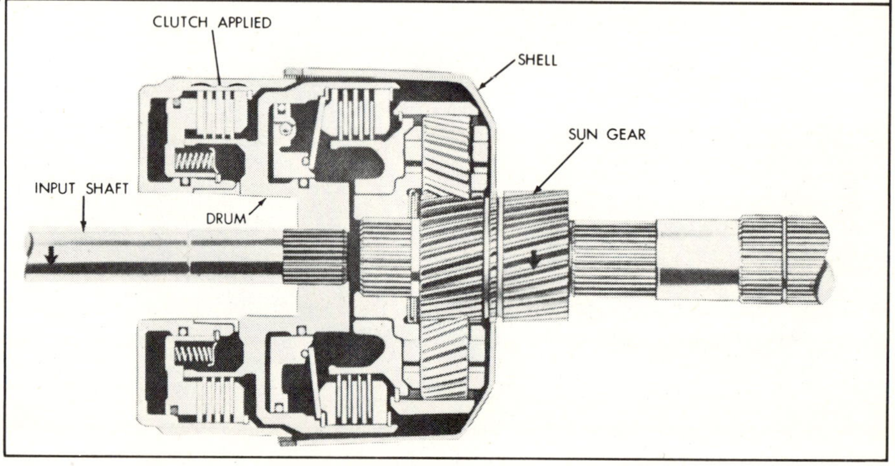

2

the radiator coolant and the transmission oil to cool the transmission oil properly. In normal driving that doesn't cause any trouble, but it can if your car is operated under conditions that cause it to run hotter than it should.

Of course if the engine's cooling system and the radiator tubes are plugged with rust and corrosion, then the transmission is really going to get a bad time from that hot coolant because the transmission oil will not be cooled at all. It'll go back to the transmission as hot as it was when it was pumped up to the radiator. One effect this has is a lowering of the transmission's efficiency. It just won't go nearly as well.

You can test that yourself sometime by checking the acceleration of your car with the transmission in a cooled condition such as you would have at cruising speed with light throttle on a level road. Also check the acceleration after you have allowed the engine to run at a fast idle while the car is standing still with the brakes set, thus heating up the transmission fluid. You will probably find that the transmission with its fluid at a normal temperature accelerates much better than when it has been overheated.

Up to this point we have talked about power flow from the engine into the torque converter. Now let's see what happens when that power is channeled from the torque converter into the planetary gear set that is used in literally all of today's modern transmissions.

PLANETARY GEAR SYSTEM

Planetary gears used in an automatic transmission are similar to the planetary gear sets used in Model T and also in an overdrive. The difference is that planetary gears in an automatic are arranged differently, usually with two planetary gear sets mounted together so that more gear reduction combinations are possible.

A simple planetary gear set has a sun gear mounted on a shaft with three or more planet pinions encircling it and meshed with it. The planet pinions are rotated on pins, and these pins are firmly attached to a cage so that all of the planet pinions operate as a unit. Around the planet pinions is a large ring gear with internal teeth that mesh with the pinions.

The gears are in mesh at all times. There is no sliding of the gears or unmeshing in the planetary system. The trick to planetary gears is that they are arranged in the transmission so that we can apply power to any one of the three members—the ring gear, the planet pinion carrier or the sun gear. We can also take power

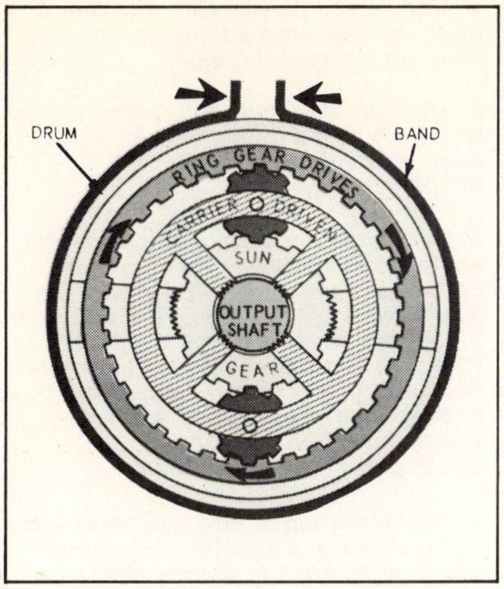

3

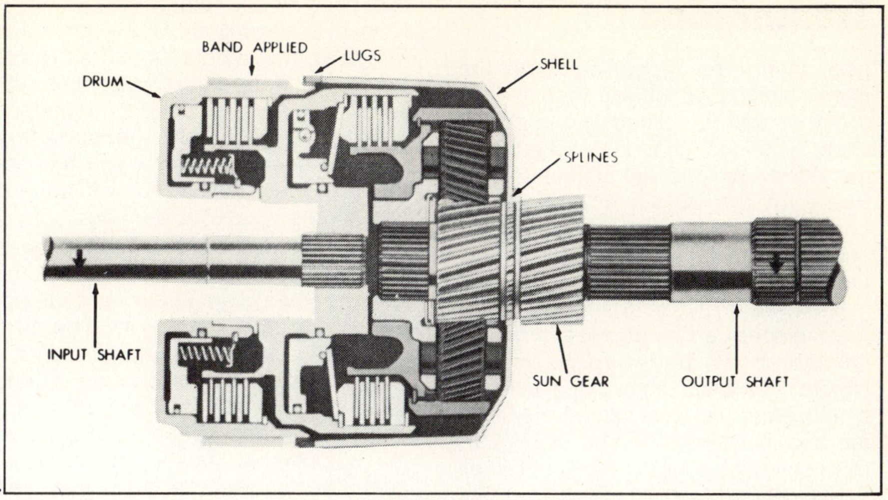

4

1. Now those ring gear projections are used to connect up with a forward clutch, which is splined to the input shaft. When forward clutch is engaged, the input shaft drives the ring gear.

2. Those projections on the forward clutch are now used to mount the reverse-high clutch. The input shell is merely a connector between the reverse-high clutch and the sun gear. When the reverse-high clutch is operating, it connects the input shaft directly to the sun gear.

3. Around the outside of the reverse-high clutch drum goes the intermediate (front) band. It is mounted on the case.

4. With the intermediate band applied, the reverse-high clutch drum, shell, and sun gear are all locked to the case.

output off any one of the members.

The third method of operating the planetary gear set is to immobilize any one of the members so that it is held firmly to the transmission case. Not all of these possibilities occur in any transmission but if they did we could get reduction in two ways, direct drive, reverse, neutral, and overdrive—all out of the planetary gear set without doing anything but changing the power input, the power output, or holding one member.

Drive to various members of the planetary gear set is obtained by clutches inside the transmission which are operated hydraulically. The holding of various members is done by brake bands which are also operated hydraulically.

To get direct drive in the planetary gear set we lock any two members together with a clutch or combination of clutches. Whenever any two members are locked together the ring gear, planetary gear carrier, and the sun gear will all rotate as a unit. There is no gear reduction and we have a 1-to-1 direct drive ratio.

If we drive the sun gear with the power from the engine and hold the ring gear stationary, the sun gear will force the planet pinions to walk around the ring gear. Our output shaft is hooked to the planet pinion carrier which rotates considerably slower than the sun gear does. With that condition we get a big torque increase which would compare to the low gear in a standard transmission.

Suppose we want gear reduction but not quite so much. Then we apply engine power to the ring gear and hold the sun gear—with our output shaft still connected to the planet pinion cage. We still get the same walking action between the pinions and the ring gear but they are also walking around the stationary sun gear. Thus we get a gear reduction that would be about the same as second gear in a standard transmission.

To get reverse we drive the sun gear, hold the planet pinion carrier stationary, and connect the output shaft to the ring gear. The planet pinions can rotate on their pins only, which causes the ring gear to turn in a reverse direction but also gives some gear reduction besides.

We could drive the ring gear and take the output off the sun gear, still getting reverse, but then we would have a reverse overdrive and it's necessary in reverse, the same as in low gear, to have gear reduction.

In all combinations where there is gear reduction we drive one member, take the output torque off another member, and hold the third. In order to go into neutral, all that is necessary to do is let go of the member that is being held. The other two members can remain hooked up to input and output but the member that is loose will spin and no power will be transmitted in the gear set.

One other function that is possible with a planetary gear set is overdrive, which we get by driving the planet pinion carrier, taking our output off the ring gear and holding the sun gear stationary. We can also get overdrive by driving the carrier with the ring gear held, if we want an even greater percentage of overdrive. The overdrive function is not ordinarily used in an automatic transmission. High gear is usually a 1-to-1 ratio.

GEAR TRAIN BUILDUP

Many different arrangements of planetary gears have been tried in an automatic transmission. Most of them have evolved into an arrangement whereby two planetary gear sets are used with a common sun gear. Thus, the ring gear and the pinions can apply power to the sun gear, which in turn transmits the power to the pinions and the ring gear of the other planetary gear set. With an arrangement of clutches and bands for holding or connecting the various members, it is possible to get the gear ratios we want for efficient operation of the transmission.

The easiest way to understand how all these gears, bands and clutches are put together is to start from the beginning with the sun gear and add the different pieces until we finally end up with the complete planetary gear assembly. The following gear train buildup describes the Ford Motor Co. C-4 transmission as used on Lincoln-Mercury cars. Much of the following information and illustrations were supplied by the Service Department of Lincoln-Mercury Division.

We start with the sun gear, add the planet pinions and then the ring gear goes around the outside of them. Then we insert the output shaft. It goes through the middle of the sun gear but it does not actually connect with it. The output shaft is splined to the planet pinion carrier. All the output shaft does for the sun gear is provide a bearing surface for the sun gear to rest on.

Outside our ring gear we add a forward clutch assembly and splined to it is the input shaft. So up to this point we have an input shaft driving the forward clutch, which is used to connect the input shaft to the ring

Transmission

gear; then the ring gear can drive the planet pinions, which are mounted on a carrier that is splined to the output shaft. The sun gear is meshed with the planet pinions but it just rotates freely on the output shaft with no connection. Study the illustrations and you should be able to see how the thing works.

Next we are going to add three more pieces—an input shell, an intermediate band and a reverse-high clutch. The input shell goes over all of the parts we have talked about so far and is splined to the sun gear. Connected by lugs to the input shell is our reverse high clutch. It operates to connect the input shaft to the sun gear; thus the power flow would be from the input shaft to the reverse-high clutch, through the clutch drum through the input shell and into the transmission sun gear.

The intermediate band goes around the reverse-high clutch drum. This band is used to hold the sun gear when it's necessary to do so. By applying the band to the reverse-high clutch drum, it holds the input shell and in turn the input shell (being splined to the sun gear) keeps the sun gear from turning. At this point it isn't necessary for you to understand when the different bands and clutches are applied. You should just attempt to get firmly in your mind the arrangement of the different pieces and the holding action or connection that they make when we want them to do so.

The pieces we have described to this point are all connected to the forward end of the sun gear with the exception of the input shell, which is splined to the middle of the gear and fits over the whole mess we have talked about so far. This is one reason why it's so difficult to understand what goes on inside the planetary gear set because even if you had the thing in your hands you wouldn't be able to see what's happening.

Now that we have all of the parts and pieces hung on the front of the sun gear, we start adding similar apparatus to the rear of it. The first items are the planet pinions of the rear planetary set, and then a ring gear goes over that, and this ring gear has a hub which is splined to the output shaft of the transmission. Just for comparison purposes you might remember that in the front planetary gear set, the planet pinion carrier was splined to the output shaft, but in the rear planetary set, the ring gear is connected to the output shaft.

One more member we have to add is the planet pinion carrier. In this case it is connected to the low reverse drum. This low reverse drum is controlled both by a band around its outside called the low reverse band and a one-way roller clutch which is attached to the transmission case. The one-way clutch allows the rear planetary carrier to revolve clockwise but if power application is such that it would tend to make the carrier revolve counter-clockwise, then the roller clutch locks it and forces the pinions to turn on their shafts, holding the pinion carrier stationary.

By this time you should be able to identify any one of the members in the illustration of the complete gear train, and you should know what con-

1. With all the pieces added to the front, we now start adding on the rear. First are the planet pinions.

2. Then the ring gear and output shaft, which are splined together. At this point we don't have any carrier for the pinions.

3. Next to be added is a planet carrier and a rear band. Note that the planet carrier is made so that it goes all the way around the outside of the ring gear It does not touch the ring gear, however. When the rear band cinches down, it holds the planet carrier from turning. In this illustration the gearset is in reverse, with the reverse-high clutch and rear band applied.

4. Our gear train is now complete with the addition of a 1-way roller clutch, attached by bolts to the transmission case, and a front band around the reverse-high clutch drum.

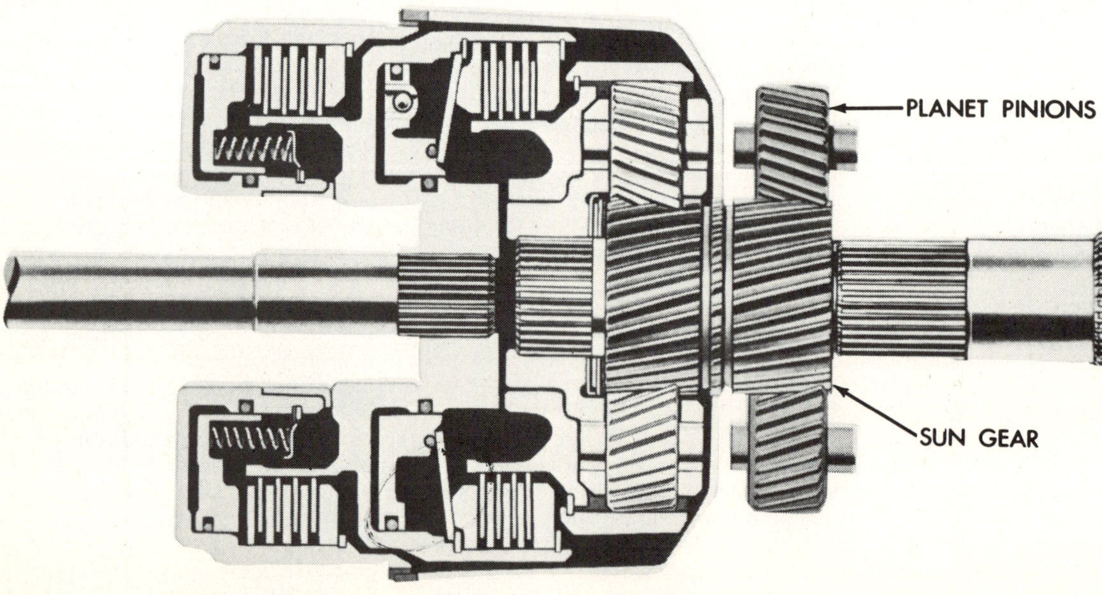

1

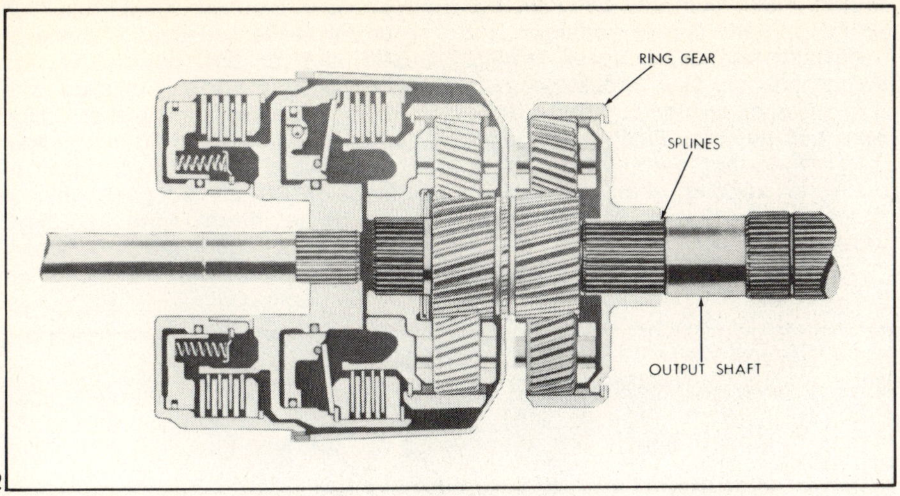

2

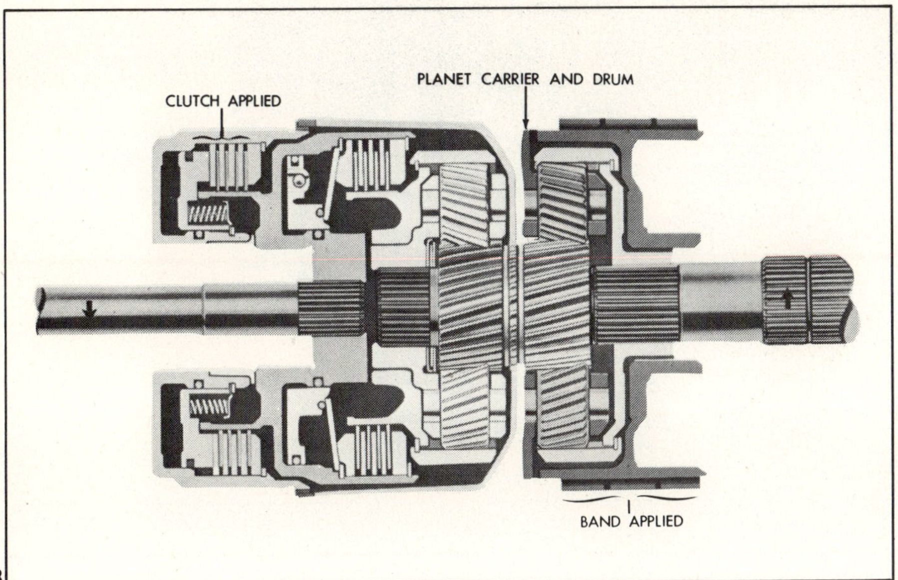

3

nections it can make between itself and the other members. All of these connections are accomplished by pressure from the hydraulic system but for now we will just accept that they are made and will study the hydraulic system later on to see specifically how it does it.

LOW GEAR POWER FLOW

In low gear the engine turning the torque converter applies power to the input shaft which is splined to the forward clutch cylinder. The forward clutch is engaged with the forward ring gear, which rotates the forward planet pinions on their shafts. Torque that the ring gear imposes on the planet pinions tries to rotate the sun gear, but this same torque attempts to turn the planet pinion cage, which is splined to the output shaft. The torque is actually divided here, with part of it going through the sun gear and part through the planet pinion cage of the mechanism.

From the sun gear the power flow is through the rear planetary pinions to the rear ring gear, which is splined to the output shaft. Power flow through the rear planetary gear set has to go through that path because the low-reverse band is holding the rear planet pinion carrier. The carrier cannot move, but the pinions can still rotate on their shafts, so they force the rear ring gear to walk around them, which applies torque to the output shaft.

Torque is actually being applied to the output shaft from the front plane-

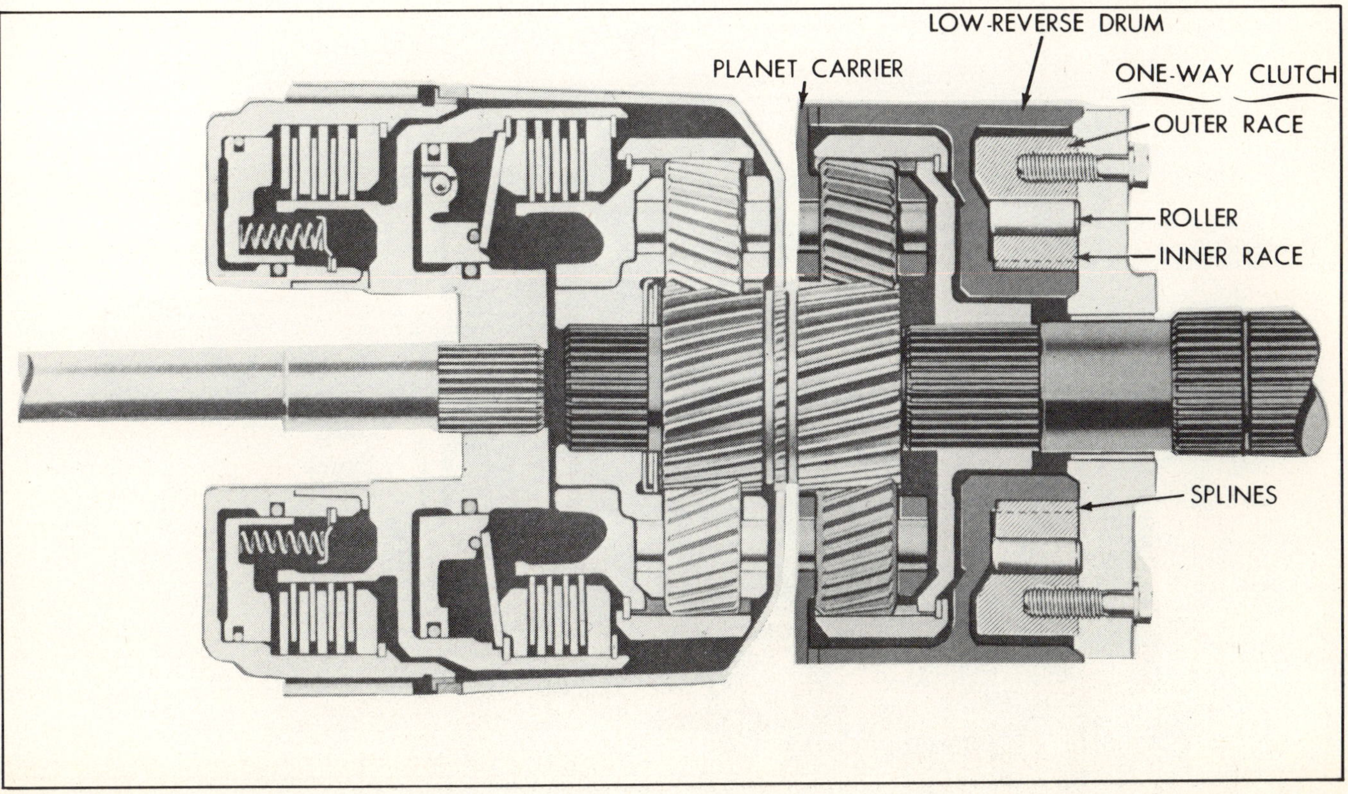

4

Transmission

tary pinion carrier and the rear ring gear. Admittedly, it is difficult to see how torque can be split between two planetary gear sets with all the members going at different speeds, and still be joined to the same output shaft where the speeds have to be identical. You can't have two parts of the same shaft going at different speeds, unless you want to twist the shaft in two.

What actually happens is that the engine is driving the ring gear, which transmits torque to the front planet pinions rotating on their pins, and also to the carrier. The carrier drives the output shaft directly, and the planet pinions also drive the output shaft through the sun gear, rear planet pinions, and so on. Actual gear ratio in low is 2.46-to-1.

Throughout all this action the rear planet pinion carrier has been held stationary by the rear band, because the shift quadrant is actually placed in the low position. A band is used to hold the carrier so that the rear wheels can drive the engine. That allows engine compression to slow the car down while decelerating and going down steep grades.

If the driver starts off with the shift quadrant in drive, the transmission is also in low gear, but the rear band is released. The rear carrier is still held from turning by the one-way roller clutch which operates between the carrier and the transmission case.

Reaction from the sun gear, turning the pinions on their shafts, tries to turn the rear carrier counter-clockwise, but the roller clutch won't allow that. When the automatic shift is made to second gear, the roller clutch allows the rear carrier to freewheel.

We have to have the roller clutch because in the intermediate and high gear ratios the sun gear is going to either stop or slow down. The rear planet pinions are still meshed with the sun gear, and the planet pinion carrier is still splined to the output shaft. In order for the output shaft to turn faster than the sun gear, which it must do in intermediate and high when the sun gear stops or slows down, we need some method of al-

1

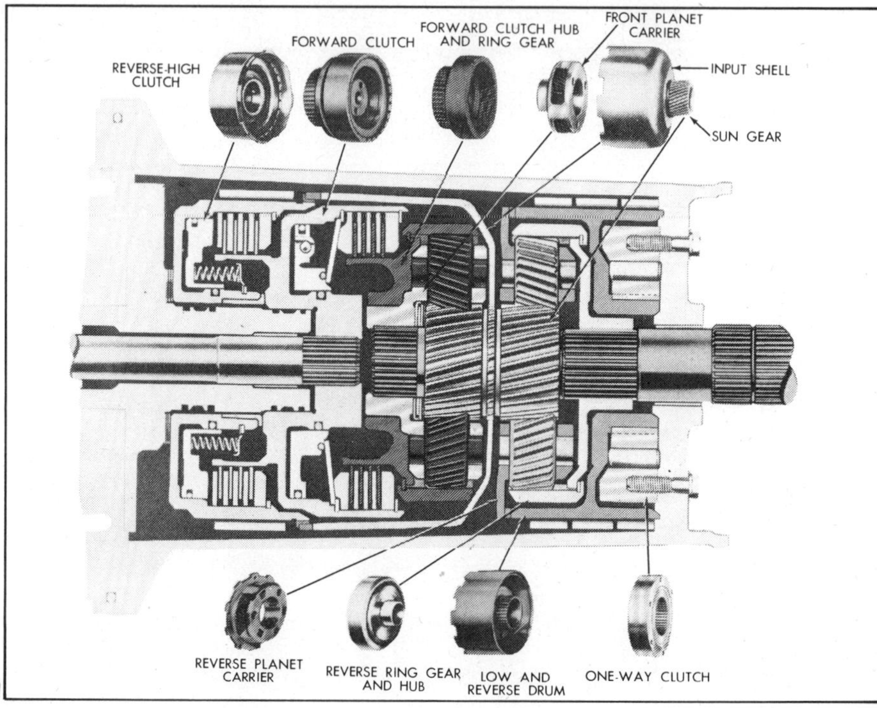

2

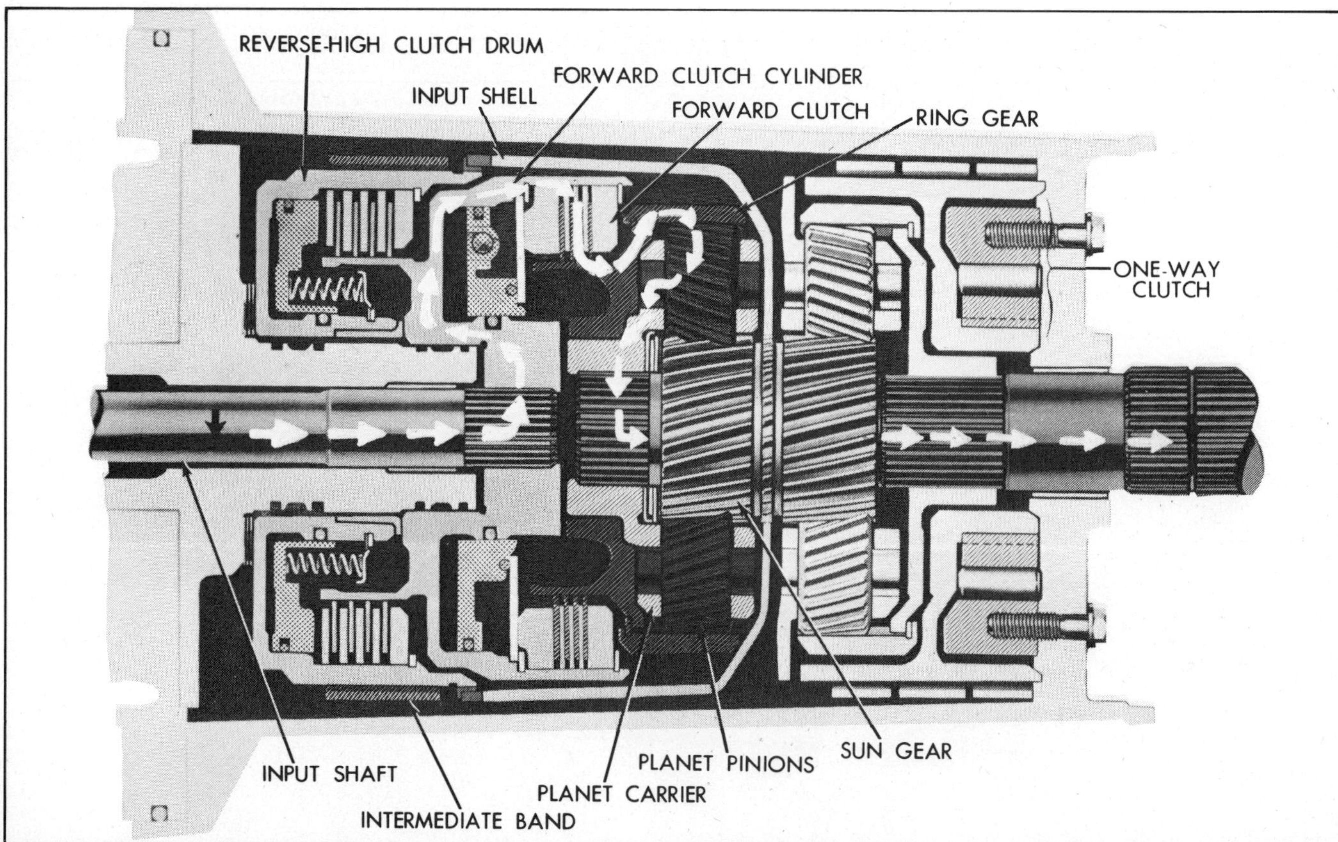

lowing the rear carrier to go faster than the sun gear. The one-way roller clutch does this. What we have, in effect, is a rear carrier and planet pinions that are not transmitting any torque to the sun gear. In intermediate and high gear ratios the rear carrier acts just as if it were disconnected from the driving force.

Now that everything is so clear about what happens when the transmission is in low gear ratio, let's take a quick trip through the power flow in intermediate (2nd) gear.

SECOND GEAR POWER FLOW

In all forward speeds the input shaft from the torque converter drives the forward clutch cylinder, and the forward clutch is engaged so that it drives the forward ring gear. From the ring gear, torque is transmitted to the forward planet pinions.

In low gear we transmitted the torque from the planet pinions to the sun gear, but in second the sun gear is locked to the transmission case by application of the intermediate band. The force applied by the ring gear causes the planet pinions to walk around the stationary sun gear, which rotates the planet pinion carrier. The planet pinion carrier is splined to the output shaft, so our torque is transmitted to the rear wheels.

We said that the sun gear was locked to the case in second. That peculiar-looking input shell that covers everything up is what does the job. It is splined to the reverse-high

1. This is the completed planetary system in the gear train.

2. C-4 gear train in intermediate (second) gear. The intermediate band is commonly called the front band. When it is applied, it firmly locks the sun gear, input shell, and reverse-high clutch drum to the transmission case.

3. Reverse gear in operation. This is the only power situation where the forward clutch is disconnected. The sun gear is turning the same direction as the input shaft, but opposite to the output shaft.

4. Low gear ratio on the C-4. Note how the power flow splits at the forward pinions, part of it going into the sun gear and the other part going through the carrier direct to the output shaft. The output shaft runs through the middle of the sun gear, but the two parts are not connected. Low gear has the most complicated power flow of any ratio. All the parts indicated in this illustration are used to produce low gear.

5. High gear operation. Everything forward from the middle of the sun gear is locked up solid. The arrows show that the power goes to the sun gear and the rear planetary pinions, but stops there because the rear pinion carrier is just spinning without transmitting any torque. Power applied through the front carrier turns the output shaft.

3

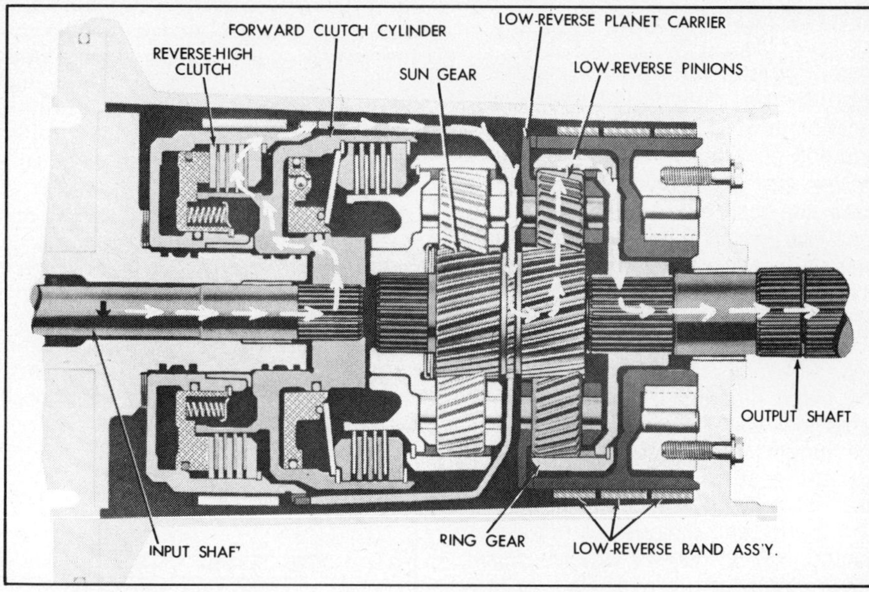

4

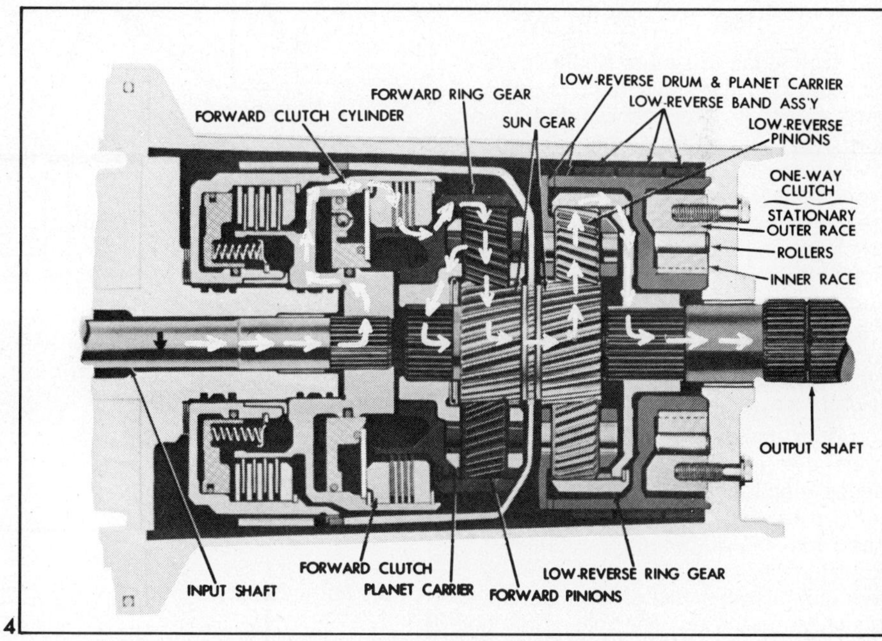

5

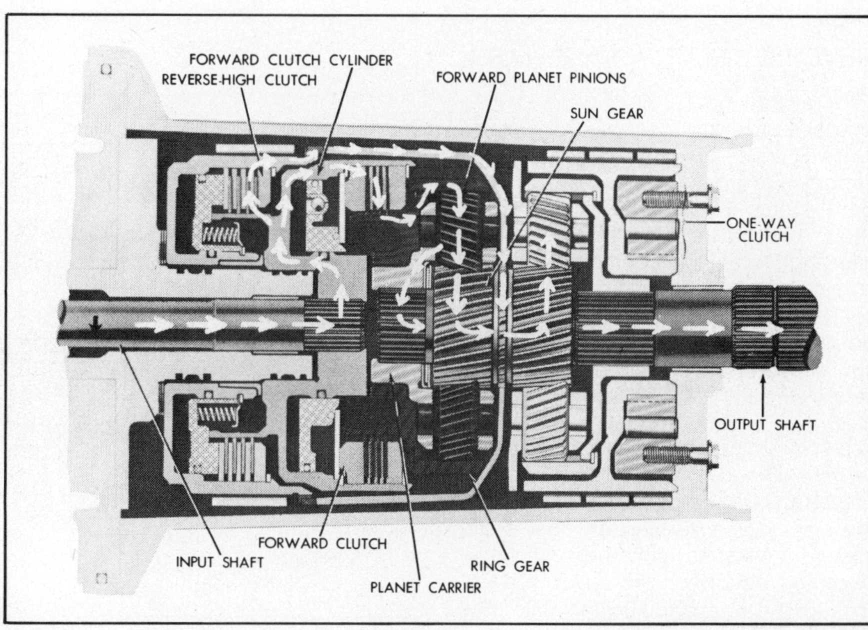

clutch drum. The intermediate band around the outside of the drum cinches down to stop the drum, shell, and sun gear.

With the sun gear locked to the case, the roller clutch is active in the rear planetary gearset, because the rear ring gear is splined to the output shaft. If you remember your planetary gear ratios, anytime the sun gear is held, and torque applied to the ring gear, the planet carrier will rotate in the same direction. The output shaft turns the rear ring gear, which makes the pinions walk around the stationary sun gear, spinning the rear carrier on the freewheeling roller clutch. The carrier and pinions are not transmitting any torque to anything, just going along for the ride. Gear reduction in intermediate gear is 1.46-to-1.

HIGH GEAR POWER FLOW

In high gear the input shaft drives the forward clutch cylinder. The forward clutch is engaged, so the forward ring gear is transmitting torque to the forward planet pinions. However, the pinions cannot rotate on their pins because the reverse-high clutch is engaged, and this is what locks the sun gear to the forward clutch cylinder.

If you trace it out on the illustration, you will see that the input shaft, foward clutch cylinder, reverse-high clutch, input shell, sun gear, front ring gear, pinions, carrier, and output shaft are all locked up solid. All pieces mentioned turn as a unit, just as if the input shaft were in one piece with the output shaft. The same conditions exist in the rear planetary gear set. It goes along for the ride, freewheeling through the roller clutch. Gear reduction in high is nonexistent, with a 1-to-1 ratio.

REVERSE GEAR POWER FLOW

In reverse gear, our good old dependable forward clutch is released, and the reverse-high clutch is engaged. Torque comes into the forward clutch cylinder from the input shaft, through the reverse-high clutch into the input shell, and drives the sun gear at input shaft speed. Of course the sun gear spins the forward pinions, and the forward pinion carrier turns with the output shaft, but because the forward clutch is disengaged—freeing the forward ring gear—the ring gear just spins away without doing anything at all.

So far, we have established that the sun gear is turning at input shaft speed. It meshes with the rear (low-reverse) pinions, spinning them so that torque is applied to the rear ring gear. That is possible because the rear (low-reverse) band is holding the rear planet carrier. It is locked solid to the transmission case. Thus, the torque of the sun gear, acting through the rear pinions, turns the rear ring gear in a counter-clockwise direction. The rear ring gear is splined to the output shaft, so the car backs up. Reduction in reverse is 2.20-to-1.

In the explanation of the various gear ratios, we told how the intermediate band clamps down on the reverse-high clutch drum and, through the connection with the input

1

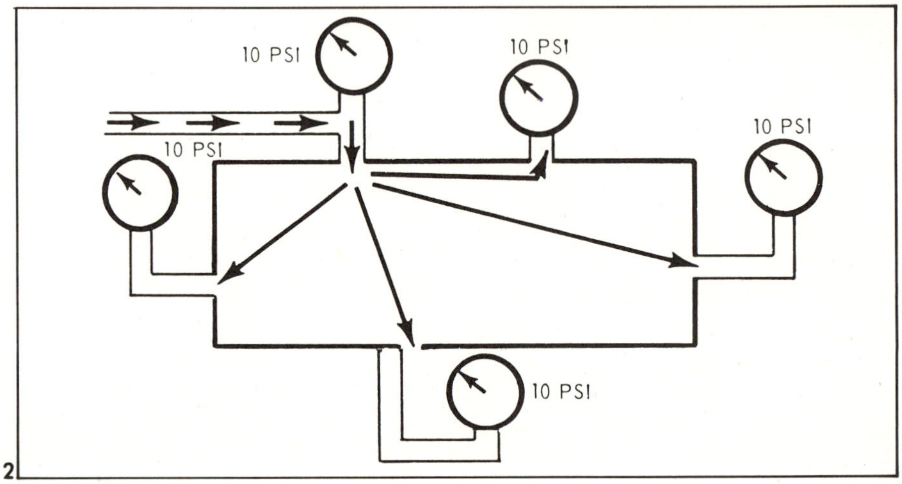

2

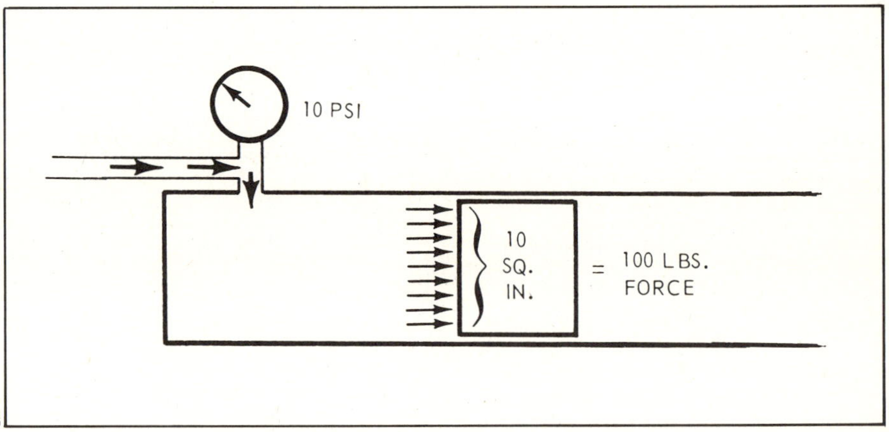

3

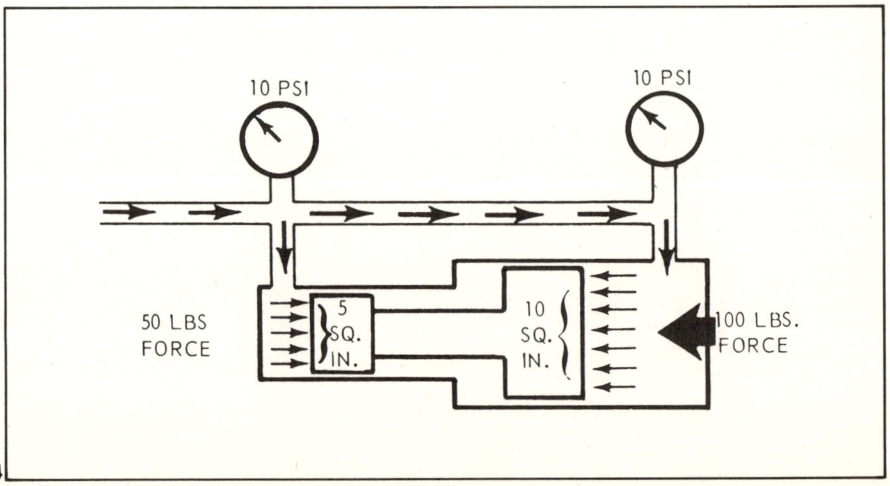

4

shell, holds the sun gear stationary. In low and reverse gear, the low-reverse band acts on the low-reverse drum to hold the rear planet carrier stationary. These bands are nothing more than a split ring of steel that can be tightened down onto the drum by closing up the split. The end of the band on one side of the split is attached to an adjustment screw which is used to bring the band close to the drum but not touching it. On the other side of the split is a rod or strut which leads right to a hydraulic servo.

If you remember any part of the hydraulic theory that some teacher or professor tried to drum into you, you will recall that pressure—whether it is hydraulic pressure, oil pressure, air pressure or any other kind of pressure is measured in so many pounds per unit area. This unit area can be cubic centimeters, square inches, square yards, square miles or anything you want to work with.

You may also remember that pressure is the same throughout any system. If you have a pump filling a pipe with oil and a relief valve which maintains the pressure at five pounds per square inch, then it doesn't matter if the pipe is 10 miles long. Any place that you want to tap in and measure the pressure along the pipe you will get a measurement of 5 psi.

You can connect side pipes onto the main pipe with fittings and build a kind of a tree or even a road map so that pipes run in all directions as if the set-up was created by a mad plumber. As long as the pump is putting out 5 pounds of pressure and the relief valve is set to release at 5 pounds, then no matter where you measure the pressure in the system you will get 5 psi.

If on one of the branches of the pipe you install a piston with a seal so that the oil can't leak out around

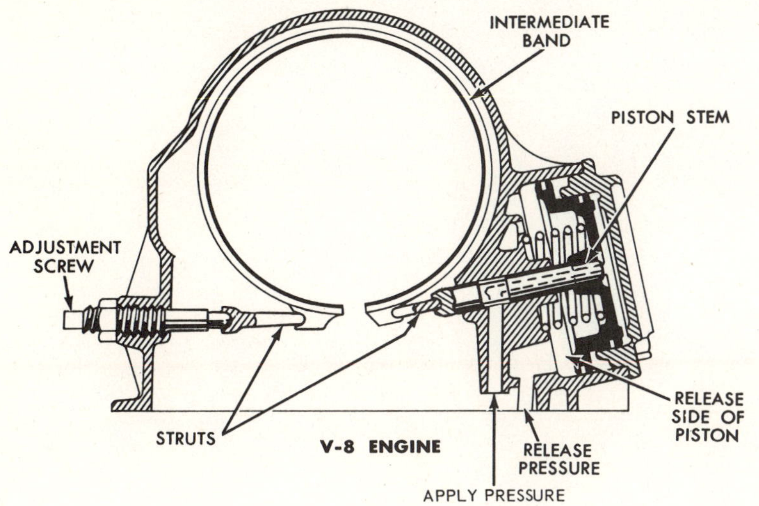

5

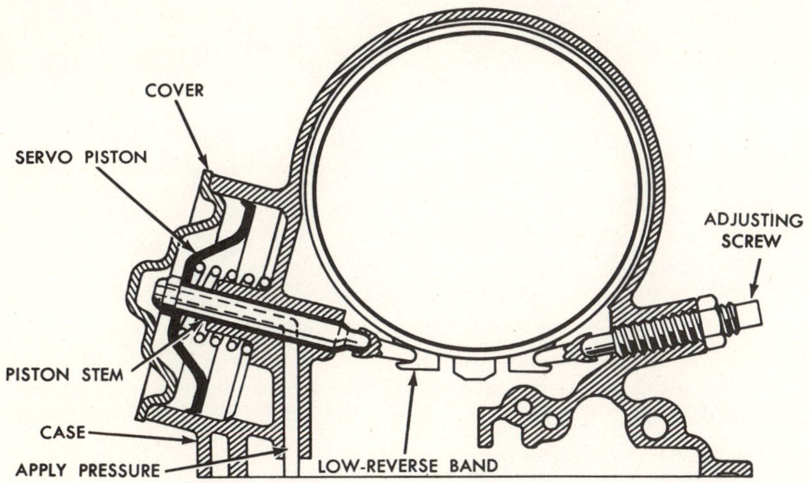

6

1. This is the means of identifying the rear band apply pin on a Turbo Hydra-Matic. Pins with one ring are shorter than this; those with three are longer.

2. If the pressure is 10 psi, it will be 10 psi anywhere in the system, assuming no leaks.

3. With the same amount of pressure, the bigger you make the piston, the more force you get.

4. Pressure is the same everywhere, but the piston will move in the direction of the large arrow because force is greater.

5. C-4 intermediate (front) band. The adjustment screw is accessible from outside the transmission case. Some transmission bands do not have a screw adjustment, as in the Turbo Hydra-Matic. When those bands wear enough to require adjustment, a longer apply pin is used, or the band is replaced.

6. Low-reverse (rear) band on the C-4. Other transmissions have the parts arranged differently, but the servos and bands are recognized by their shape.

7. A slightly different situation, but the same reasoning applies. The large piston head will pull the small piston head along with it, because of the greater force.

8. Moving inside a housing, it is the valve lands that control fluid pressure by covering or opening oil holes.

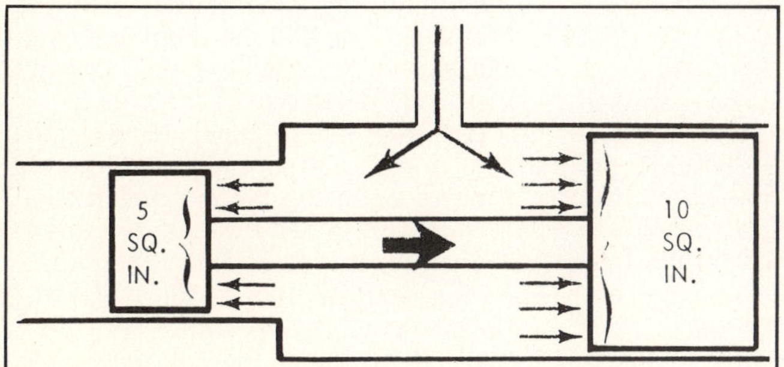

7

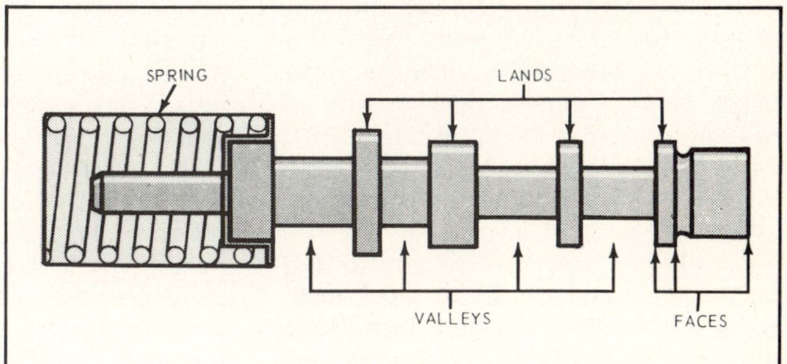

8

it, you will have 5 pounds per square inch pushing on the head of the piston, but the total force pushing on the piston will be 5 pounds multiplied by the number of square inches that are on the head of the piston. If the piston has 4 square inches on its surface where the oil is pressing, then the force on the piston is 20 pounds.

Our total push on any area such as a piston is called the force, but the pounds per square inch are called by the name pressure, so that we don't mix up the two actions. If you take a look at the illustration of the band and servo, you will notice that there is a passage admitting hydraulic fluid under pressure to the side of the piston opposite from the split. This hydraulic pressure pushes the piston against the spring pressure and tightens up the band so that the drum underneath the band is stopped completely from moving. Of course the drum does not stop immediately, and there's a lot of friction and heat buildup, and some wear too. The servo piston can only move so far and then it runs out of travel, so you can see that if the band wears enough it isn't going to stop the drum no matter how hard the servo pushes. This is why bands must be adjusted periodically or at major overhauls in order to keep them close to the drum so that the servo can act efficiently.

Servos have fairly large pistons, some of them three inches in diameter. They are made large purposely so that enough force will be exerted on the band to stop the drum quickly with little wear. Release of a servo can come in two ways. You can shut off the hydraulic fluid which is pushing on the piston and allow the spring to push the piston back and exhaust the fluid through the same passage it came in. If you want a quicker release you can run fluid under pressure to the underside of the piston so that the servo will relax that much faster. The spring is there to be sure that the servo stays retracted.

Both the spring-only release method and the hydraulic release method are used in various transmissions. In the C-4 Ford transmission the hydraulic pressure is used with the spring to release the intermediate band, and the spring only is used to release the low reverse band. The valve that admits hydraulic pressure to the servo has a double function when it comes time to release the servo. It not only has to cut off the hydraulic fluid activating the servo, but it has to allow the fluid to exhaust from the servo piston so that the servo can retract. Hydraulic pressure is used not only to operate the servos and bands but to make the clutches engage or disengage properly.

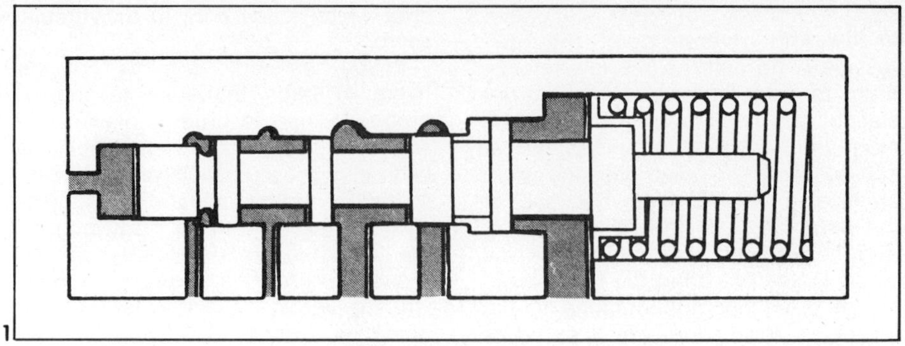

1

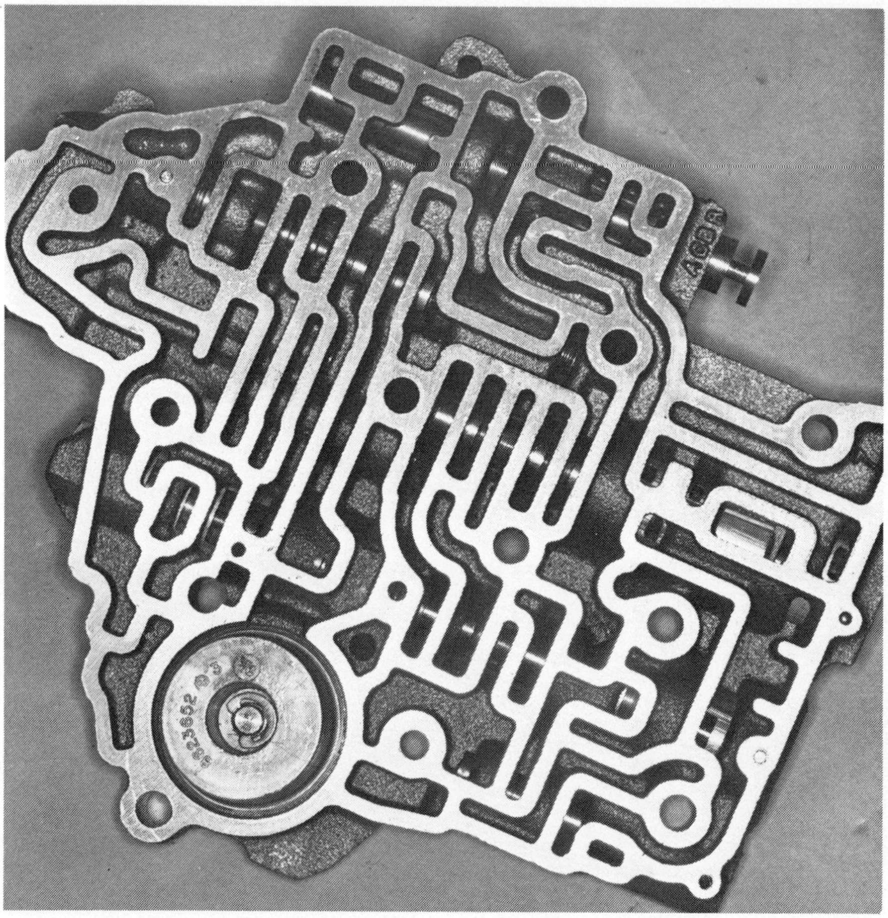

2

CLUTCH OPERATION

The C-4 transmission has two clutches: the forward clutch and the reverse-high clutch. We have already covered the action of these clutches in the dIfferent gear ratios but now we want to find out what it is that makes these clutches work. By looking at the illustrations you can see that the clutch assembly is a series of plates with some plates joined to the hub and the others joined to the drum or cylinder. What we have is a multiple-disc clutch similar to the type used on some race cars today. They were also used on passenger cars in the early days when the automobile industry was just beginning.

In a multiple-disc clutch the plates are free to float back and forth because they are just held by grooves in the hub or the drum that accept tangs on the disc. When the clutch is disengaged the set of discs that is connected to the hub is spinning away while the other set connected to the drum is stationary. It could be the other way around too, with the hub stationary and the drum spinning. But all motion is relative so it doesn't make any difference to the clutch.

With one set of clutch plates spinning and the other stationary, there's a bit of bouncing around and friction in the clutch when it is released. Transmission fluid lubricates the clutch so that this friction doesn't bother anything. A clutch is much more dependable than a band. Because of this extra dependability,

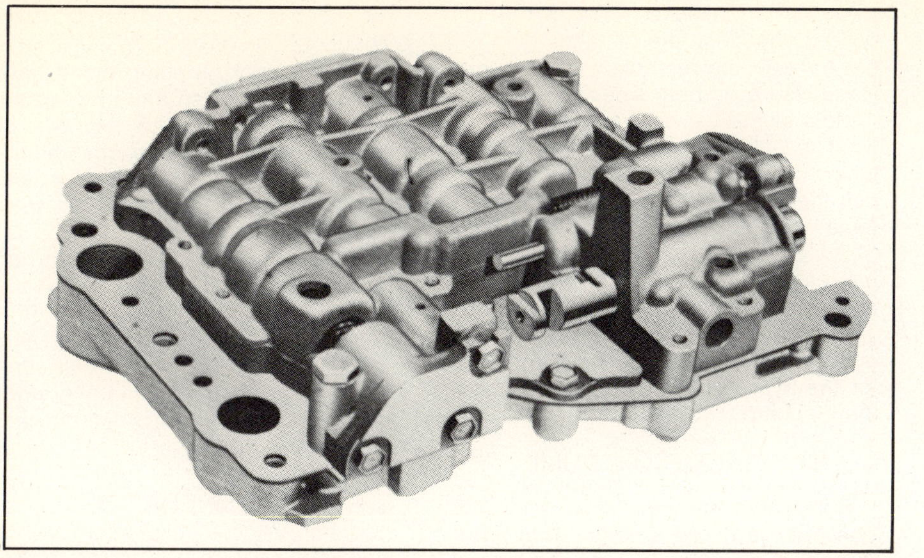

3

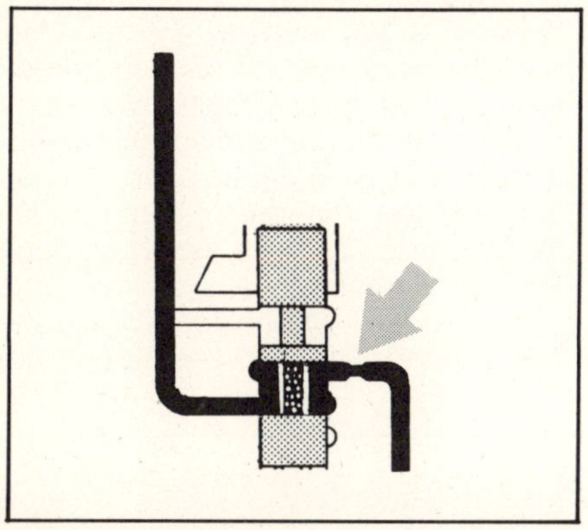

4

1. One of the many possible arrangements of passages around a hydraulic valve.

2. Underside of Turbo Hydra-Matic valve body. With all those passages and sliding valves, you can see that cleanliness is important. The valves can easily stick in their bores if a speck of dirt gets into the wrong place.

3. Outside of the valve body on a Ford C-4 transmission.

4. Put a restriction in a passage and it will slow down the build-up of pressure, thus delaying the action of the valve. A shift can be made softer, easier that way.

some transmission designers have tried to do away with bands entirely, using clutches instead.

Built into the clutch cylinder is the clutch piston, which is right next to the plates and rotates with the whole assembly. In order to apply the clutch a valve admits hydraulic fluid under pressure to the clutch piston and it squeezes all of the clutch plates together. It does a pretty good job of squeezing, locking the plates up tight into one solid mass so that the hub is locked to the drum.

When we want to release the clutch a valve releases the hydraulic pressure and a spring pushes the piston back. There's nothing holding the clutch plates together so they float apart and the hub starts moving independently of the drum. The release spring can be a flat washer or coil springs.

There has been quite a variation in the number of plates used in clutches through the years. One of the problems in reassembling an automatic transmission is in getting the right number of plates back in. Of course if too many are put in, then the clutch will not release and if too few are put in, the clutch will not engage lightly enough to lock the hub and drum. To be sure of putting in the right number of clutch plates, the transmission man has to have a book for the particular year and make of transmission he is working on, or he has to be so familiar with the transmission that he is positively sure what's supposed to go in it.

To keep the clutch piston from leaking oil it is sealed on both its inner and outer diameters with O-rings. When the mechanic takes it apart he will not only check the clutch plates for wear and replace them if necessary, but he should in every case replace the O-ring seals.

In relation to diagnosing trouble in an automatic transmission, you may have seen charts that show the gear positions and tell which of the clutches and bands are engaged in the different positions. These charts are useful in determining which band or clutch may be responsible for slippage in the transmission. For example: If you get slippage or harsh jerky application in manual low, it could be the forward clutch or the low-reverse band but if the same condition occurs in reverse, then you know the trouble must be in the low-reverse band because this is the only band that is applied in both manual low and reverse gear positions. Common sense goes a long way toward diagnosing the ills that may occur in an automatic transmission. A lot of these ills can be caused by dirt or wear in the hydraulic control system.

HYDRAULIC CONTROLS

We know what the terms pressure and force mean and that it is possible to run hydraulic fluid through a line into a large cylinder to act on the head of a piston and move it. If we want to move the piston faster or with more force, we just make the area that the hydraulic pressure can act on considerably larger. Designers of automatic transmissions have used this principle of hydraulics to make pistons do some pretty tricky things when they are mounted in the cylinders in different arrangements.

For instance, suppose we had a piston that was double headed, with one end of it being 10 square inches in area and the other end in a smaller cylinder that was only five square inches in area. If we admit 10 pounds hydraulic pressure to the small end of the piston, we will have 50 pounds of force acting against it, which will move the piston in the direction of the larger end.

Suppose we want the piston to move in the other direction against the pressure that acts on the small end, but we don't want to shut off the pressure on the small end. All we have to do is admit the same pressure to the large end of the piston and because it is larger, it has more force acting on it therefore the piston will move in the direction of the small end.

By admitting pressure into different areas of cylinders that have different size pistons we can do some pretty tricky things, causing pistons to move back and forth wherever we want them to go. Suppose we have the same large and small piston and we admit pressure between the two pistons. Of course the large piston is going to move and drag the small piston along with it because the force on the large piston is greater.

These pistons that we have been talking about are used in an automatic transmission as valves. Most of the valves have a spring at one end but all the spring does is hold them in one certain position so that there's no hydraulic pressure applied anywhere on the valve. That is done to ensure that vibration will not cause the valve to move and turn something on when it's supposed to be off.

The valves are really peculiar looking things—like a bunch of rings stacked on a pole. And if you didn't know they were valves you would never consider them as such. The exterior surfaces of the rings are called lands, and the spaces between the

Transmission

rings are the valleys. The flat surfaces of the rings next to the rod are called the faces.

Actually the valve is not a valve until it is installed in the body, a housing that is machined to close tolerances so that the valve will not leak. You can see from the illustrations that the lands seal against the housing and the valleys allow the fluid to enter from a passageway, go around the rod and exit through another passageway. When precise control is needed the land will have a shallow groove cut in it which actually amounts to a very small valley. This groove allows oil to enter on one side of the land, go around both sides of it and exit on the other side into another passage.

Another trick the designers use is passage restriction. If a servo, clutch or a valve is operated too quickly by hydraulic pressure, instead of changing the size of the surface where the hydraulic fluid is pressing it is sometimes easier just to put a restriction in the passageway. The total force is still the same to make the component work but it takes a while for it to build up because the oil takes time to get through the restriction and equalize pressure on both sides of it.

Most of the valves are contained in what is called the valve body. It's a lumpy thing that you see when you pull the pan off the bottom of an automatic transmission. It has all kinds of valves, check balls, and other pieces and looks like a plumber's nightmare. If you have an exploded view it's really amazingly easy to take apart and put back together. Pressure for the hydraulic systems comes from the oil pump.

HYDRAULIC OIL PUMP

Sometimes the impeller in a torque converter is called a pump. That term is correct because the impeller actually does pump fluid but it is not the pump we are referring to here.

If the oil pump quits then everything stops, so it really might be considered the heart of the automatic transmission. The pump keeps the torque converter full of transmission fluid, supplies this same fluid as lubrication to the various parts of the transmission, and also maintains pressure with this same fluid in the main control system which consists primarily of the valve body.

As if the poor oil pump didn't have enough to do, it also has to keep the pressure up enough so that the oil from the torque converter goes all the way up to the radiator, is cooled, and comes all the way back to the automatic transmission.

Oil supply for the pump is carried in the oil pan which is attached to the bottom of the transmission. A fine screen or filter keeps most wear particles out of the pump. It's really one heck of a pump, with much greater capacity than the transmission will ever need.

VALVE OPERATION

To show their mechanics what happens in the hydraulic control system of the C-4 transmission, Lincoln-Mercury provides over 20 full page drawings of hydraulic circuits. We don't want to go into the subject that deeply here but in order to give you an idea of how the pump can operate the valves all by itself simply through hydraulic pressure, let's examine the control pressure regulator valve in the C-4 transmission and see how it not

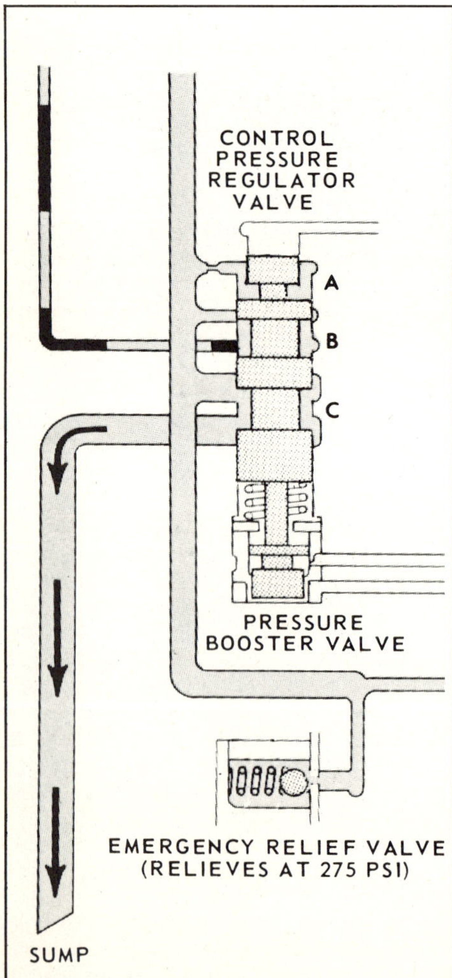

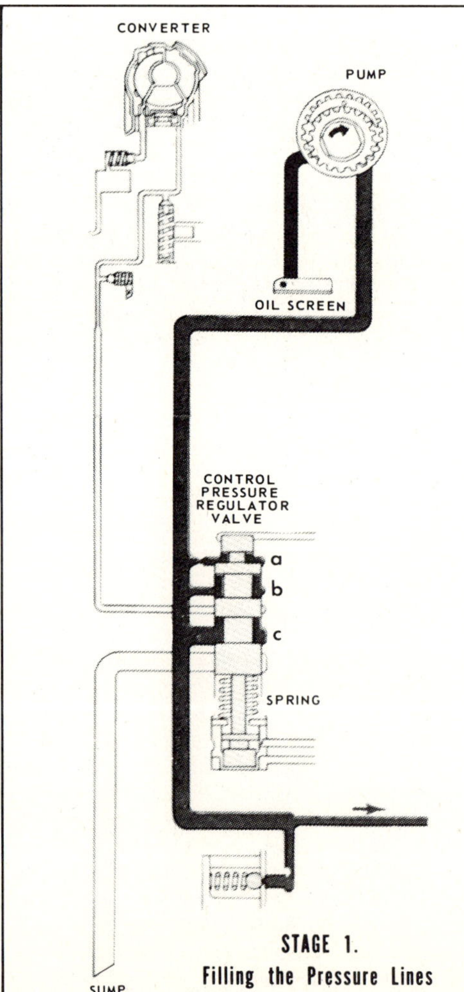

STAGE 1.
Filling the Pressure Lines

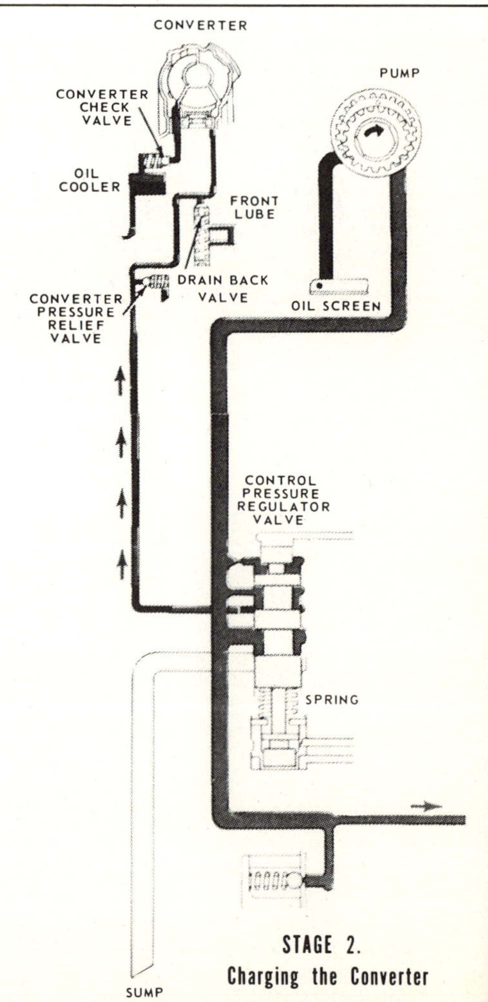

STAGE 2.
Charging the Converter

only controls the pressure that the pump puts out but also supplies pressure to keep the converter full.

The pump is usually mounted just behind the converter in the front of the transmission, and it runs at any time that the converter is turning, which means any time the engine is running. In the illustration showing stage 1, the engine is running and the pump is supplying automatic transmission fluid through the control pressure regulator valve. A spring holds the valve up and there are three valleys into which hydraulic fluid is admitted—A, B and C.

Pressure of the fluid in valleys B and C will not move the valve in either direction because the faces of the valve on either side of those valleys is equal in area. But take a look at valley A. The face on the bottom side of valley A is greater in area than the face on the top side. We know from our study of hydraulic pressure that the larger area will have more total force applied to it. The upper part of the valve including the valleys A, B, and C, and the lands surrounding them is all one piece, so the force acting in valley A causes the valve to move downwards against the pressure of the spring below it. The rod in the center of the spring fits into a hole in the valve like a slip joint and does not affect the movement of the valve.

In the illustration showing stage 2, you can see that as the valve moves downward it uncovers the passageway to the converter, so that the converter is pressurized. The pump does not fill the converter because it stays full of oil at all times. However, this oil is not under pressure until the pump starts pumping and the control pressure regulator valve begins to move downward.

In stage 3 you see that the passageway leading into the oil pan or sump has been opened by the downward movement of the valve. When that passageway is opened the pressure bleeds off from the pump and falls enough to allow the spring to push the valve back up. Of course the instant that the spring pushes the valve back up, the pressure builds up again and forces the valve right back down.

Because the pump pressure is a constant force and the spring is a constant force, there is no pulsation of the valve. It just stabilizes itself at a point determined by the force of the spring. The spring is calibrated so that the pressure will hold between 55 and 62 pounds per square inch.

Other valves in the system work in a similar manner, admitting or releasing pressure to perform the various functions of controlling the clutches and bands in the planetary gear unit.

Control of the hydraulic system for various engine conditions is determined by four units in the transmission: the governor, the vacuum diaphragm unit, the downshift valve, and the manual valve.

1. Can you see why the valve will move down against the spring when the fluid is admitted under pressure into the three areas, A, B, and C? Passage A is the one that moves the valve, because the force is greater on the side nearest the spring. The emergency relief valve is a safety feature in case some freak happening should cause an abnormal pressure rise. The pressure booster valve acts with the spring pressure to close up the exit hole to the sump, thus raising the control pressure when needed.

2. A basic hydraulic system. Note that the pump has three jobs, keeping the converter full of oil, lubricating the various parts of the transmission, and giving oil under pressure to the main control. Both the governor and the primary throttle control (vacuum modulator) tell the main control what to do and when to do it. The governor senses car speed; the modulator senses the load on the engine. The main control activates the clutches and bands so that the transmission will go into the various gear ranges.

MANUAL VALVE

There isn't much difference in the looks of the manual valve from any other valve in the transmission. Its unique feature is that it is the only valve controlled by the shift lever. For all normal driving a manual valve stays in the drive position. The only time that it is necessary to move it is to go into neutral, reverse, or to shift the transmission manually instead of allowing it to go through its full cycle of automatic shifts.

Also connected to the shift lever is the parking pall, but this is not a hydraulic function. It is simply a pall that is forced into a toothed wheel connected to the driveshaft. When in park, the driveshaft is locked so that

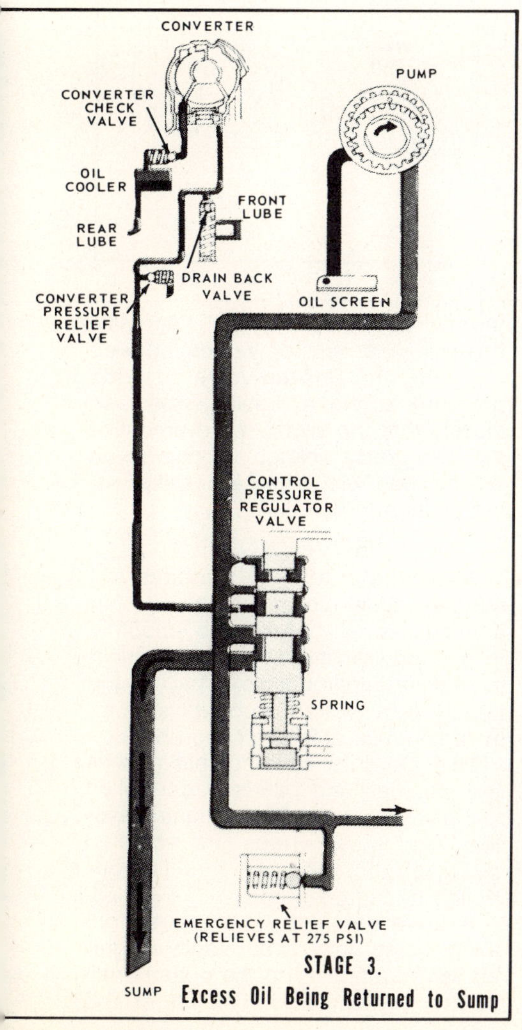

STAGE 3.
Excess Oil Being Returned to Sump

2

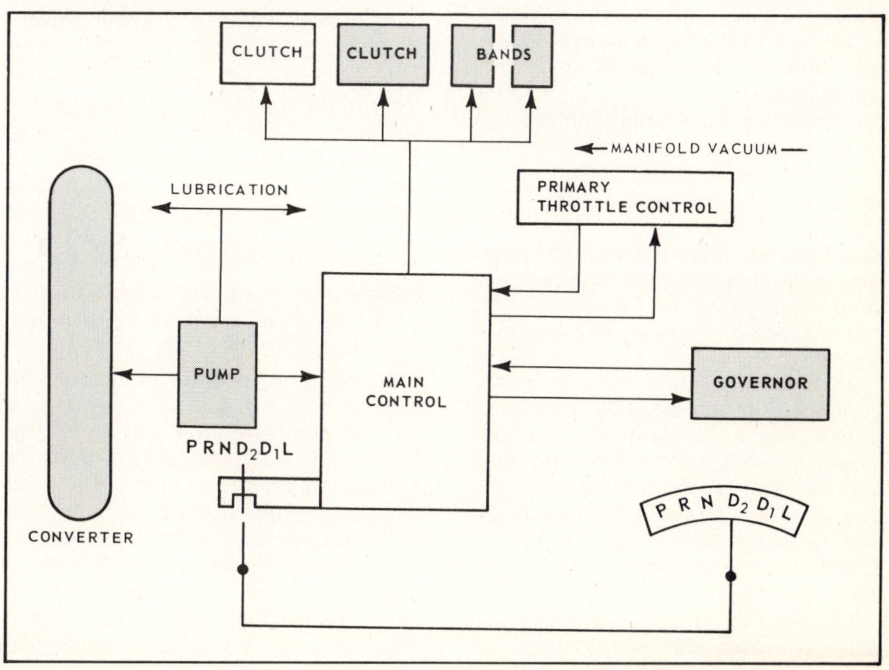

Transmission

the car will not roll out of control while it's unattended. Park position is only used while the car is standing still. If you have ever shifted into park while the car was moving even a half a mile an hour you know from the horrible sound why it's necessary that the car should be motionless before this particular shift is made.

Mounted on the carburetor or somewhere on the throttle linkage there may be a plunger-operated electric switch that sends a signal to a solenoid in the transmission to operate the downshift valve. The downshift valve can also be operated mechanically from a rod which is connected to the carburetor linkage. Whichever way it's activated, it amounts to the same thing. When the throttle is opened—wide open—the downshift valve is activated, and it channels hydraulic pressure to the proper parts of the transmission so that the gear ratio will change from whatever gear it is in to the gear which is the next lower gear.

On some early models of various makes of transmissions there was no separate downshift valve although there was a rod going down to the transmission. In that case the rod was connected to the transmission throttle lever. Now on most transmissions the throttle valve is controlled by the vacuum diaphragm unit.

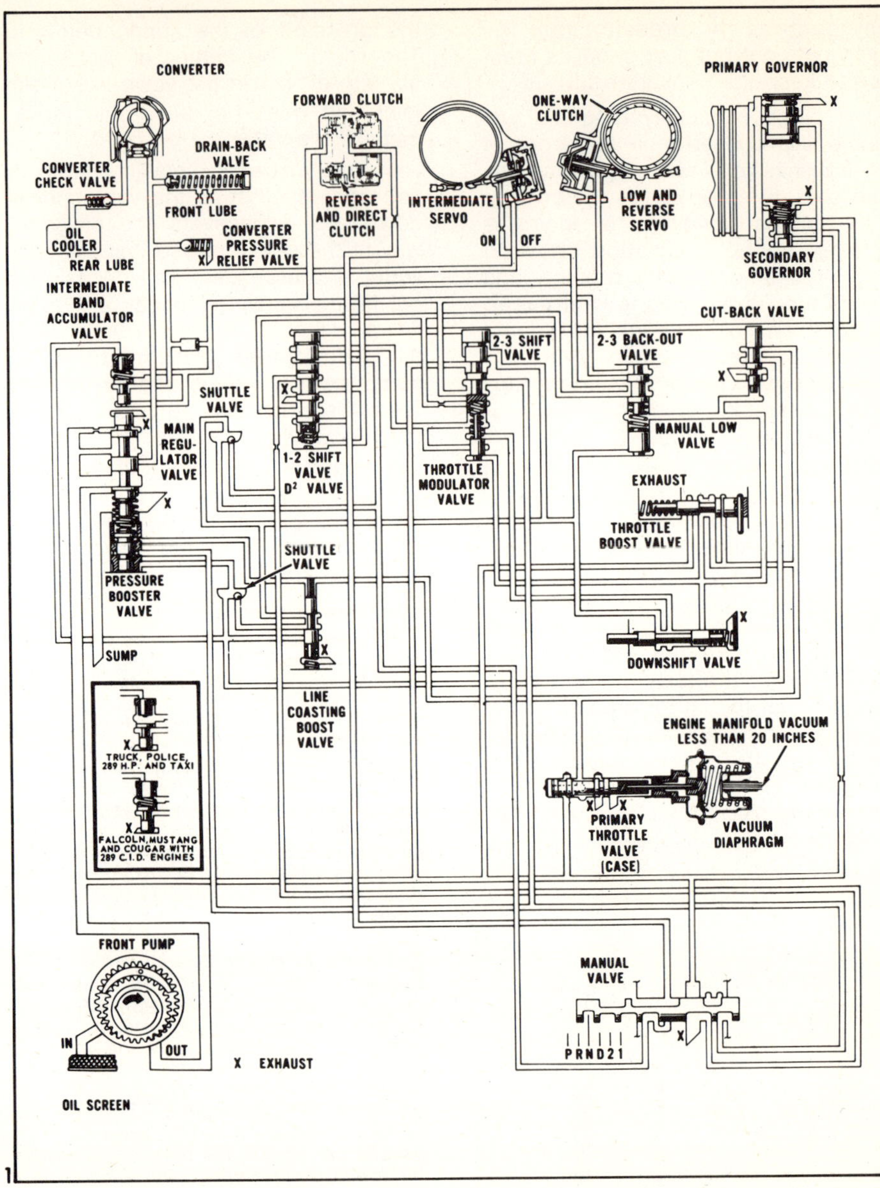

1

VACUUM DIAPHRAGM

One of the major control points in the hydraulic system is the primary throttle valve. Other manufacturers call it by different names, such as modulator valve or just plain throttle valve. It is important because it controls the amount of pressure available in the hydraulic control system to operate the clutches, bands, and various control valves.

Hydraulic pressure that controls the clutches and bands, and even the valves, must be regulated so that shifts are smooth without harshness, and so that there is no splippage. You can see that if hydraulic pressure is inadequate, the clutches and bands will not be firmly applied, and that unwanted slippage may occur in the transmission.

Most of the older transmissions controlled the throttle valve with a rod connected to the throttle linkage. When the carburetor throttle was opened, the TV rod would move the throttle valve in the transmission so that extra pressure would be applied to the clutches and bands to prevent undue slippage.

The old TV rod also had a lot to do with controlling shift points. Many an adjustment was made on the TV rod by shortening or lengthening it to change shift points or again prevent slippage.

The TV rod knew what the carburetor throttle opening was because it was connected directly to the linkage. But the TV rod could not sense the load on the engine. At one particular throttle opening it is possible to have the car operating at any speed, just by varying the load. A hill or a trunk full of luggage can be enough to confuse the transmission so that it doesn't shift at the right time.

To get more positive control of the throttle valve, designers went to a vacuum diaphragm control on the throttle valve. The diaphragm is hooked directly to engine manifold vacuum. High vacuum closes the throttle valve and a spring opens it. Thus, when the carburetor throttle is floored, vacuum drops off, and the spring opens the transmission throttle valve so more pressure is applied to the hydraulic control system. Actually, through a series of valves hooked together, the throttle valve controls pressure in the hydraulic system so closely that the slightest movement of the carburetor linkage or change in engine load will cause a corresponding change in hydraulic pressure.

GOVERNOR

When the automatic transmission reaches a certain speed, it will shift automatically. It happens so smoothly in a good transmission that the shift is almost imperceptible. That is the beauty of a hydraulic control system: Instantaneous pressure build-up.

This pressure build-up that applies the clutches and bands is controlled by valves in the hydraulic control system, and those valves are actuated by hydraulic pressure that is controlled by the governor.

A governor is a relatively simple thing, being nothing but some weighted valves attached to the output shaft of the transmission. The faster the output shaft goes, the more the valves uncover holes and let the hydraulic pressure through to activate the various other valves that control the clutches and bands. Governors are usually driven by a gear on the output shaft, and can in most cases be removed while the transmission is still in the automobile.

VALVE FUNCTIONS

We mentioned several times the various valves that operate in the hydraulic control system. To give you an idea of what they are and what they do, here is a list of the valves in a C-4 transmission and their function.

Manual Valve: Linked to the shift lever, it is used by the driver to select gear ranges.

Control Pressure Regulator Valve: Regulates pump pressure for basic transmission control.

Pressure Booster Valve: Increases control pressure if needed.

Control Pressure Reducer Valve: Reduces control pressure if needed.

Primary Throttle Pressure Valve: Accepts control pressure and changes it to primary throttle pressure.

Secondary Throttle Valve: Accepts primary throttle pressure and acts to reduce it.

Throttle Modulator Valve: Reduces throttle pressure for the 2-3 and 1-2 shift valves.

Throttle Booster Valve: Increases throttle pressure.

Cutback Valve: Reduces line pressure above 10 mph.

1-2 Shift Valve: Provides the 1-2 shift in D1, and prevents manual shift into low above 10 mph.

2-3 Shift Valve: Provides the 2-3 shift in D1 and D2.

2-3 Backout Valve: Acts as a timer to prevent simultaneous application of reverse-high clutch and front servo during a closed throttle 2-3 upshift.

Manual Low Valve: Closes the 2-3 backout valve during the shift into manual low.

3-2 Coasting Control Valve: Times release of the reverse-high clutch and application of intermediate band, in D2 coasting and 3-2 downshift.

Downshift Valve: Functions during kickdown shifts.

Primary Governor Valve: Controls the governor pressure below 10 mph to prevent certain shifts.

Secondary Governor Valve: Changes control pressure to governor pressure.

Emergency Relief Valve: Keeps control pressure from exceeding 275 pounds.

Converter Pressure Relief Valve: Keeps converter pressure from exceeding 70 pounds.

Drain Back Valve: Holds oil in converter when engine is off.

Converter Check Valve: Holds oil in converter when engine is off.

As you think back over what you have read in this chapter, don't feel bad if you have the idea that maybe you aren't ready to rush right out and overhaul your automatic. To do intelligent work on an automatic, you must have hydraulic circuit diagrams, exploded views, troubleshooting charts and a few special tools.

A lot of mechanics take automatics apart every day but many of the jobs they do are only simple seal replacements or band adjustments. To be a really good automatic transmission repair man, you have to admit that you don't know anything about the darn thing except how to take it apart and put it back together. Intelligent diagnosis of transmission ills and the ability to fix them the first time without having to take the transmission out and in like a yo-yo, requires a reliance on the training materials supplied by the manufacturer.

Show me a man who never reads a transmission book or looks at a troubleshooting chart, and I will show you a man who doesn't fix many transmissions. He may do a lot of removing and replacing, but he is either replacing too many parts unnecessarily, or he is unwittingly fixing it by the disturbance he creates when he takes it apart.

So if you want to work on an automatic, by all means do it. But first be sure you have the official manual put out by the manufacturer of the transmission. It might be a separate manual, or be included in the factory shop manual of the particular car you are working on. Independent publishers put out excellent manuals, too, but read and compare them first, before you buy. Some are abbreviated and condensed to a degree that their usefulness is limited. That would naturally be expected, because they cover all transmissions in one book, whereas the manufacturer puts out a single large volume covering only one transmission, and in some cases only one model year.

If you decide to take on your automatic, you will probably find it's a lot easier than you thought, and certainly a lot cheaper than paying someone else to do it.

1. Complete hydraulic circuit diagram for the Ford C-4. Plugged passageways or stuck valves can be diagnosed with a diagram like this. Some manufacturers supply a separate diagram for each condition in the trans, showing oil flow to the units that are operating. When a mechanic is stumped, it really helps to study a hydraulic diagram to see where the trouble might be.

2. When oil goes around a valve in a groove like this, it goes around both sides, so that the pressure is equalized on all sides of the valve for equal wear and smoother operation.

3. Turbo Hydra-Matic parking pawl in disengaged position.

4. When the driver shifts to Park, the lever (removed in this photo) moves the pawl down into the space between teeth.

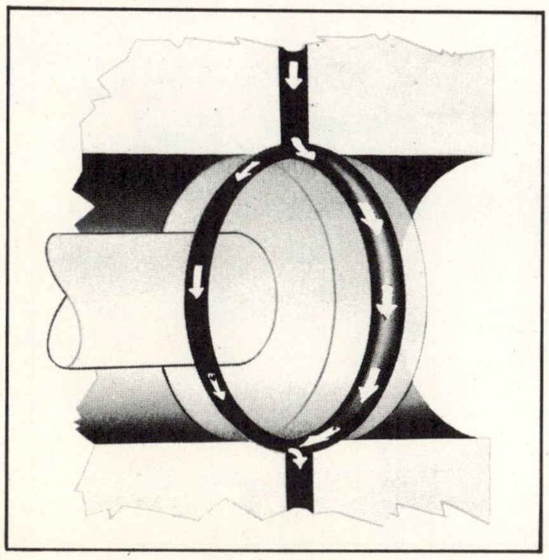
2

3

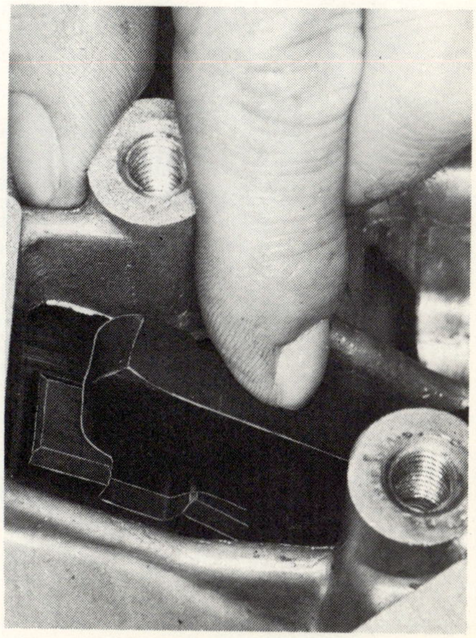
4

Chrysler Corp. Automatic Transmissions

The now-famous TorqueFlite was originally introduced by Chrysler Corp. in a limited number of 1956 Imperials. The first one was designated as the A-466 and was of cast iron, with a torque converter housing and rear extension of die-cast aluminum. Later versions are entirely aluminum and cast as a single unit, contributing greatly to the low overall weight of about 160 lbs. with the torque converter installed.

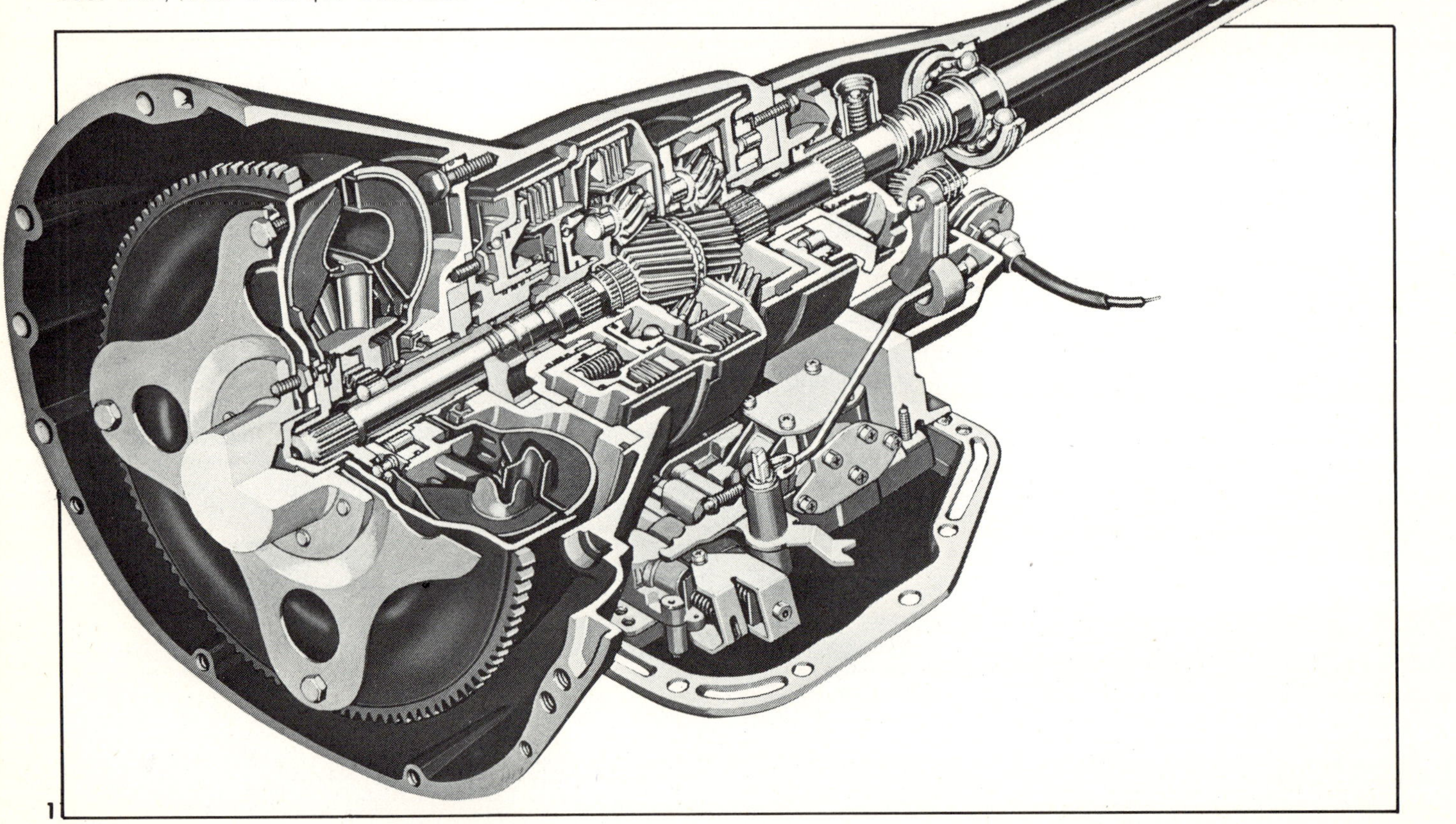

1

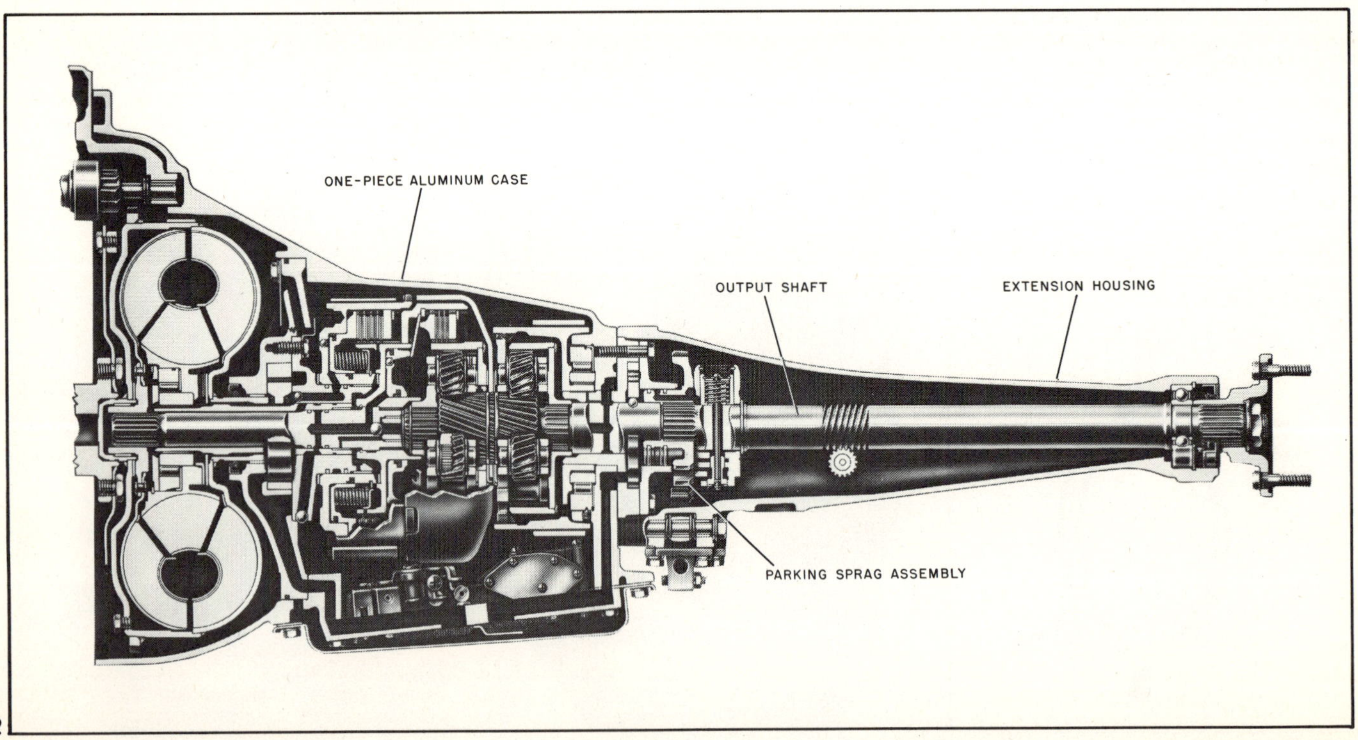

2

3 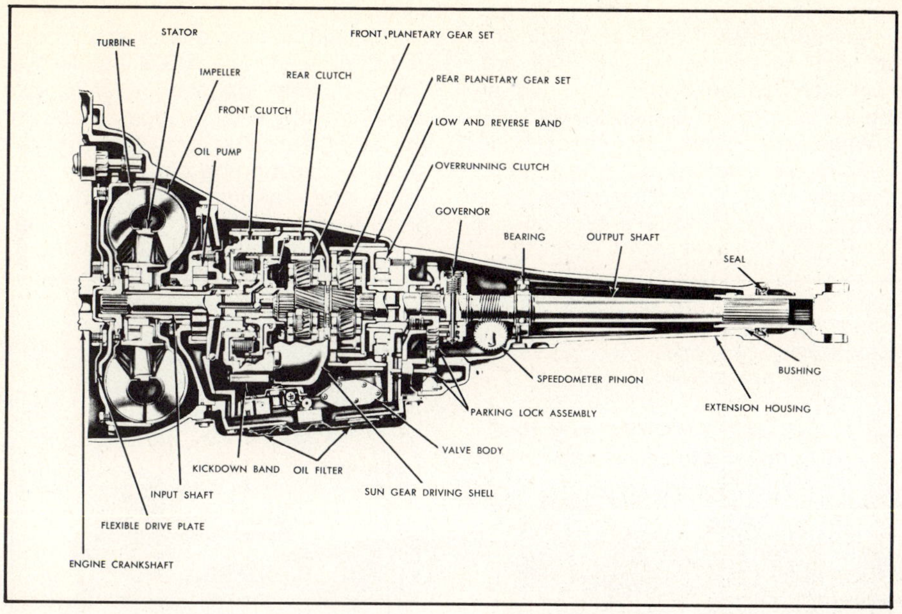

4 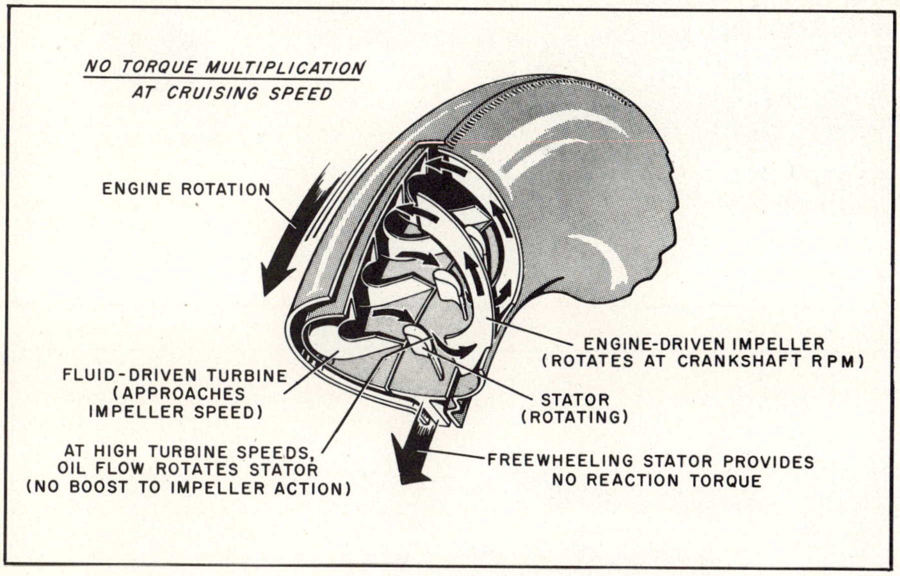

5 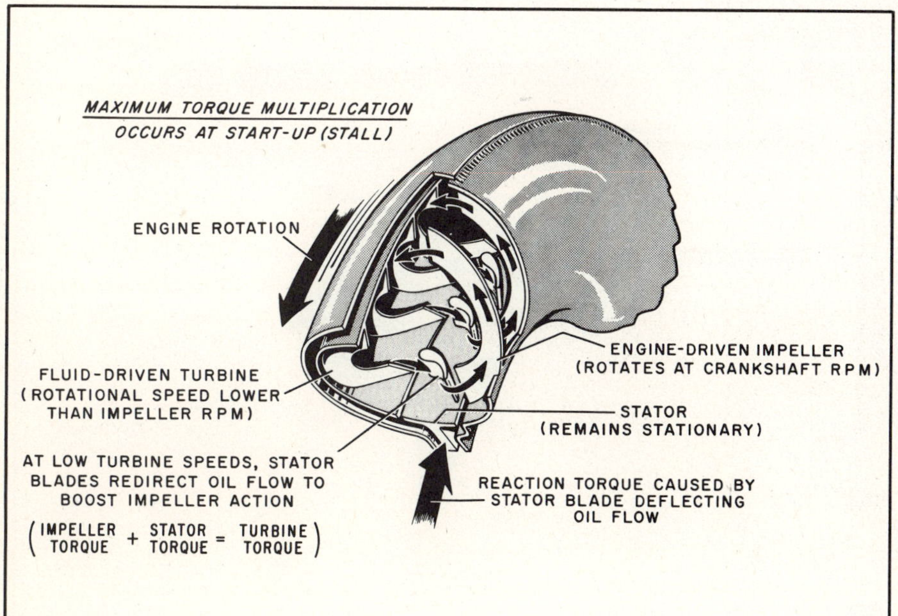

The A-466 had gear ratios of 2.45:1 for low, 1.45:1 for second, and 1.00:1 for direct, with 2.20:1 in reverse. With the exception of a reverse that is now 2.21:1, these ratios are still in use after 18 years, and are found in the latest units designated A-904 for the light-duty and A-727 for heavy-duty use.

Numerous improvements have been made since 1956, including the replacement of the original propeller shaft parking brake with an internally actuated sprag and lock; new paper-type organic friction material on the rear clutch discs to cushion neutral-to-drive shifts; a high-accuracy speedometer pinion located behind the governor on the output shaft; and standardization of extension housing lengths and mounting pad locations.

For 1967, both front and rear clutches in all TorqueFlites (except those used with the 383 and 440 engines) were faced with paper-type friction material. The input shaft diameter was increased and the torque converter shaft has an added spline. TorqueFlites used with the 426 and 440 engines had four planetary gearsets instead of the usual three.

The light-to-medium duty A-904 underwent major changes in 1968 to increase its torque handling capabilities, but the construction features do not differ materially from the larger A-727's. The output shaft bearing on

1. The A-727 Chrysler TorqueFlite is generally used in conjunction with the larger displacement engines. The housing is a single aluminum casting—for lightweight and better heat dissipation—and uses a 4-planetary gearset when attached to either the 426-in. Hemi, or the 440-in. wedge.

2. Chrysler's continuing upgrading of the transmission caused the '66 version of the A-727 to be far quieter. Trans featured noise-free pump, inside-acting parking sprag and control rods replacing previously used cables.

3. Chrysler continues to refine even the refined, and for '74 the A-727 looks like this.

4. When the converter reaches cruising speed, the rotation of the turbine and impeller becomes nearly identical. When this happens, the oil flow across the stator blades is changed, ending the force retaining the stator and allowing it to freewheel on its overrunning clutch at the same rotational speed as the impeller and turbine. When this occurs, the converter acts as a simple 2-element coupling.

5. The torque converter has 3 elements—impeller, turbine, stator—sealed in a doughnut-like housing which are then interconnected by transmission fluid. Drawing depicts converter at stall.

Chrysler Corp.

current models of both transmissions has been moved from the rear of the extension housing to a position just behind the revised speedometer pinion, with a seal and bushing taking its place. Identification marks on TorqueFlites were once cast in raised letters about ⅜-in. high on the lower left side of the bellhousing, but are now stamped on a pad provided at the left side of the oil pan flange.

While early production models for 1974 were unchanged from 1973, changes made during the model year to the A-727 include a ball check valve added to the rear clutch feed circuit and a new output shaft front clutch oil feed. The A-904 received a stamped steel planetary gear pinion carrier, along with changes made in the front annulus gear support assembly, overrunning clutch cam, reverse drum and reverse band one-piece lever.

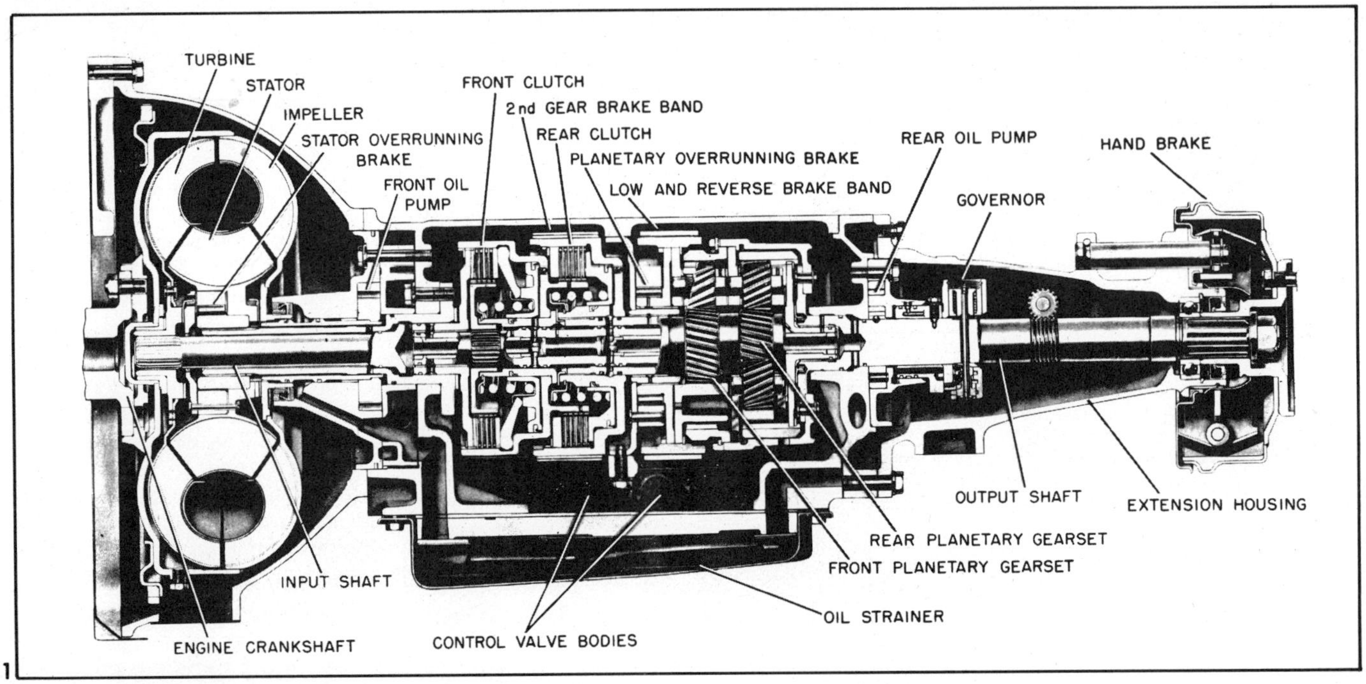

1

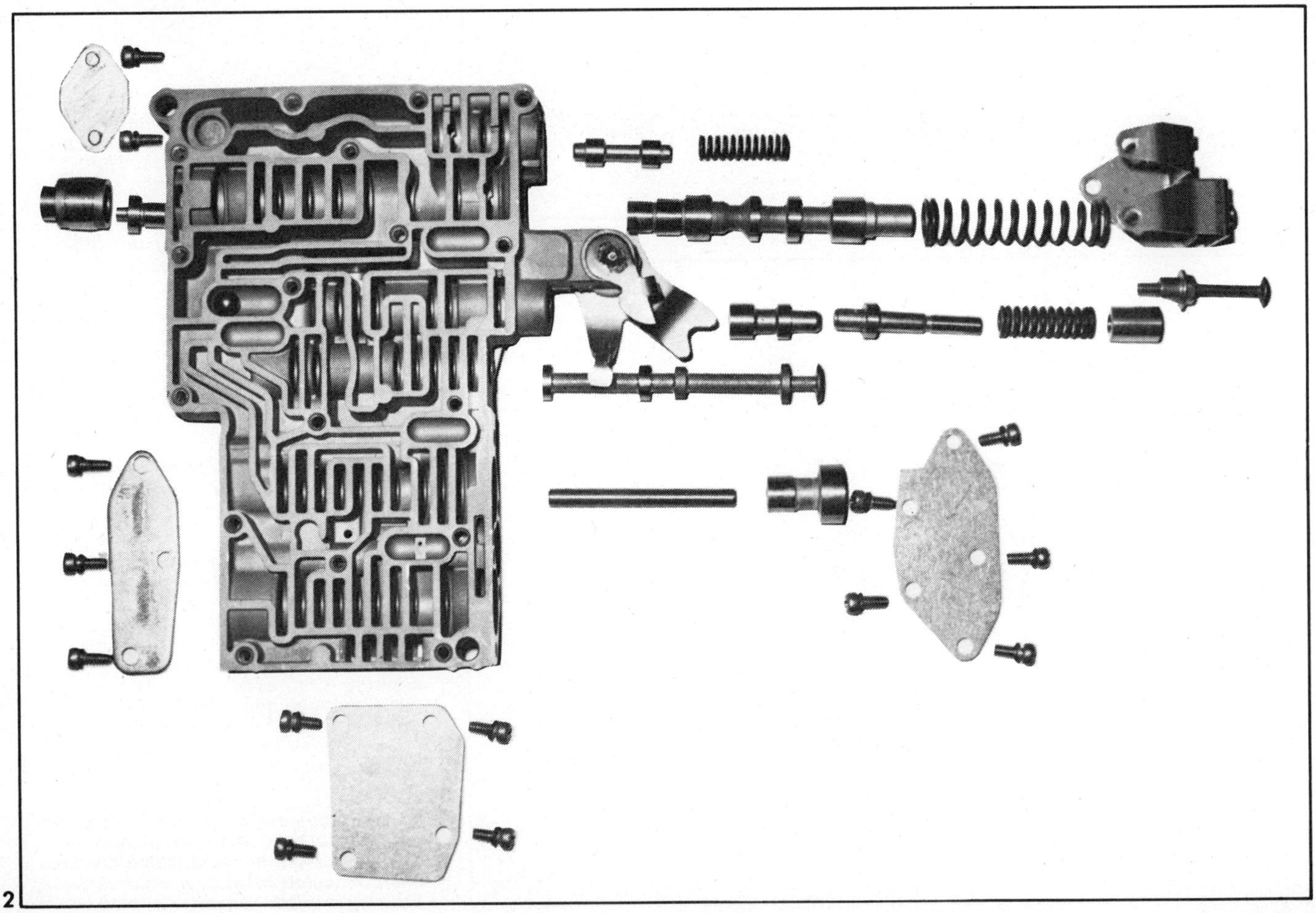

2

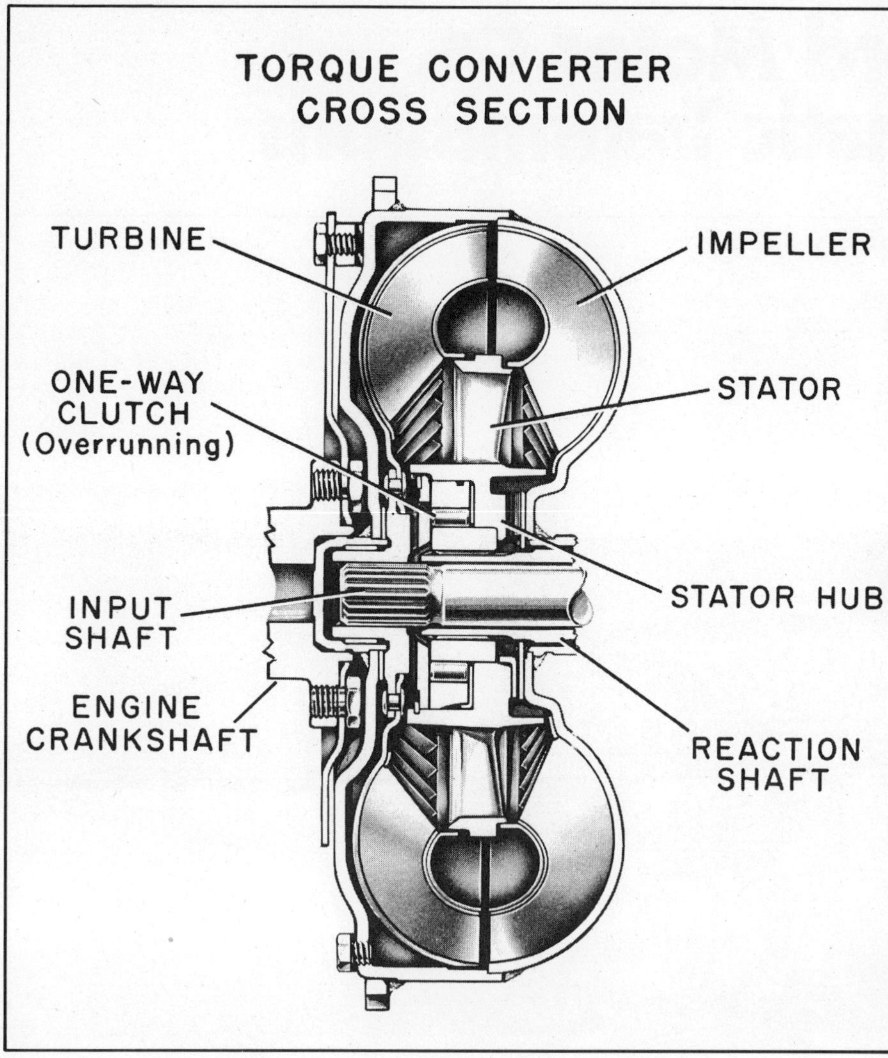

3

5

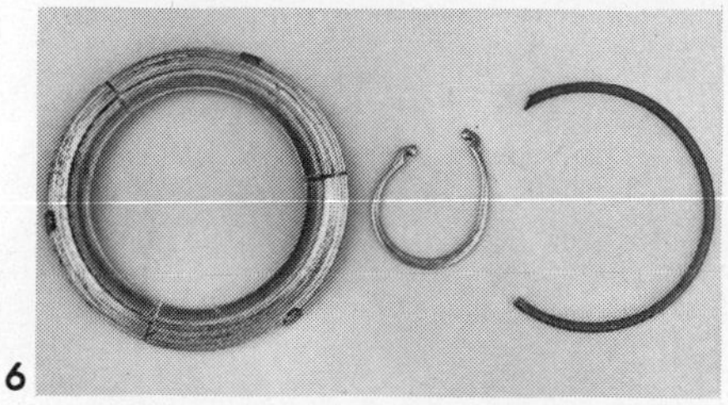
6

7

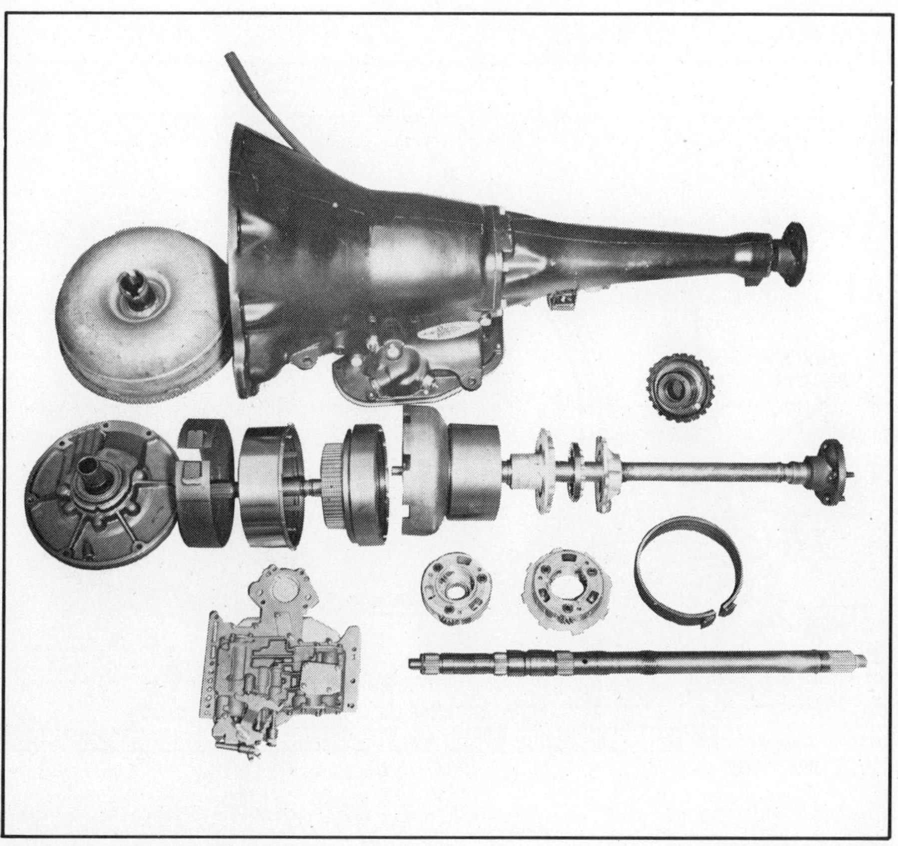
4

1. Going backwards, here's the original TorqueFlite as installed on just a few Imperials for '56. By '57, though, it was available in all Chrysler Corp. V-8-equipped cars.

2. The valve body on the smaller TorqueFlite can be modified when heavy-duty, manual operation is desired.

3. Cross section of the torque converter illustrates the "heart" of a TorqueFlite automatic transmission. The interaction of turbine, impeller, and stator provides the needed slip when a car is at idle, torque multiplication during acceleration and full "lock-up" when cruising.

4. The exploded view of the modern TorqueFlite shows the relative complexity of the unit. Its tailshaft can be replaced with a shorter truck housing if space limitations necessitate its use.

5-7. Part failure is a rare occurrence on a stock TorqueFlite, but it does happen with some frequency on some highly modified units. Snap ring, thrust washer, plus planetary carrier can be "lunched" if subjected to hard use, part misalignment, case distortion.

Ford Motor Co. Automatic Transmissions

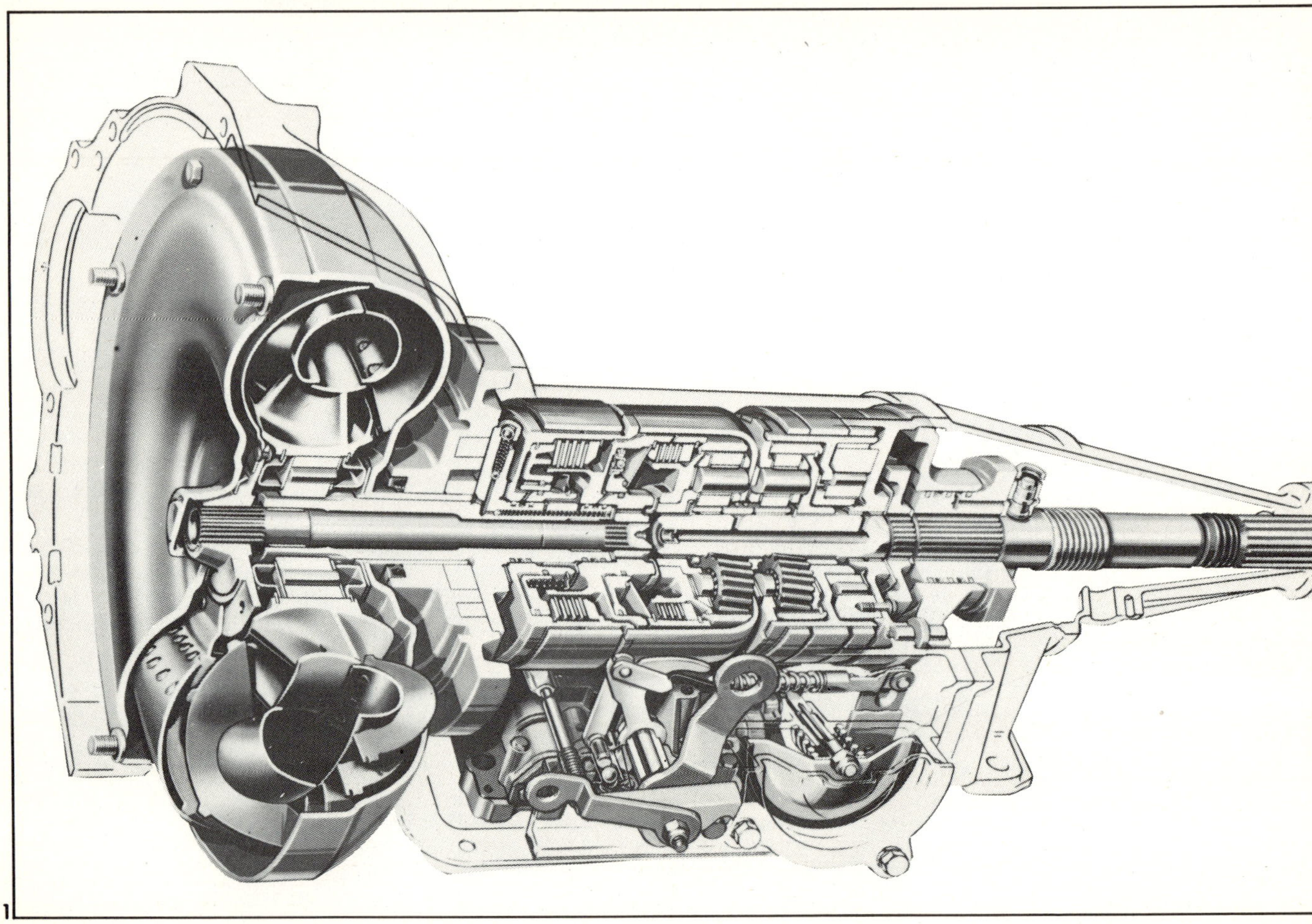

1

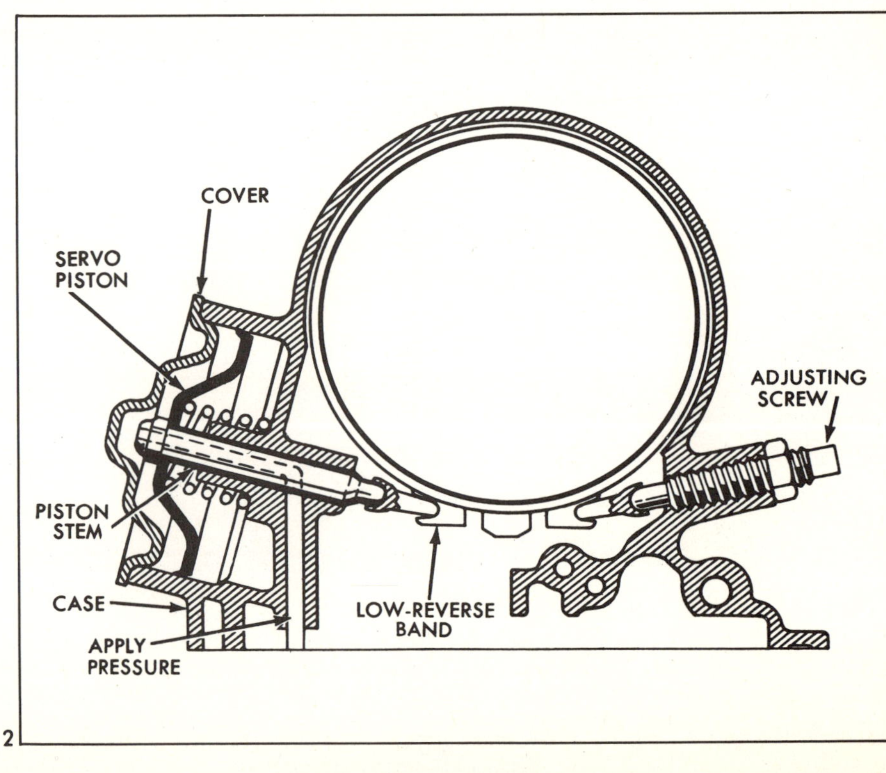

2

Although Ford products have had a variety of names for their automatic transmissions, they are now reduced to three basic units—the C-4, for use with engines through 289 cu. ins., and the C-6 and FMX for the larger engines.

Over the years, there have been the 2-speed Ford-O-Matics used in the Comet, Falcon and Ford cars, and the Merc-O-Matic used in Mercurys; 3-speed versions were the Cruise-O-Matic used in some Fords and in Thunderbirds, the Turbo-Drive used in Lincoln Continentals, and the Multi-Drive used in some Mercurys. In 1964, a 3-speed Ford-O-Matic was introduced for some Ford cars.

All three of the present models are called SelectShift Cruise-O-Matics and have similar gear ratios to the older 3-speed Ford-O-Matic. The C-4 and C-6 both use a 2.46:1 low, 1.46:1 second, and 1.00:1 high, but the C-4 reverse ratio is 2.20:1 while the C-6 uses a 2.18:1.

Through 1967, some Fords and Mercurys had a 3-speed automatic designated FX-MX, with gear ratios identical to the old Cruise-O-Matic. This was dropped for 1968 and brought back in 1969 as the FMX. While variations in construction exist among the C-4, C-6 and FMX, their operation and performance features are quite similar. The FMX now carries gear ratios of 2.40:1 in low, 1.47:1 in second, 1.00:1 in high, with a reverse of 2.00:1.

The SelectShift type allows the driver to manually choose the gear that he wants, and the transmission will stay in that gear until the selector lever is moved. This is a desirable feature for many, as the transmission can be operated as an automatic by leaving the selector in Drive, or shifted manually through the gears.

The C-6 closely resembles the C-4,

1. The C-4 Select Shift Cruise-O-Matic consists of a converter and extension housing, with a servo mounted on each side of the transmission case. As you can see, the C-4 and C-6 are similar in concept, but not in application.

2. The C-4 trans is equipped with a low-reverse servo band not found on C-6.

3. The smaller C-4 tranny can be converted to manual operation by modifying the torque converter, valve body, low-reverse servo piston and housing.

4. Ford's old Cruise-O-Matic has become its new FMX. The trans is still a 3-speed unit with capabilities for automatic up- and downshifting in all gears, plus manual selection in 1st and 2nd.

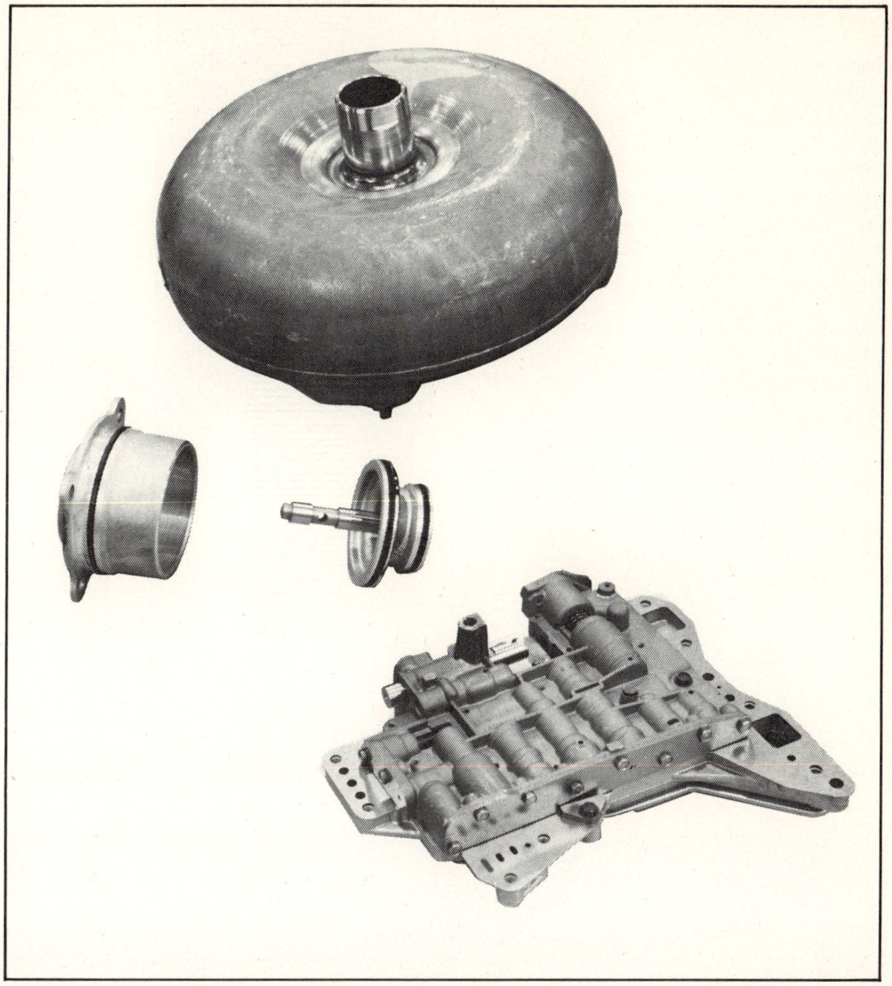

3

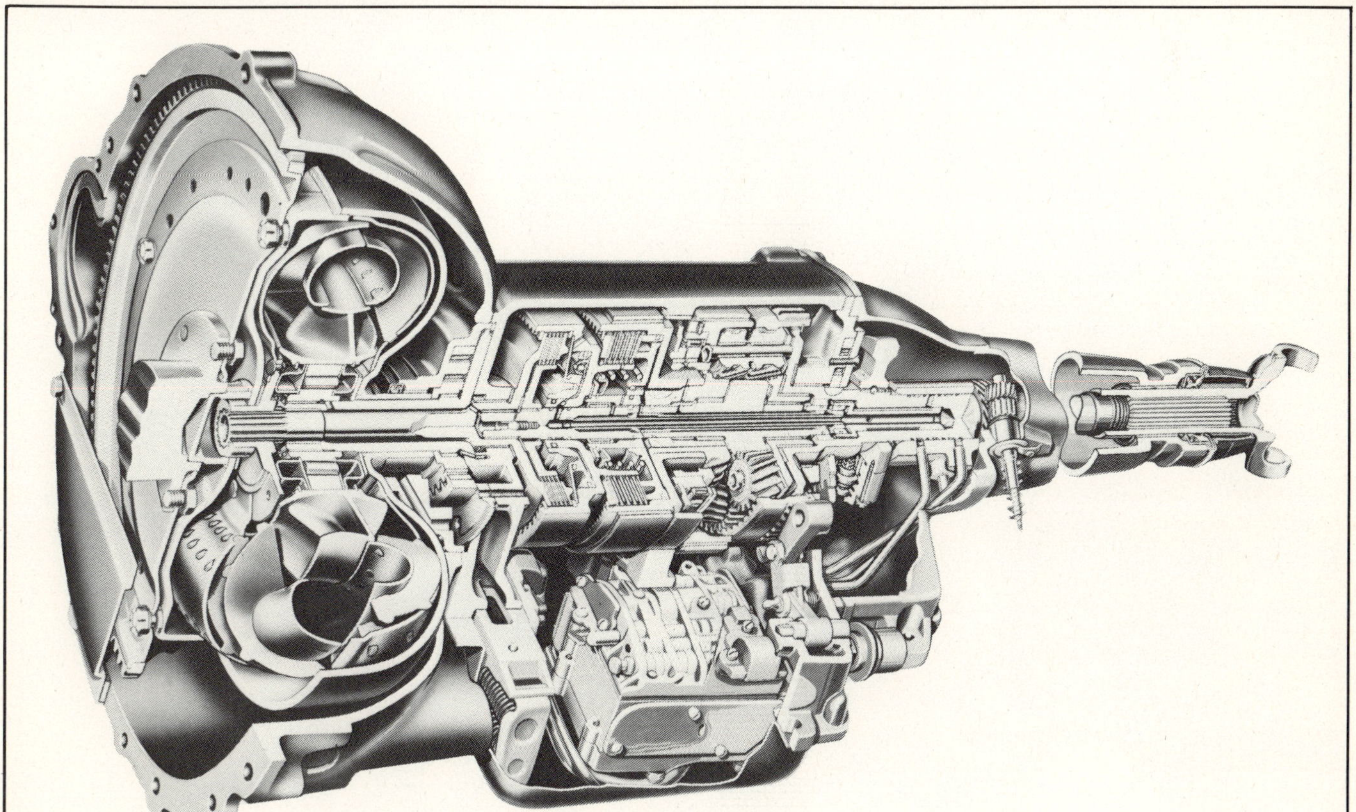

4

Ford Motor Co.

which was introduced in 1964. It is, of course, a larger version of the same to handle the higher torque engines. The gear trains are similarly designed, and so are the clutch and band combinations with the exception that the C-6 uses a low-and-reverse clutch in place of the C-4's low-and-reverse band. Valves in the hydraulic control system perform identical functions, although there are some minor differences in design, as well as some additional valves in the C-6 used to ensure shift quality.

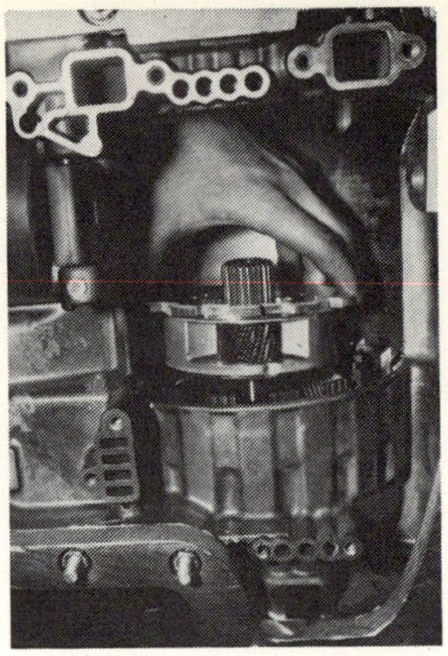

1

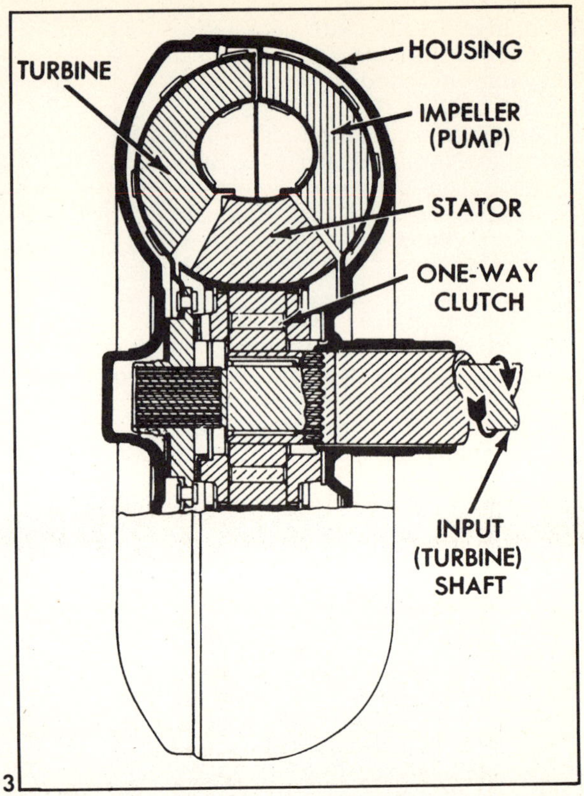

3

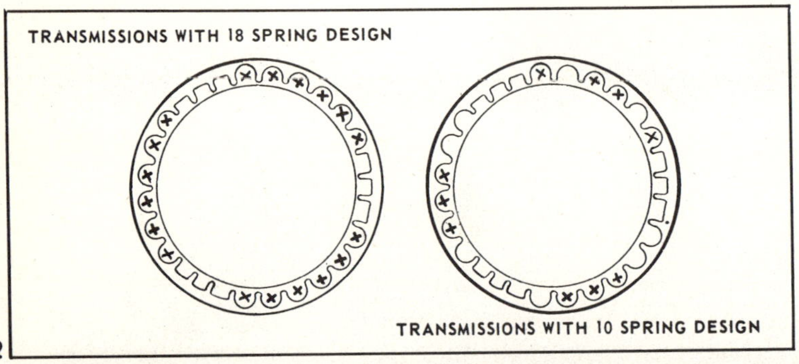

2

4

5 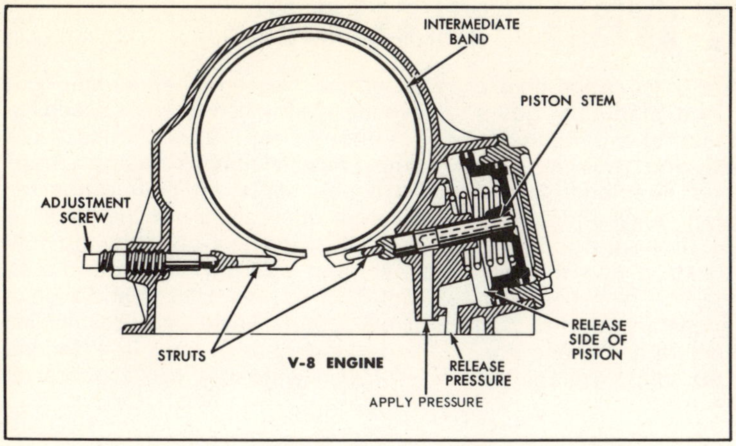

6 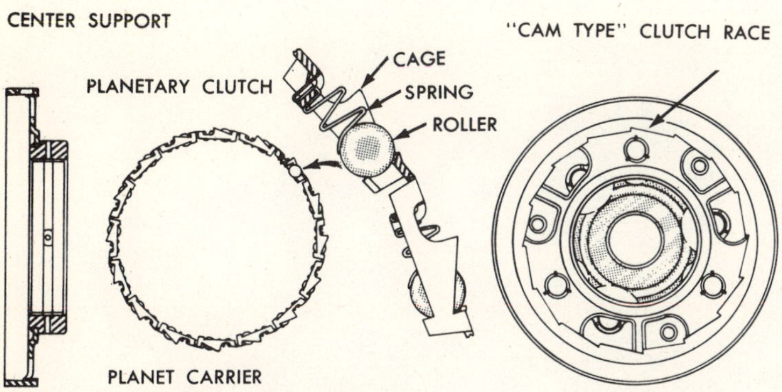

7 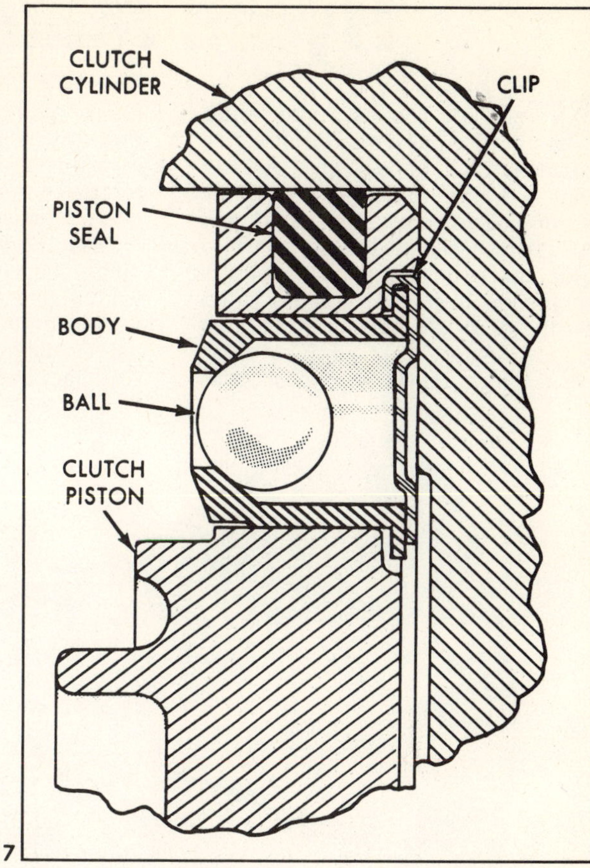

8 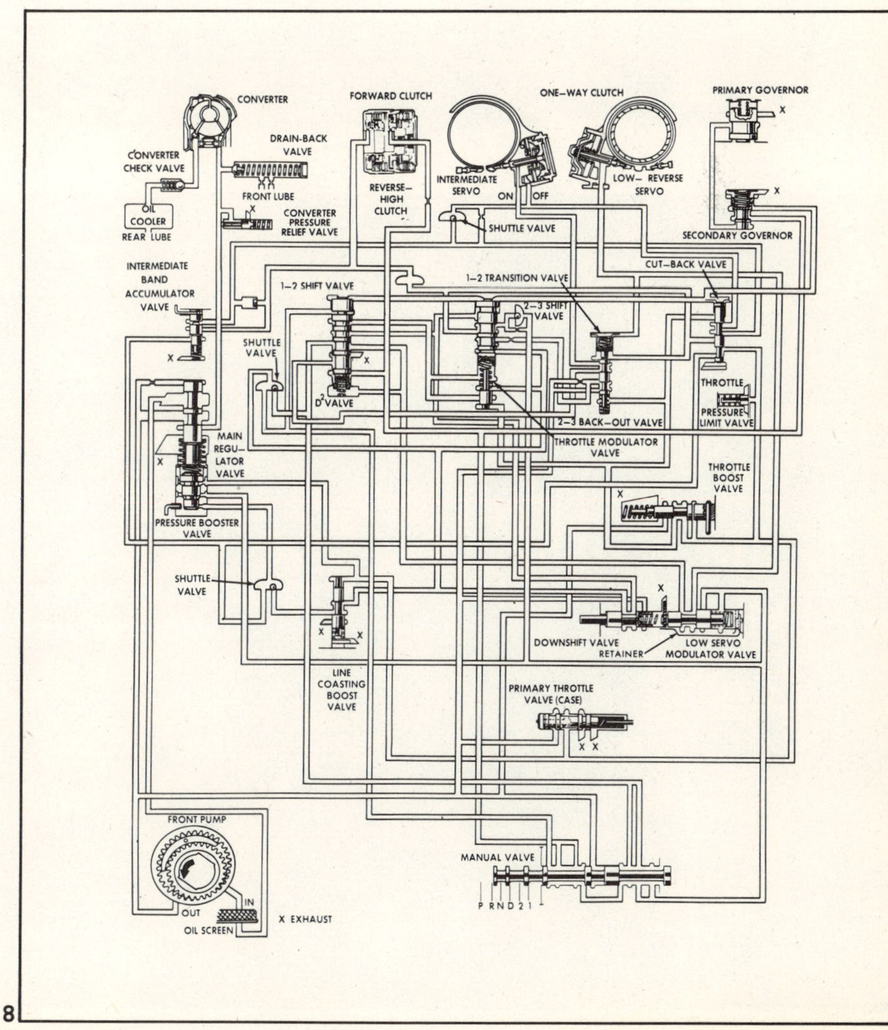

1. Ford's "big tranny" is finally feeling the magical touch of the rodders. Here the rear carrier is placed on output shaft.

2. The mechanic may discover minor differences between otherwise (and supposedly) identical transmissions. In the C-6, for example, he may find either 10 piston return springs, or 16. Whatever, here's the way they go back in with X's indicating the respective pockets.

3. The torque converter consists of: housing, pump, turbine, and stator.

4. Ford's big, 3-speed, dual-range C-6 is really upgraded version of old Cruise-O-Matic. This trans was needed (1966) when small C-4 units couldn't take the punishment dished out by high-riser 427-in. Ford engines. Special Comet and Fairlane GT models have manually selected 1st and 2nd gear ranges; another version (Merc's Police Interceptor) prevents the misengagement of reverse gear.

5. The intermediate servo and band on both the C-4 and C-6 have an adjusting screw on outside of the aluminum case.

6. Schematic drawing of the center support, planetary clutch and the planetary carrier in the FMX trans. Clutch rollers should be lubricated with petroleum jelly when transmission is being reassembled after a teardown.

7. Cross-sectional drawing of the clutch piston check valve is identical for both C-4 and C-6.

8. The hydraulic control system for the C-4 used in Pinto and Mustang II maps out to look like this. Version for larger Ford cars is slightly different.

Hydra-Matic Transmissions

Now known throughout the entire GM line as Turbo Hydra-Matics, all versions are fully automatic units consisting of a 3-element hydraulic torque converter and a compound planetary gear set. Multiple-disc clutches, a sprag unit, a roller clutch unit, and bands provide the friction elements required to obtain the desired functions of the compound planetary gear set.

The torque converter, the clutches, the sprag and the roller clutch couple the engine to the planetary gears through oil pressure, providing the forward speeds and reverse. The torque converter supplements the gears by multiplying engine torque.

When it was first introduced in 1939, the Hydra-Matic had four forward speeds. In 1961, a 3-speed version made its appearance in the Pontiac Catalina, Ventura and Grand Prix. Cadillac and the big Pontiacs retained the 4-speed for some time, but eventually switched to the 3-speed.

Although the various units have been very similar in design, there have been different gear ratios for different applications. In the Cadillac and Oldsmobile 4-speeds, the low gear ratio was 3.97:1, second was 2.55:1, third was 1.55:1, and high a direct 1.00:1. Some late Oldsmobiles had a low gear of 3.51:1, a second of 2.93:1, a third of 1.57:1 and a di-

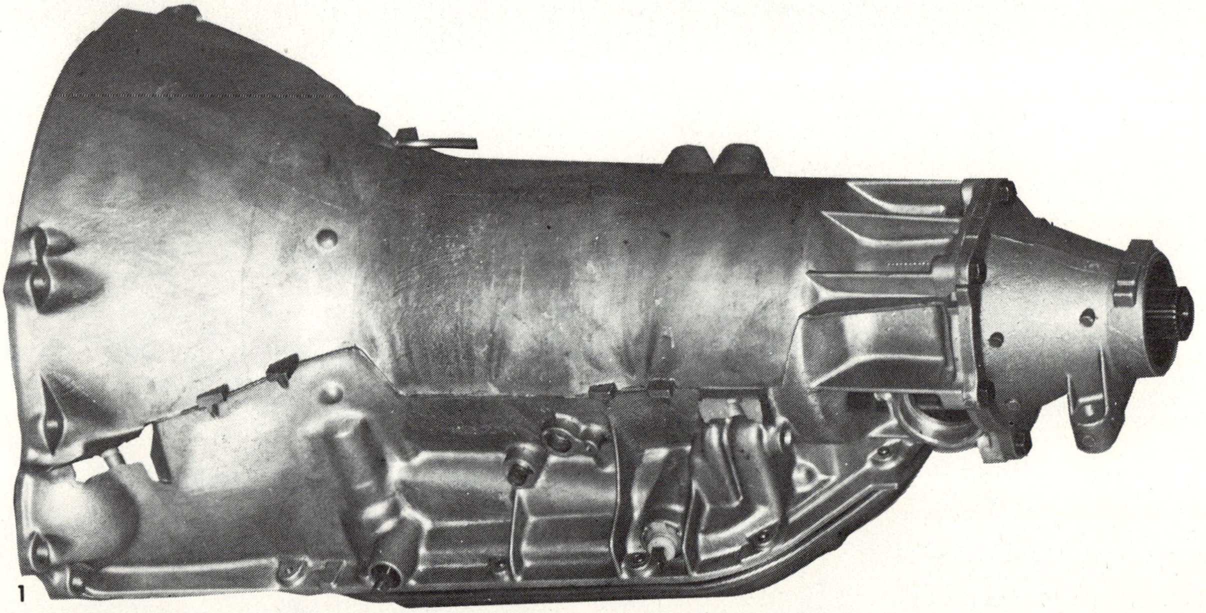

1

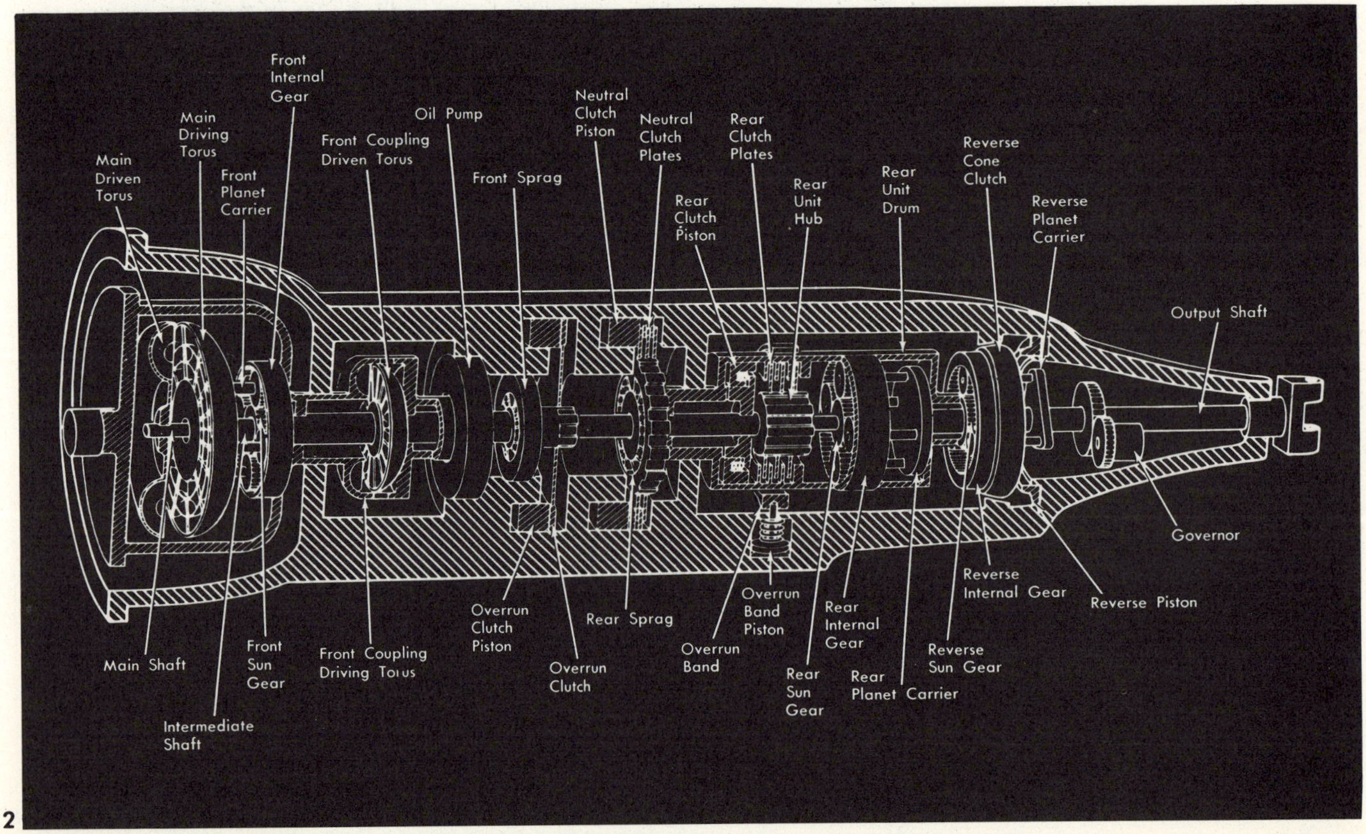

2

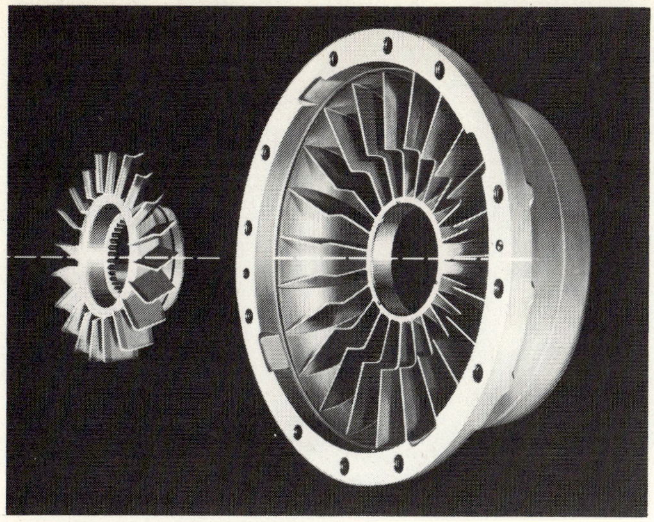

3

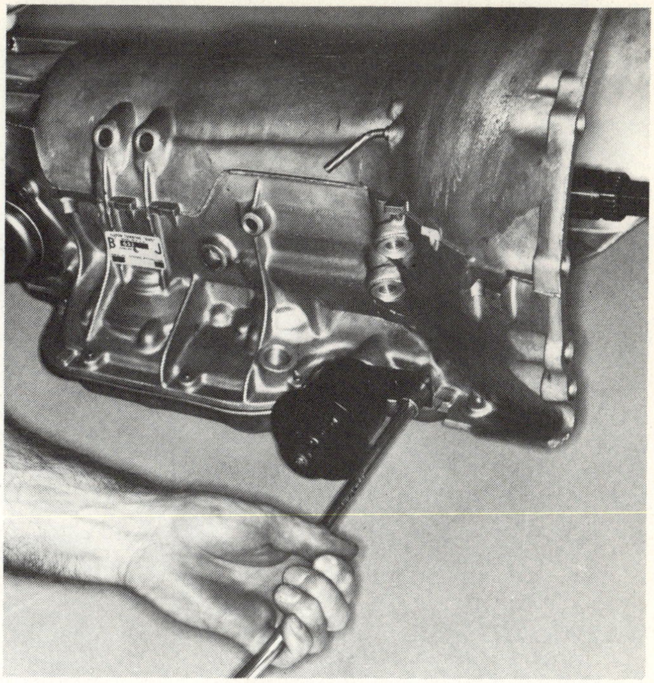

4

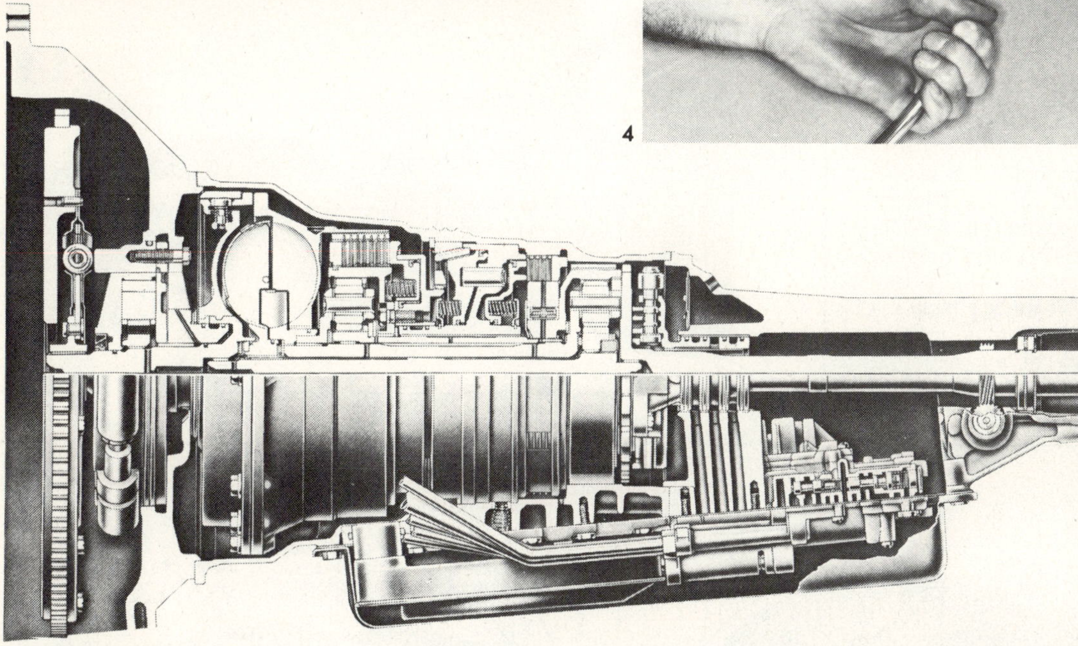

5

1. GM's Hydra-Matic transmission has undergone numerous changes since it came on the scene in '49—though GM put an automatic in the Olds 10 years earlier. This is the popular Turbo 400 model.

2. Here's a much simplified drawing of a typical Hydro showing graphic relationship of various components.

3. The driving torus used in the "roto-rooter"—at right is the stator or multiplier; it bolts to flywheel and allows coupling to multiply torque 1.20 times.

4. Vacuum modulator is on the right side of the Turbo Hydra-Matic. It connects to engine vacuum with a hose. That hose is a source of lots of unnecessary overhauls. If it's leaking or plugged, the trans may act like it's worn out, when all it needs is a new hose.

5. Pontiac's "Roto" Hydra-Matic was a unit used by the company on its Catalina and Ventura models during the mid-'60's. Trans multiplied power with three forward speeds and a semi-torque converter.

6. Disassembled Turbo-Hydro shows off its inner workings. This design is well suited for hard usage, with clutch plates instead of bands.

6

rect high. But the majority of applications used a 3.32:1 low, a 2.93:1 second, a 1.56:1 third and a direct high.

Three-speed versions of the big Hydra-Matic have either a low gear of 2.97:1, a second of 1.56:1, a high of 1.00:1, or, as in current Cadillacs and Chevrolets, a low of 2.48:1, a second of 1.48:1, a direct high and reverse of 2.08:1.

In 1969, yet another version of the 3-speed Hydra-Matic was installed in F-85 Oldsmobiles and was later made available to Vegas. Designated the Turbo Hydra-Matic 350, this transmission is similar in design to the existing Hydra-Matic 375/400, but was tailored to match the needs of smaller engines. This brought forth the Turbo Hydra-Matic 250 in 1974, which is a still more elemental refinement of the 350, using three multiple disc clutches instead of the 350's four, and one roller clutch instead of two.

The low gear ratio on these latest models is 2.52:1, second is 1.52:1, high is direct at 1.00:1, and reverse is 1.93:1. In conjunction with a low rear end ratio, the relatively high internal gear ratios are intended to provide better economy at the expense of performance.

1

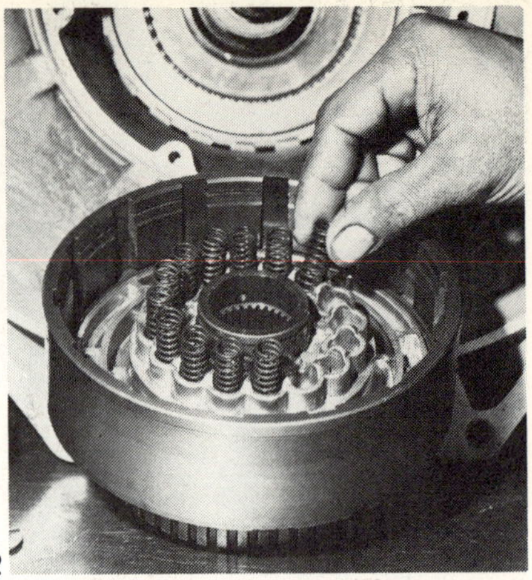

2

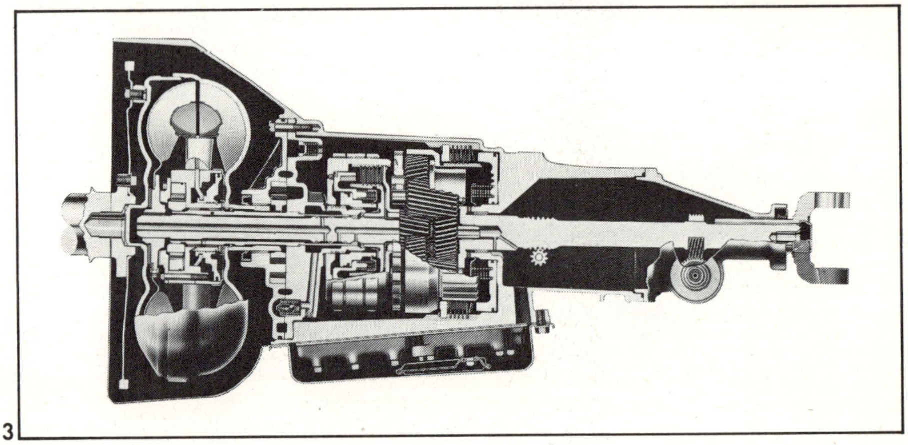

3

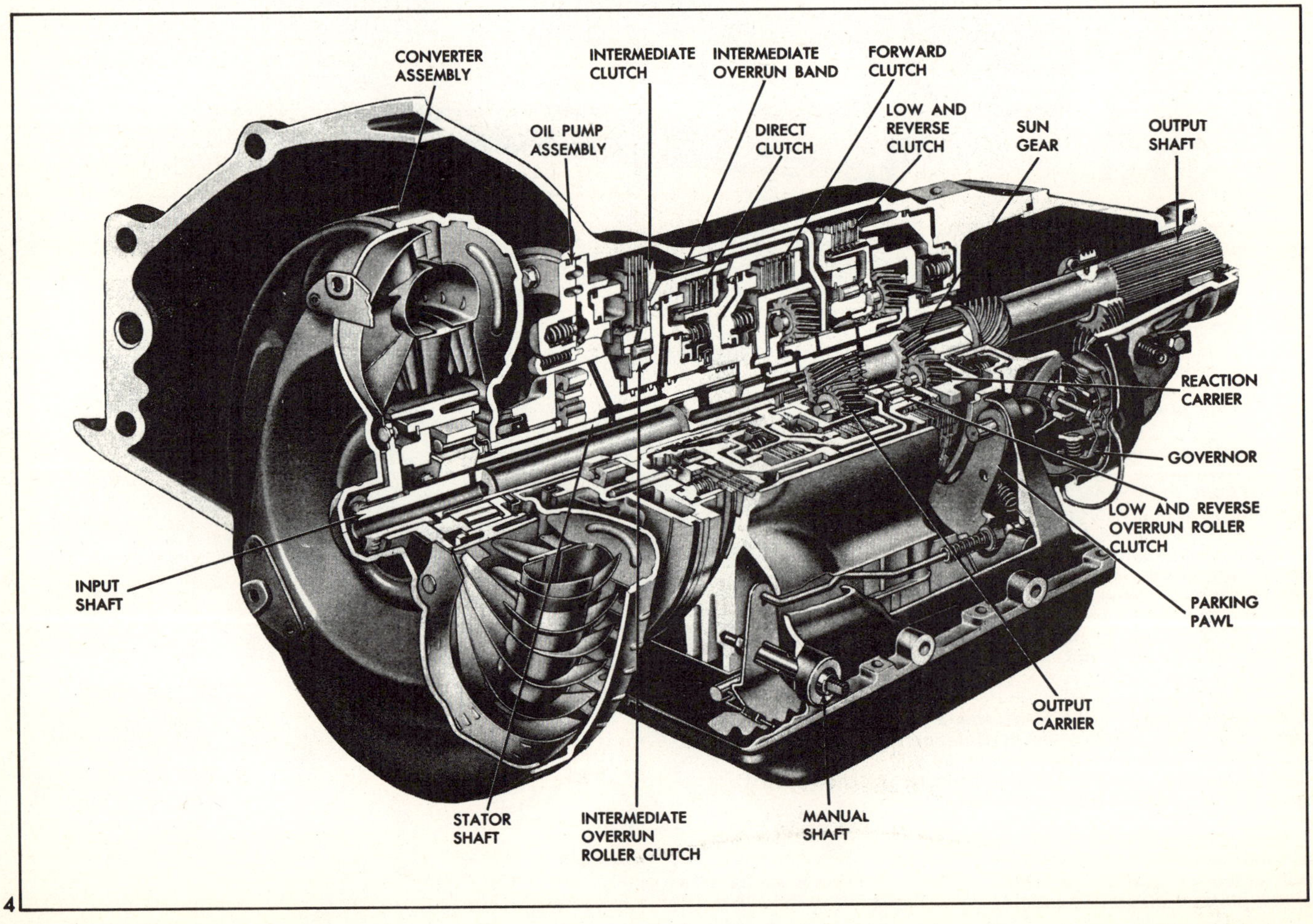

4

1. Extension housing on Turbo Hydra-Matic is made in different lengths to fit various cars. In the model shown, it is the attaching point for the rear mount, and also contains the driveshaft seal.

2. On the Turbo 400, the 16 clutch release springs sit in pockets around the clutch release piston.

3. Buick's mid-'60 Super Turbine 300 was offered as an option on many of their division's smaller models. The trans was a 2-speed, driven by a 3-element converter; its stall ratio was 2.45:1; gear ratios of 1.77:1, not much for competition.

4. Turbo Hydra-Matic 350 was designed to work in conjunction with GM's intermediate models' small-block powerplants.

5. This pump in converter was destroyed because it was not tightly bolted to back of converter. Leave mods up to expert.

6. During reassembly of a 400 guide, pins or screwdriver may be used to align the valve body so that screws can be installed.

7. A '64 Cadillac Turbo Hydra-Matic trans was used on 60 and 63 series cars and convertibles. It consisted of a 3-element converter, compound planetary gearset.

8. GM's Hydra-Matic 400 is the 350's big brother. Either as an option or standard equipment, the 400 is used on all large displacement, high-performance models.

5

6

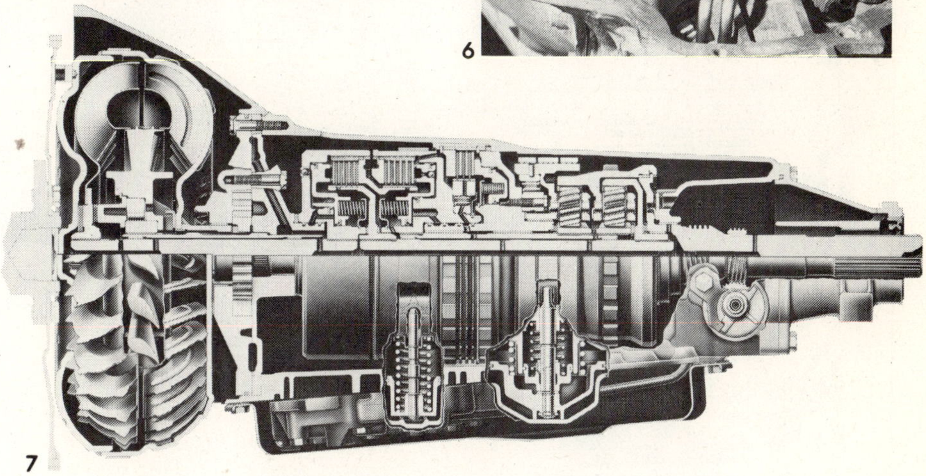

7

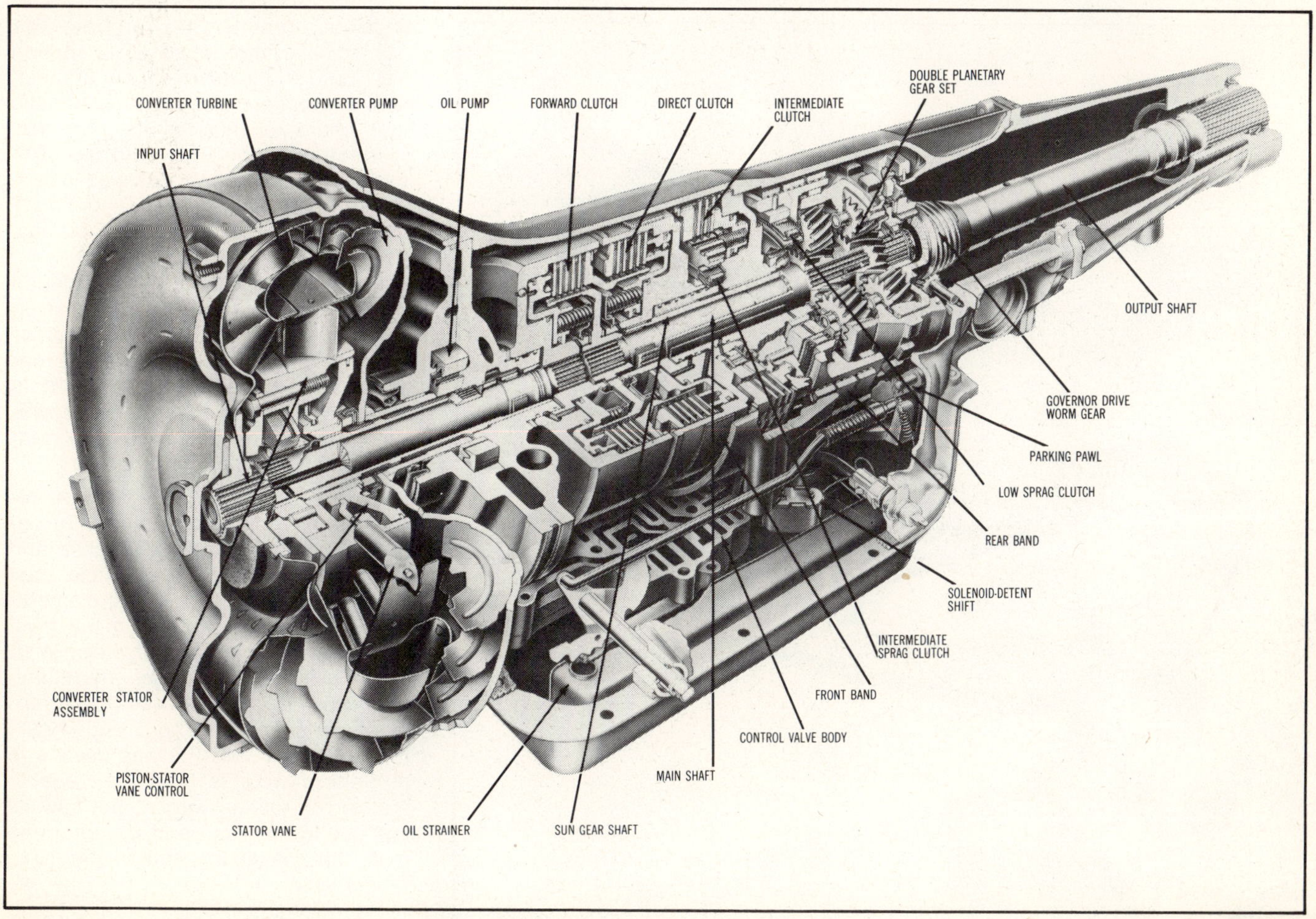

8

Powerglide Automatic Transmissions

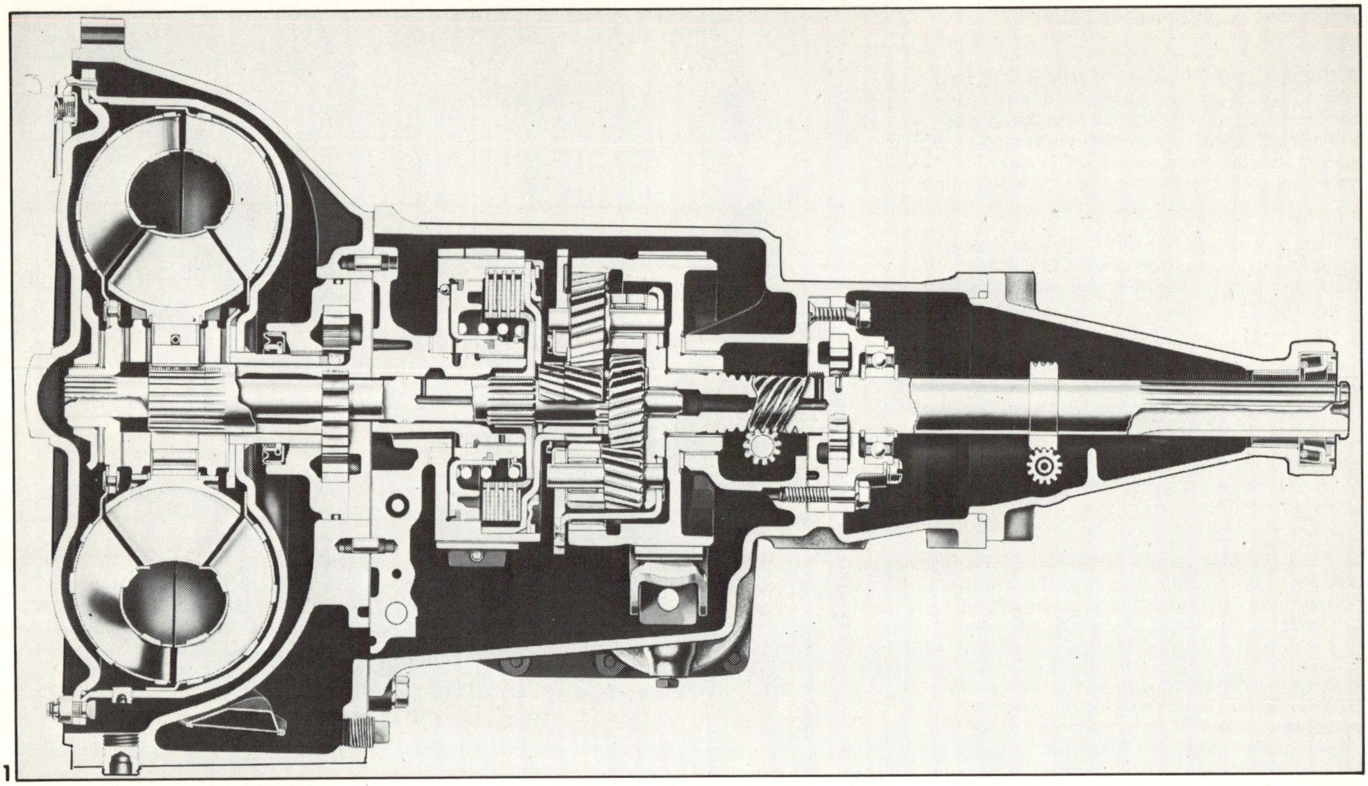

1

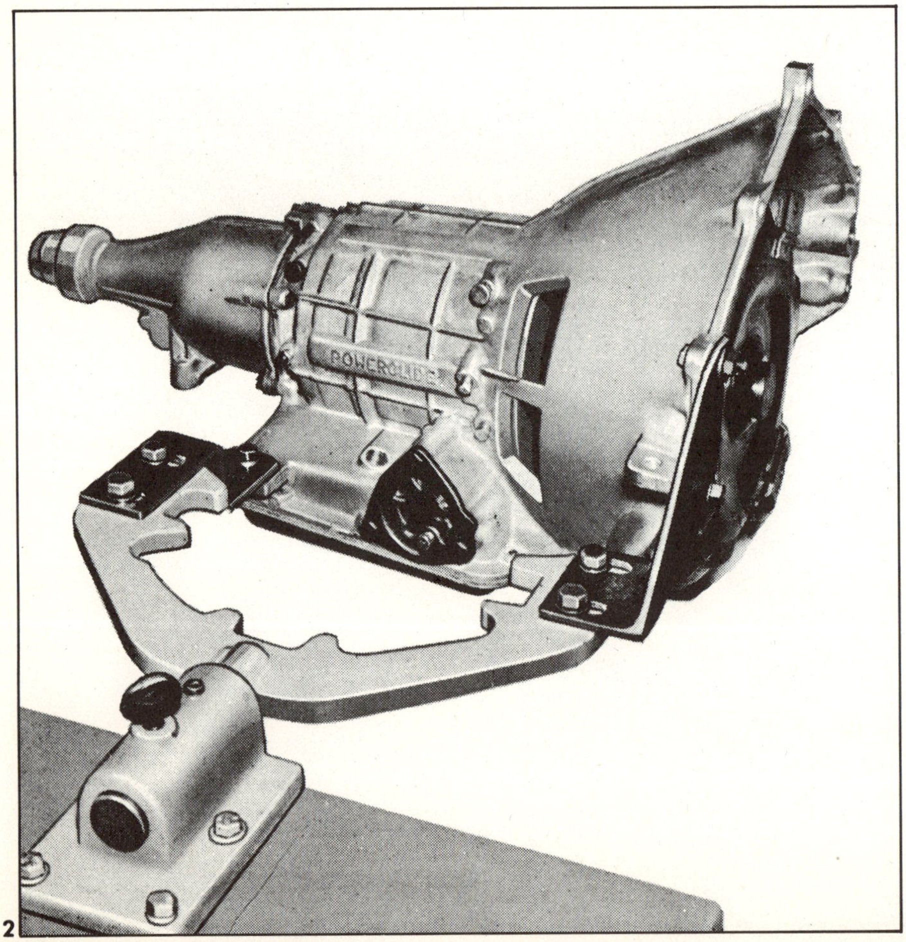

2

The aluminum Powerglide was initially introduced as an option on 1962 Chevrolets. A 2-speed with the same manual and automatic shift characteristics as its cast iron predecessor, the unit weighed 30% less at 150 lbs. Low gear was 1.82:1 for light-duty applications, and 1.76:1 when used with V-8's of 327 cu. ins. or larger.

When the manual control is placed in the drive position, the transmission automatically shifts to low gear for initial vehicle movement. As the car gains speed, an automatic shift is made to high gear. A forced downshift feature provides a passing gear by returning to the low range.

General appearance of the case resembles the long-discontinued Turboglide in many respects. The case and converter housing are a single cast aluminum unit with a large-diameter oil pump housing acting as the front bulkhead of the transmission. With minor component changes, the aluminum Powerglide was used with engines ranging from the 90-hp 4-cyl., to the 400-hp V-8. PG was used with all vehicle weights from the Vega to the largest station wagon, but saw its last use in the Vega and was dropped from the line at the end of the 1973 model year.

Intentionally designed as a basic

3

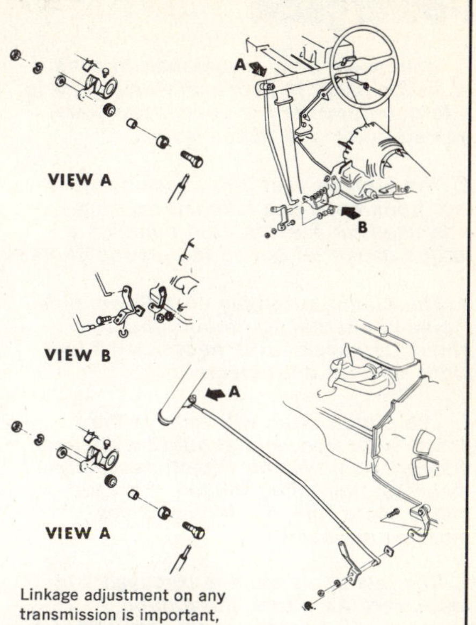

Linkage adjustment on any transmission is important, and the Powerglide is no exception. Several points of adjustment are provided on all models equipped with this automatic and must be kept properly adjusted if good shifting qualities are to be retained.

transmission for the entire Chevrolet line, it had a high degree of parts interchangeability. All models shared a common case as well as identical reverse clutch plates and springs, high clutch driven plates and many gaskets, seals and component parts. In addition, most remaining parts had a common usage on at least two versions of the transmission.

In 1967, the rear oil pump was eliminated and replaced by a governor support that is a casting similar to the rear pump housing. The main control valve body is bolted to the lower surface of the transmission and covered by an oil pan in a manner similiar to the Turboglide and the Corvair Powerglide.

1. Inside the not-too complex Powerglide is Chevy's own automatic which was offered, as an option, first in 1962. It underwent various design changes and refinements until—alas—it enjoyed its last days, in the Vega, at the end of the '73 model year.

2. When Chevy shifted to the alloy-cased Powerglide they sort of combined most features of the defunct Turboglide, cast iron Powerglide and Corvair Powerglide. Aluminum saved 70 lbs., aided cooling.

3. It's easy enough to pull the converter housing out of the way, if the transmission is out of the car. Just stand the trans on its tail, and lift.

4. One feature of the Powerglide that repairmen appreciate is that aluminum models are some of the lightest units to take out or install, and the internals are simple enough to make repairing them an easy task.

4

Powerglide

1. Main valve body is attached to back of bellhousing. In essence, Powerglide is a torque converted, 2-speed, planetary-geared automatic transmission.

2. The 3-piece cast iron case—bellhousing, trans case and tailshaft housing—was used for 8 years. Unit contains a torque converter bolted to 2-speed gearset.

3. The clutch assembly on a Powerglide includes alternating friction and steel plates; replacement is necessary if any signs of wear are detected.

4. The clutch drum will wear at the points indicated, and should be replaced if wear is significant. Clutch assembly includes alternating friction and steel plates. As a rule, the friction discs must be replaced.

5. The low apply band is removed and inspected. As a rule, the friction material will be worn excessively on older transmissions, so replacement is almost mandatory.

6. The low band servo spring may be broken; ring seal should be replaced.

1

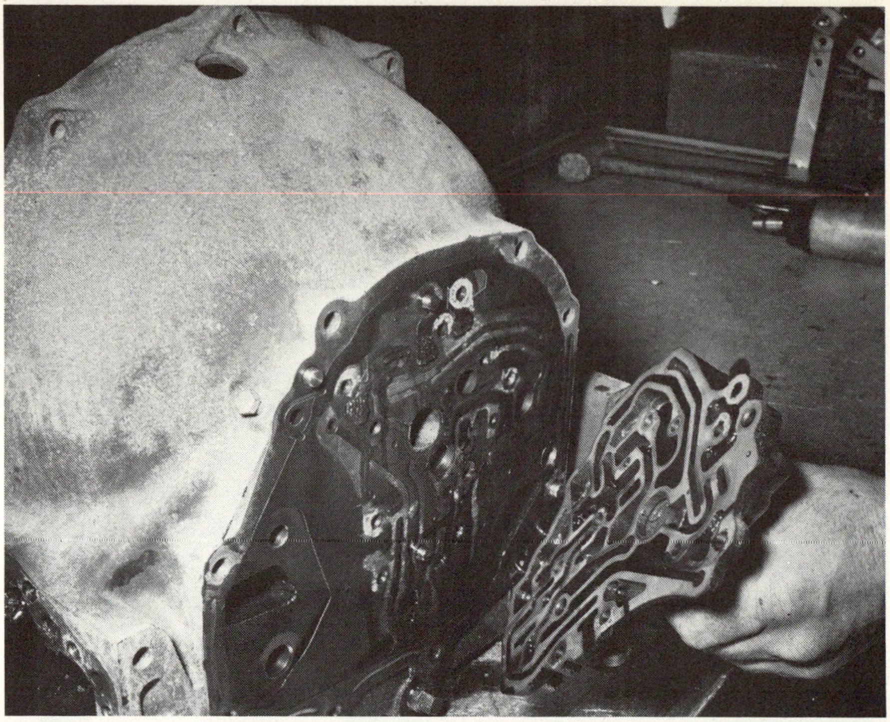

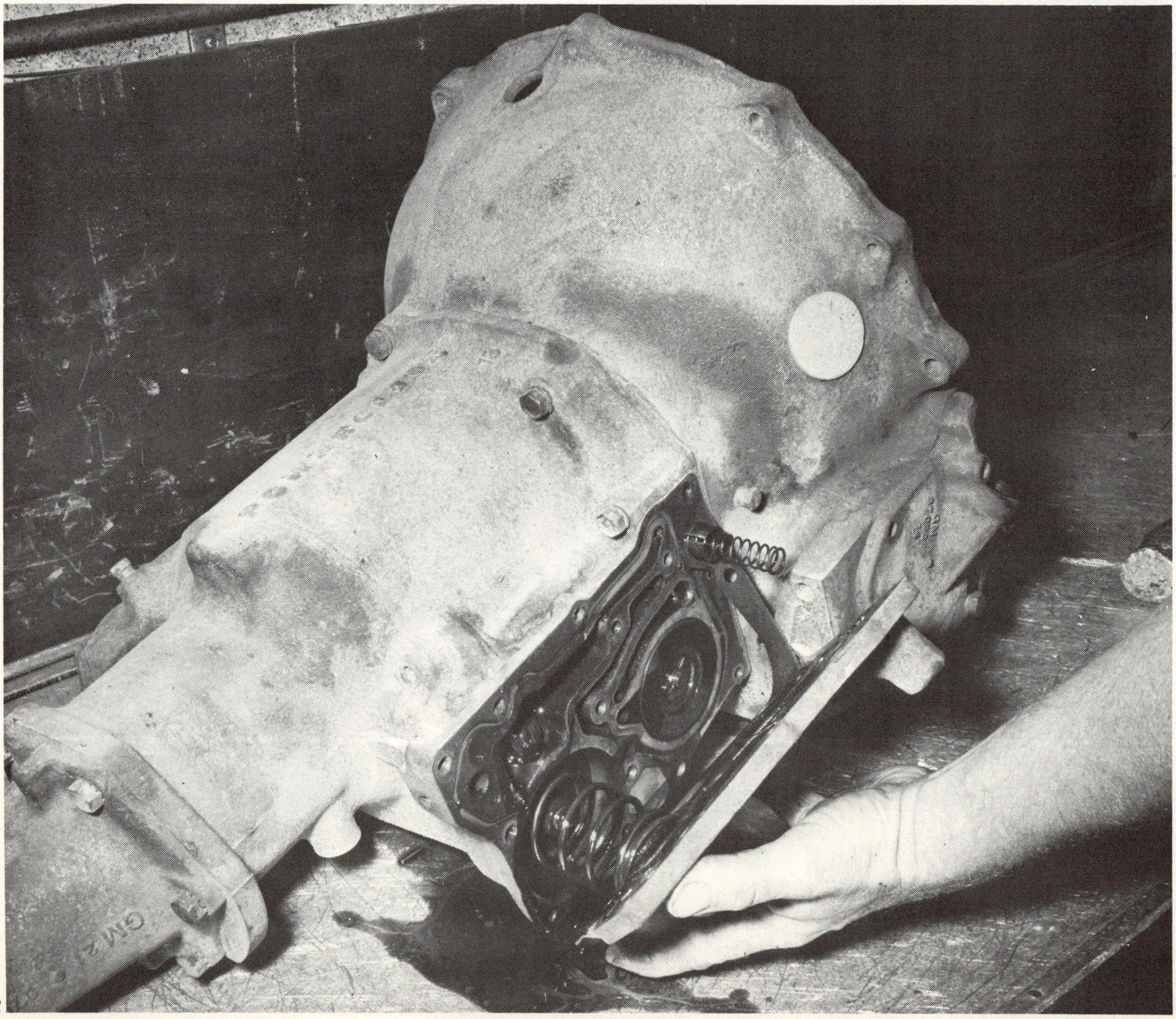

2

3

4

5

6

Warner Automatic Transmissions

The Warner Gear Division of Borg-Warner Corp. supplied the automatic transmissions used in American Motors products for a number of years until AMC changed over to use of the Chrysler TorqueFlite in 1972 cars. The last version was designated as "Shift Command," but was preceded by a unit known as the American Motors "Flash-A-Matic."

The "Flash-A-Matic" was a 3-speed housed in a cast iron case, and was used in the 10- and 80-series vehicles. The low gear ratio was 2.40:1, second gear was 1.46:1, high gear was a direct 1.00:1, and reverse was 2.00:1.

The "Shift Command" versions used two types of units; the model 36, 37 and 40 transmissions had the 3-speeds encased in an aluminum

1

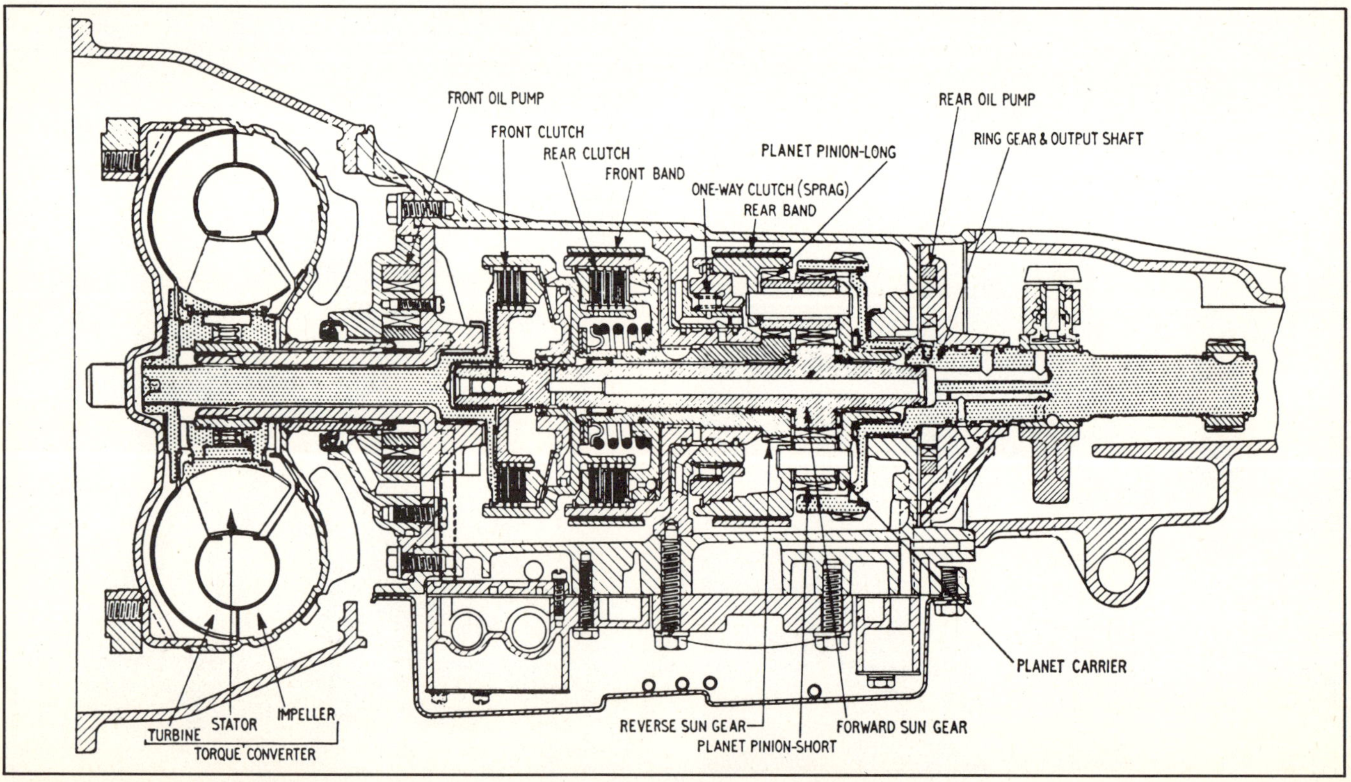

2

3

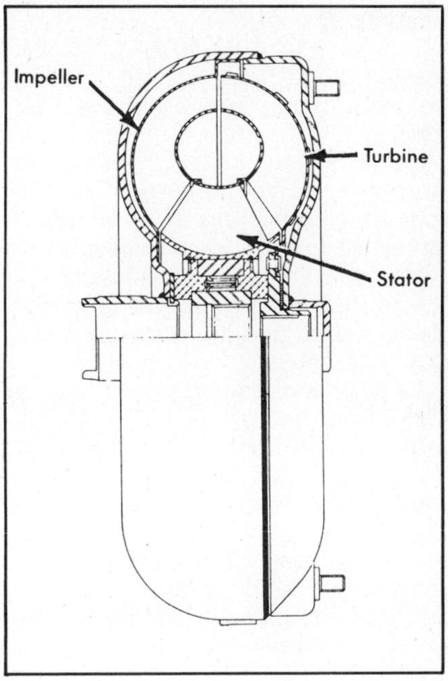

4

5

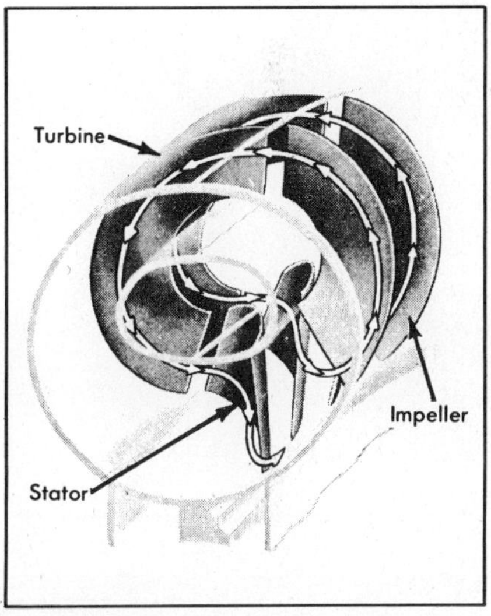

6

1. No tricks here, the B-W automatic is pretty much straightforward; standard torque converter design, 3-speed forward, reverse planetary gearset.

2. The cross-sectional view of American Motors Borg-Warner auto reveals the interaction of the various clutches and bands required to transmit engine power to forward reverse movement. Shown is the Borg-Warner #35.

3. Removing an M-40 with a garage jack. The later M-40's, or "Shift Command" have an aluminum housing. Trans ID tag is mounted on left-hand side of the case will show model, serial, part number.

4. Borg-Warner's torque converter is just like all others. In its doughnut shape lies a turbine, impeller and fixed stator.

5. A special puller is needed to remove oil pump seal; should be attempted only by factory-trained personnel.

6. Oil flow in torque converter is illustrated here as it flows from impeller to turbine then back to impeller through the stator. Power flow is transmitted through two sets of planetary gears, two clutches and two bands.

Warner

housing, while the model 11 retained the heavier, but less expensive cast iron unit.

Engine power is transmitted through a conventional-type torque converter consisting of an impeller connected to the engine crankshaft, a turbine splined to the transmission input shaft, and a fixed blade stator connected to the transmission case through a one-way free wheel unit.

A compound planetary gear train is used to transmit power to the wheels and is controlled by a series of clutches, bands and servos. The transfer of power through the transmission for the various ratios is dependent on the combination of the gears being held by these multiple disc clutches and the bonded linings on the servo-operated bands.

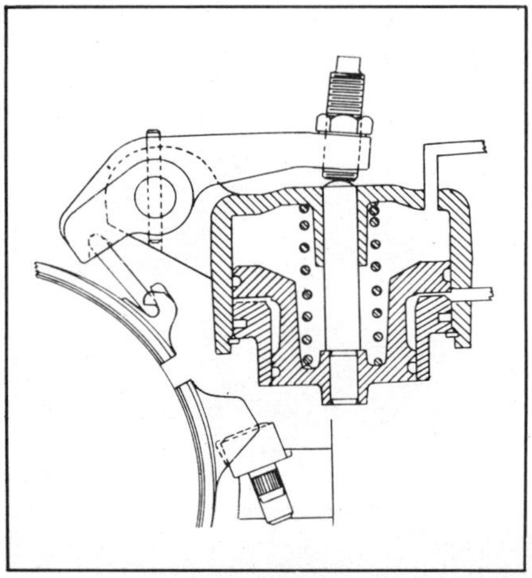

1

2

3

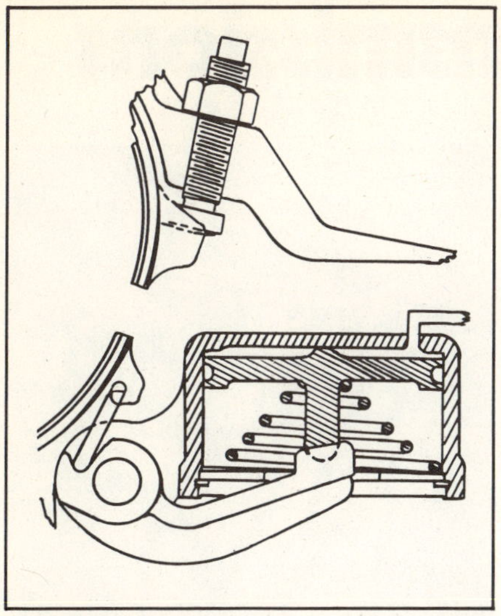

4

5

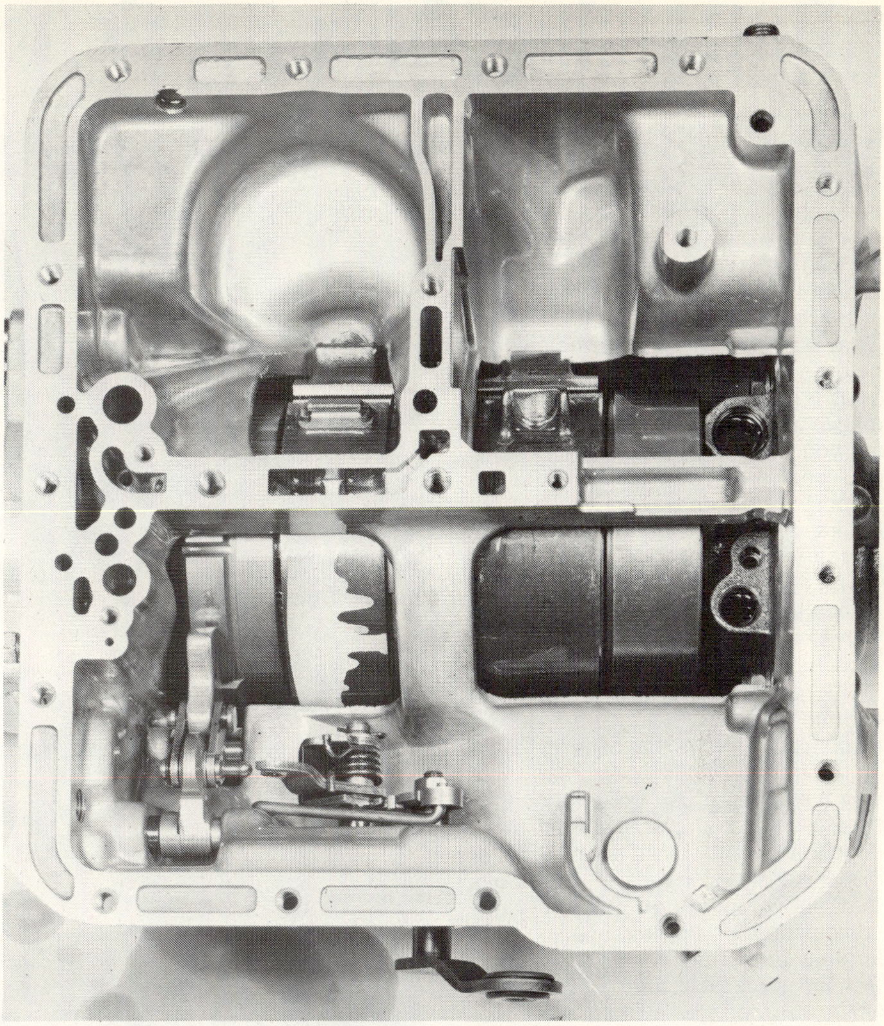

6

7

8

1. This servo is released when hydraulic pressure is applied to the reverse side of the piston. Spring augments return.

2. On B-W's "Shift Command" trans, oil tubes are placed here, underneath unit. From left, tubes are rear servo apply tube, rear clutch apply tube, front servo apply tube, front servo release tube, and converter out tube.

3. Pump is removed through case's front.

4. Cross section illustrates workings of rear servo on a Warner transmission. Unit applies band force when hydraulic pressure is forced against piston.

5. AMC's trans pressure can be adjusted with screw inside modulator. One turn will vary pressure approximately 18 psi on M-36, 37, 40 models (used on small inlines and V-8's); 10 psi on M-11 and M-12's (used on big V-8 models).

6. Oil passages in trans case must be checked for cracks, stripped threads, nicks on mating surfaces, scoring on bushings, elongation of holes during overhaul.

7. To remove center support, sprag clutch, and planet carrier as an assembly, first take off the pair of cap screws located from the sides of the case. These hold and secure the center support.

8. Transmission components are "R-'n R'ed" through front of housing. Planetary gears, rear clutch are being installed.

Repairing Automatic Transmissions

When looking at an exploded view of an automatic transmission, probably the first thing that you might worry about is how you would get all those pieces back in the right place. That actually should be the least of your worries. A much bigger problem is finding out what's wrong with the transmission before you take it apart.

Of course nobody can wave a magic wand over a transmission and tell you exactly what is wrong with it and what is needed to fix it. But it's surprising how much you can find out before you even lay a wrench on any part of the car.

The first thing to consider is whether there is actually anything wrong with the transmission. Hardly a man alive has not heard a mechanic say, "They all do that." It sounds like an excuse for not being able to fix something. But in the case of automatic transmissions, particularly the way they shift, it can be unfortunately true. Some of the smoothest, most dependable and most efficient automatic transmissions have little faults that show up under just the right combination of throttle opening, car loading and road conditions. Some transmissions will shift up or down when you don't want them to. Other transmissions won't shift when you wish they would. And still others have annoying hesitations that make you think the engine just went to sleep for a second or two.

The only way you can determine if a transmission fault is part of the design or an actual malfunction is by talking with mechanics, other car enthusiasts or someone who really knows your make of car and model of transmission. It doesn't do any good to talk to a TorqueFlite man about a Turbo Hydra-Matic, and a Powerglide expert wouldn't necessarily know the finer points of a Cruise-O-Matic.

Once you have determined that the problem is actually a malfunction, the next thing to find out is whether it is caused by the transmission or by the engine and shows up in the transmission. An engine that is out of tune or that needs a valve and ring job is going to be a poor performer not only because the engine does not put out the power that it should but because the transmission can become confused and not know when to shift. You will have problems that might seem to be transmission troubles but in reality are caused by an engine that is not in good shape.

1

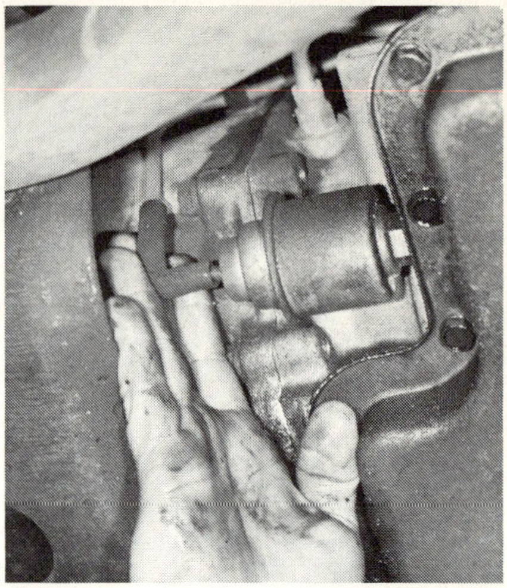
2

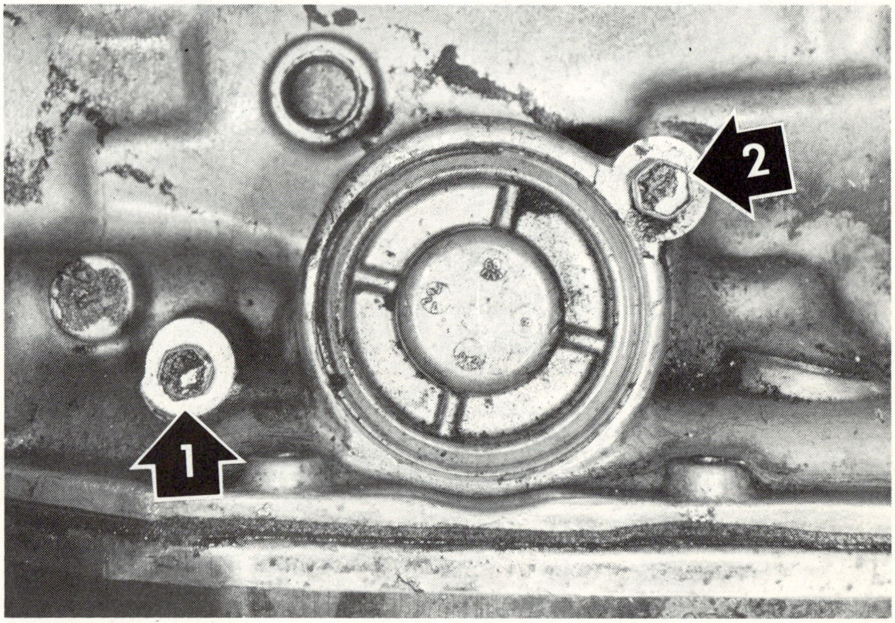

3

If the transmission has any vacuum controls, the vacuum lines to the engine must be checked for leakage. If everything looks good as far as engine operation is concerned, the first thing any automatic transmission man will do is pull out the transmission dip stick. When he finds discolored fluid with a definite burned smell, he doesn't have to go any further in diagnosing the trouble. The problem is slippage in the clutches or bands which have gotten hot enough to start burning. The only solution is to remove the transmission and replace the worn out parts.

LINKAGE PROBLEMS

If the manual linkage is not adjusted so that it falls into the detent position easily in every gear range, the lever may be between gear positions which would cause a restriction in the amount of oil pressure sent to the clutches or bands. That can ruin a transmission very quickly, so the manual linkage should be checked to make sure not only that the quadrant on the shift stick points to the gear the transmission is in, but that the transmission is actually in that gear. Checking for proper detent position is done easily by disconnecting the lever at the transmission. If the shift stick is holding the lever between two gears, the pressure on it will be felt when it is disconnected.

Oil leaks in an automatic are probably one of the most common reasons for either teardown or removal. If you can't keep oil in it, you can't go anywhere—and it gets expensive because automatic transmission fluid is not cheap. Some oil leaks are easy to diagnose, but others are not so obvious. What it amounts to is run-

4

5

6

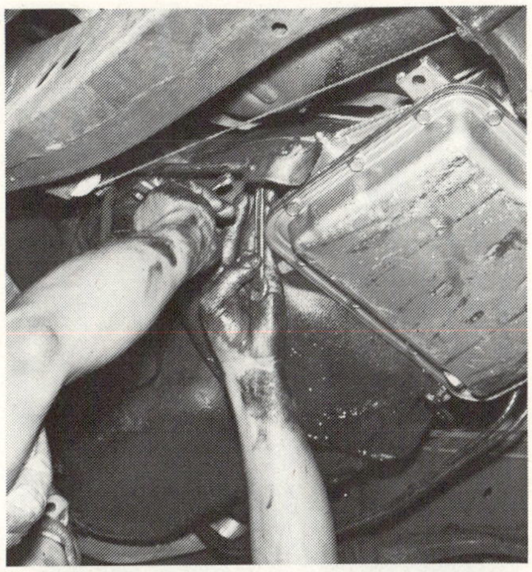

7

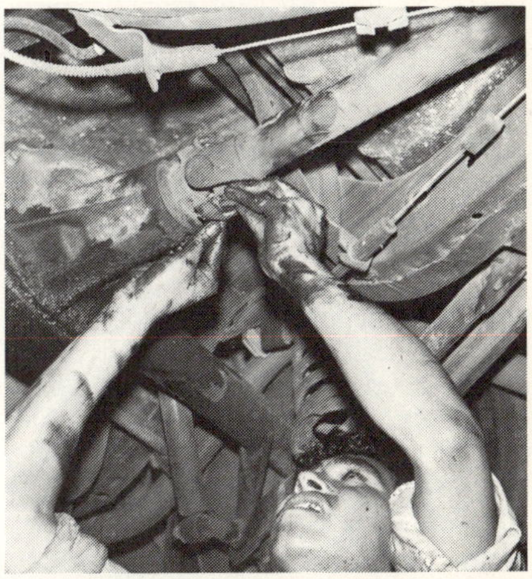

8

1. Some auto transmission problems can be solved without removal from the car. If you suspect a trans leak, cleaning the converter housing, pan, etc., is the first step in locating it.

2. Leaks can develop in vacuum lines. Your problem may be as simple as a tubing replacement.

3. If you know where to look, you'll find plugs in the case for checking pressure in the drive ranges. On the Turbo Hydra-Matic 350, you can connect a pressure gauge to (1) to check line pressure or to (2) to check the 1-2 clutch pressure.

4,5. Band adjustment can be done without pulling the trans. You'll usually find a front adjuster (4) and a rear adjuster (5). How easily you can get to them depends upon the particular car and trans installation.

6. When you have good equipment, it makes the job a lot easier. Instead of draining this Powerglide in the car, specialist waited till he had it out, then just tipped it up.

7. Disconnecting the linkage. Removing the transmission is a greasy job. If you work on your own, a good idea before starting would be to drive the car to a 25¢ self-service car wash. If you can get up in there with the wand, hot soapy solution will take that grease right off.

8. Disconnecting the driveshaft at the rear U-joint. Bearing cups will fall out if you're not very careful.

ning the engine while the car is up in the air and checking underneath the transmission to see if you can tell where the oil is coming from. In most cases the transmission will have to be wiped completely clean, and then you will have to wait underneath the car while the engine is run long enough for the leak to appear again. That way you are sure that you know where the oil is coming out of the transmission. When the transmission is removed and disassembled, you know that you have to look in the area of the leak to find the cause.

LEAKAGE CAUSES

Most leaks are caused by worn-out seals or by nicked or gouged surfaces. If a seal wore out from old age, then all that is necessary is to replace it with a new seal and your transmission is back in business again. But if the seal wore out because the shaft it seats on is wobbling—caused by a worn-out bearing on the shaft—then putting in a new seal will not do any good at all. In that case you have to replace the bearing so that the shaft will run true. Otherwise it will knock out the new seal very quickly.

One type of leak that is rare but can really cause trouble when it happens to you is the one that comes from a porous casting. It most often happens to transmissions with aluminum cases. But it can also occur in cast iron, either in the transmission case or the pump casting.

If you ever find water in the oil, it's pretty obvious that you have a leak in the oil cooler which is mounted within the radiator. Some shops use black light for checking to tell whether a leak coming out of the front of the transmission is engine oil or transmission fluid. Oil and transmission fluid look different under black light, and it's easy to make a comparison check to tell what is doing the leaking.

In most transmission overhaul manuals, you will find reference to an oil pressure check. Somewhere on the outside of the transmission is a pipe plug that can be removed while the engine is off. By connecting a pressure gauge to the opening, the transmission can be checked, with the engine running, for proper pressure in the various drive ranges. If the pressure is not correct, then a troubleshooting chart is used to tell what would cause the low pressure.

Troubleshooting is probably the most difficult part of working on an automatic transmission. You should always try to fix it without pulling it out. But if work is required on the innards then there is no way around it. The transmission must come out.

REMOVING AN AUTOMATIC

Almost anybody with a good pair of arms and a strong back can take out a manual transmission and lower it to the floor without any assistance. If you're in really good shape, you might even be able to do it with an overdrive which makes the manual transmission considerably heavier. But, where the automatic is concerned, there isn't any way in the world you are going to get it out of the car without help.

Help may come in the form of a couple of husky friends who can muscle it out for you, or an automatic transmission jack that bolts to the bottom of the transmission and lowers it safely to the ground. You might even be able to make a cradle that would fit on top of an ordinary floor jack so that the transmission could be lowered without losing it. Don't ever attempt to move an automatic transmission without some kind of a cradle to hold it firmly in position. If you use an ordinary floor jack and try to balance the transmission on it, you will certainly lose it and it will either land on top of you or break something personal. Or that aluminum case will go right down onto the concrete and end up with a big piece missing out of it.

Automatic transmission jacks are very expensive and they aren't used by the individual often enough to make them a good buy. If you are a good carpenter or welder, you can make a strong cradle out of angle iron or wood, but that's a lot of work and most of us don't like to take the time to do it. The best solution is to seek several friends who don't mind getting their hands dirty.

Getting the transmission ready for removal is not too difficult; it's just a case of disconnecting everything that attaches to the transmission so that it will come out without trying to drag the engine or part of the car with it. The first thing to do is to get the car up in the air so you have lots of working clearance under it. Commercial axle stands work fine on pavement but they can be very dangerous when used in the dirt. Concrete blocks, if placed the wrong way, can shatter from the weight. If you don't have axle stands, the safest method is probably to use heavy timbers about the size of railroad ties. The car should be elevated at least two feet. Three feet is better, if you can

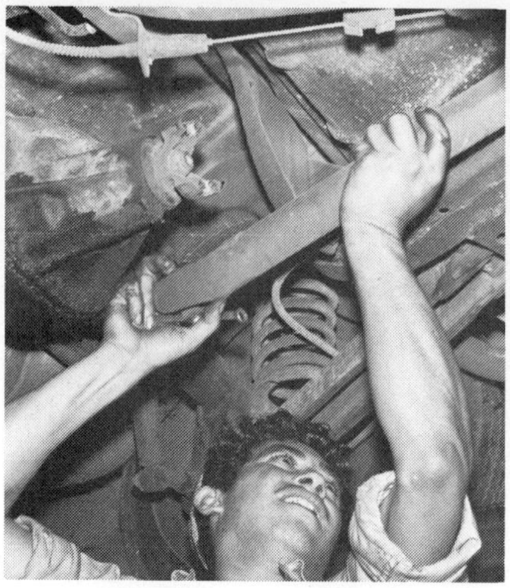
1

2

3

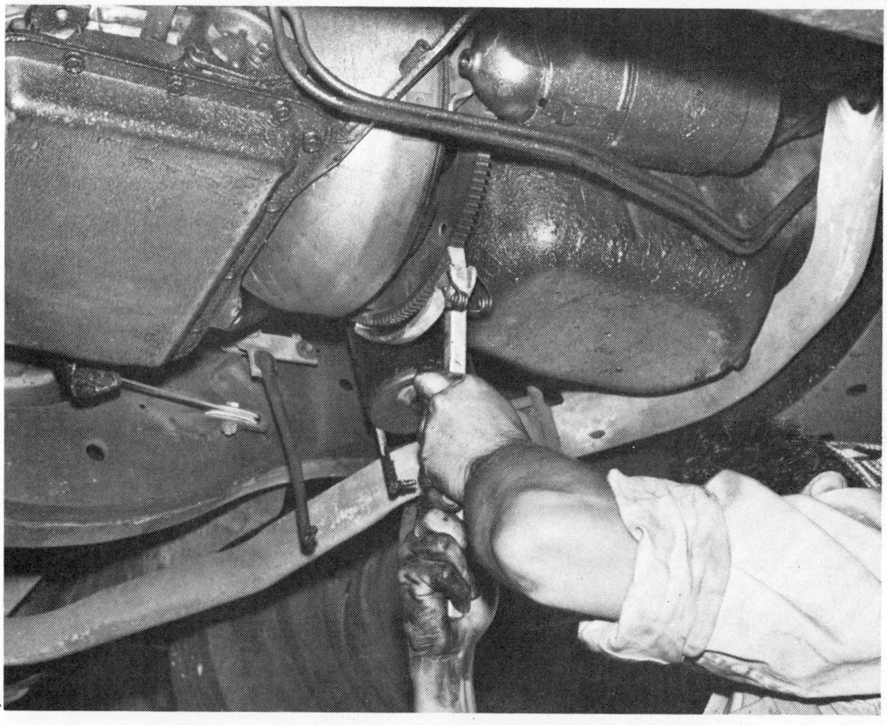
4

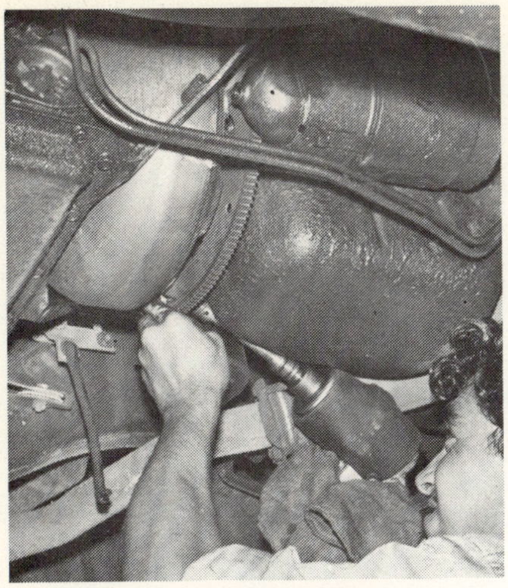

5

6

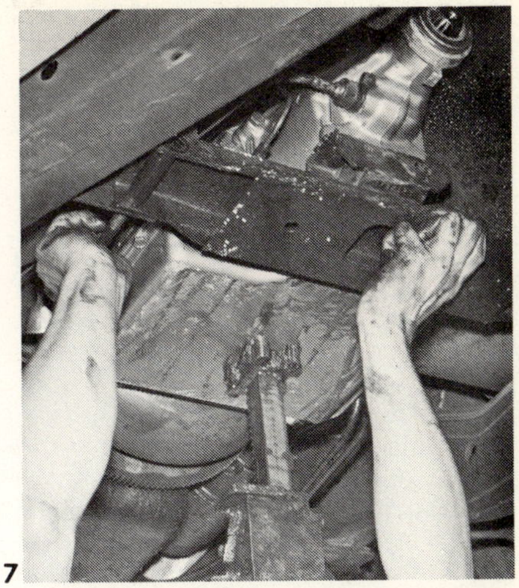

7

8

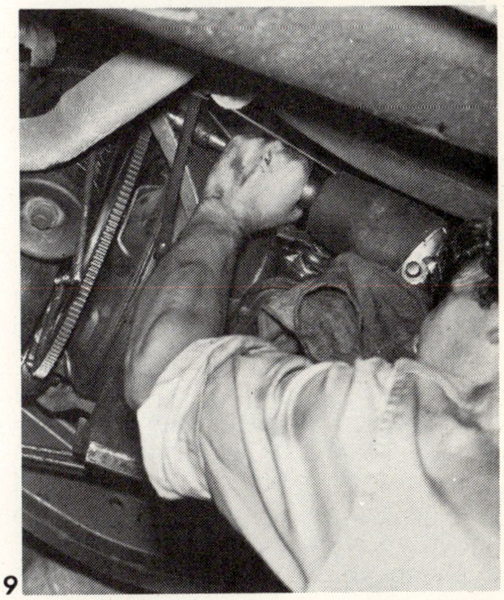

9

1. The bolts at the X-member are removed so the whole driveshaft assembly can be slipped to the rear and out of the way of the transmission.

2. A slip joint is just that. Once the driveshaft is disconnected at the rear, all you have to do is pull on the slip joint and it will come right out of the back of the transmission.

3. Holding the bellhousing cover in position while unscrewing the bolts.

4. You can turn the flywheel over with a screwdriver or wrap a rag around it and use your hands, but this special tool makes the job a lot easier. It works just like a wrench.

5. The flywheel is turned until one of the bolts is in a position where it can be reached with an air wrench. These are close quarters, but you will find an ordinary socket wrench will work just as well as an air wrench, but not as quickly.

6. Removing the rear transmission mount bolts at the crossmember.

7. The support under the transmission is to take the weight off the crossmember. After the support is in place and the transmission is raised slightly, the crossmember can be unbolted and dropped out of the way.

8. Only two bolts hold the crossmember in position. Notice the jack under the back of the transmission. On this particular engine it is only for the purpose of removing weight from the crossmember. The engine has motor mounts at its rear which will support the weight of both the engine and transmission when crossmember is removed.

9. With the crossmember out of the way the special jack under the transmission is removed and then the cradle to hold the transmission is put into place, while the bellhousing is unbolted.

do it safely, because it gives you more room to work underneath.

Every car is different and you should have either the factory shop manual or one of the independent manuals for your make and year of car. Those manuals are needed in addition to the transmission manual put out by the maker of the transmission. Ordinarily the maker of the transmission does not deal with removal or replacement of the unit, only repairs to the transmission itself.

On some cars the exhaust pipes may have to be loosened or removed so that the transmission and engine can be lowered slightly in order to reach the bellhousing bolts up near the floorboards. You may have to remove the starter motor, too. It's a good idea to remove it on any car so that the drive will not hang up on the ring gear when the transmission is moved back.

DRAINING THE FLUID

Draining the transmission is best done while it is still in the car. And, you can drain the converter there also if it has drain plugs. Remove the lower cover on the bellhousing and you can turn the converter by hand and look up to see if there is any drain plug on it. Many cars do not have any drain plug on the transmission pan itself, and the only way you can drain the fluid is to remove the pan and let it run out on all four sides at once and make a big mess.

The idea behind not building a drain plug in the pan is very good. If the fluid is bad enough that it should be changed, then the pan should also be dropped and cleaned out. Eliminating the drain plug forces the mechanic to remove the pan, and it is hoped that he will clean it once he has it off.

Everything that attaches to the transmission should be removed or disconnected. Some of the things that may have to be disconnected on your car are cooler pipes, the filler pipe, vacuum lines, shift linkage, shift cables, crankcase ventilator tube, TV rod, electric wires, and, on Chrysler Corp. cars, the driveshaft hand brake.

On cars with a one-piece driveshaft, the shaft is usually disconnected at the rear joint. Then the shaft is lowered and slipped out of the back of the transmission. If the rear U-joint is of a type that can go back together several different ways, it's a good idea to mark the parts so you can get them back the same way they were. On many driveshafts, it won't make any difference; but on the one you or I work on, the shaft will probably be out of balance if we don't put it back the same way it came apart.

On 2-piece driveshafts the technique is a little bit different. You may have to remove the center support in order to free the shaft, or you may have to disconnect it at both the rear and the center support. If you remove the center support on any driveshaft, make a careful note of exactly where any shims go because these shims determine driveshaft angles which can cause all kinds of vibrations to come in at highway speeds if they aren't correct. On the old Chrysler products with the driveshaft hand brake, disconnecting the driveshaft is done at the front simply by unbolting it. Whenever you disconnect a driveshaft at one end only and it's not necessary to remove it from underneath the car, it's a good idea to tie it up against the floorboards so that it won't get in your way as you work.

The rear of most transmissions sits on a crossmember that goes between the frame side rails of the car. This crossmember usually has to be removed in order to get the transmission out. First, you have to raise the transmission with the cradle and jack we talked about previously in order to take the weight off the crossmember. Don't get too wild on this raising procedure. The transmission only has to go up a few thousandths of an inch in order to remove the weight. If you go too far, you may put the fan into the radiator.

Removal of the crossmember is more or less obvious. It will be bolted to the frame in several places and may have all kinds of accessories attached to it. If you can get the crossmember down and shift it over to one side without disconnecting emergency brake cables and things like that, then more power to you. However, you may find that it is better to take everything off the crossmember so that you can get it out from under the car for more clearance.

When you lower the crossmember the first time, do it gently because there may be some shims between it and the transmission mounting. These shims are used to determine driveshaft angle, the same as shims used under a center support on cars that have one. Save the shims and be sure you get them back in the same place.

WEIGHT SUPPORT

After the crossmember is out of the way, removal procedure gets a little tricky. The weight of the engine and transmission is still supported by the cradle and jack. Before you can go any further, you will have to support the back of the engine because the crossmember that you just removed is in many cases not only a transmission support, it happens to be the rear motor mount also. In some instances, the back of the engine has its own motor mount, and it is strong enough to hold it in position while you remove the transmission. But, in many cases, there is nothing but open space between the front motor mounts and that one mount that is on the crossmember at the back of the transmission.

Support for the back of the engine must be just as secure and safe as the supports that you have holding up the car. In garages they use special supports that are made for the job, consisting of a bar that goes across underneath the back of the engine and bolts to the frame rails. You can make one of these bars from a piece of channel iron and a couple of bolts, or you can use a good heavy chain wrapped and bolted around the frame rails and running underneath the engine. You can also use concrete blocks or wooden blocks, but they must be built up so that they really support it and not let it wobble all over the place.

Once you have the support in position under the back of the engine, you can lower the cradle and jack under the transmission slightly and see if your engine support is going to work. On some cars the engine support must allow the engine to settle about two inches below its normal position so that you can have room to get to the bolts on the bellhousing that are up near the floorboards. On other cars, there is a hole under the floor mat and you can go through that hole in the floorboards to get the top bolts out of the car's bellhousing.

If your engine support is too low, and you let down on it without taking a careful check on clearances in the engine compartment, you may find, for example, that you have just broken the distributor cap against the firewall. If this is your really unlucky day, you may even bend or break the distributor body itself.

In case you don't have the lower cover off the bellhousing by this time, now's the time to do it. Then go in and remove the bolts that hold the converter to the flywheel. The location of the converter on the flywheel should be marked so that you can get them both back together in the same relative position. On a few

1
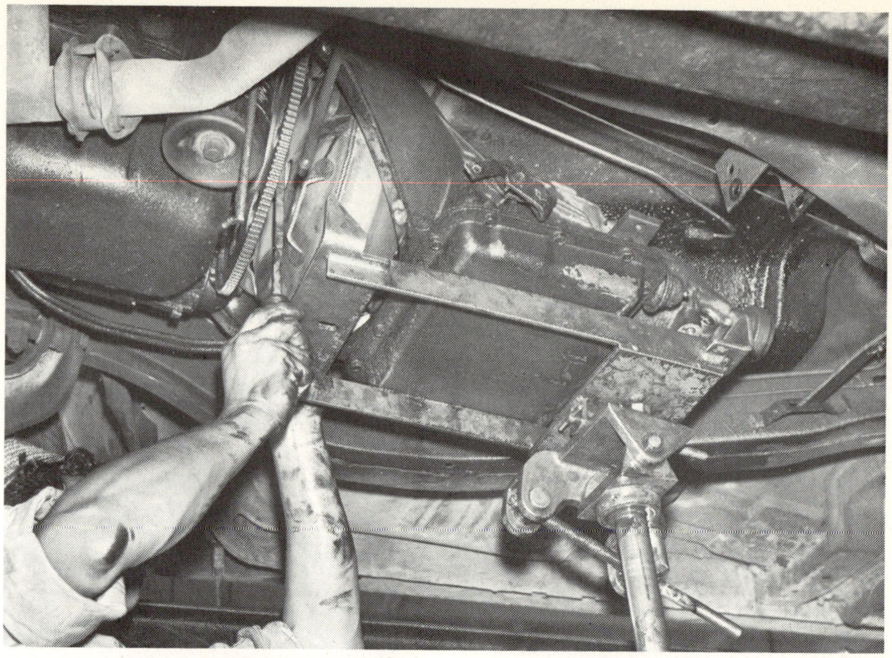

2
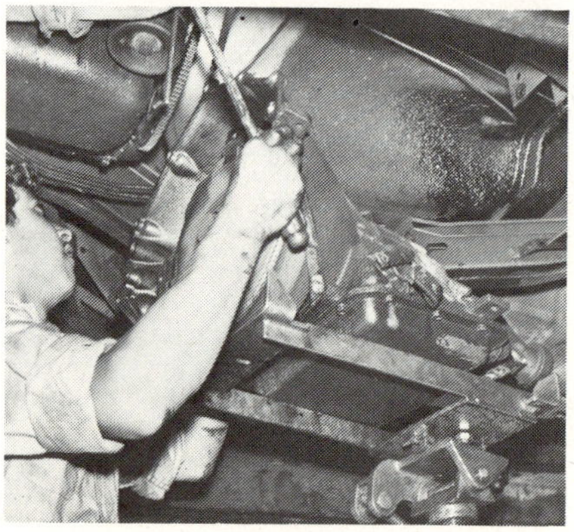

3
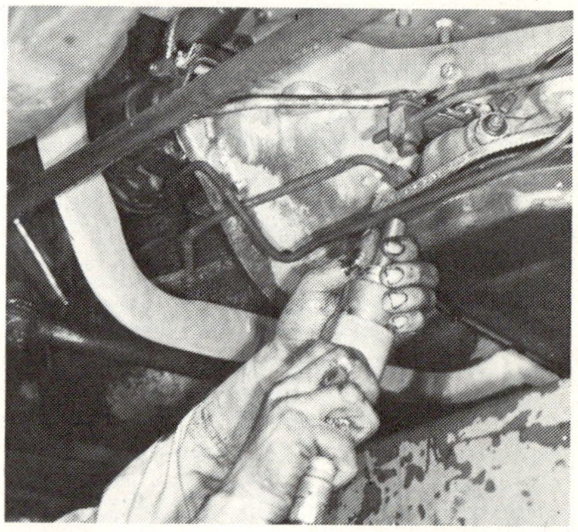

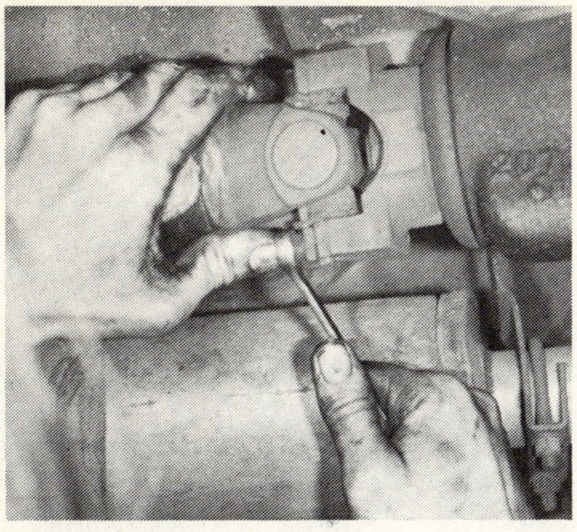

4

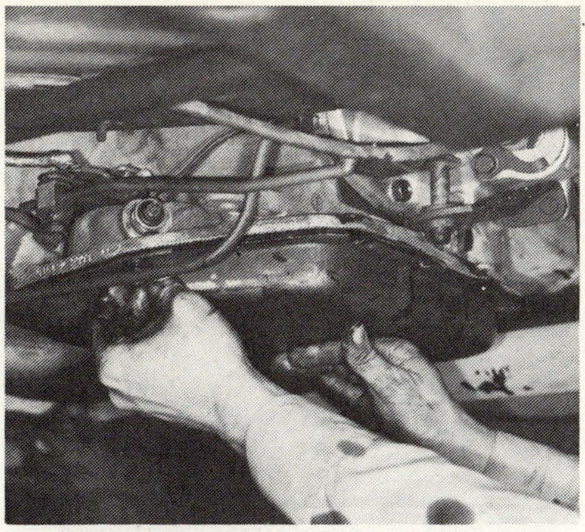

5

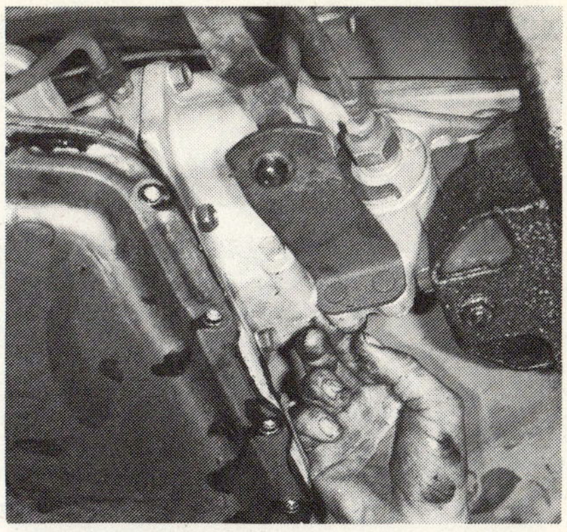

6

1. The transmission weight is supported by the special transmission jack and cradle, and the bellhousing bolts have all been removed. The large screwdriver is used to gently pry the transmission to the rear to free it from the engine.

2. And that's all there is to it! We have just removed a Powerglide transmission. It isn't obvious here but the man pictured is holding the torque converter in to keep it from falling out of the front of the transmission.

3. Transmission designs and mountings differ widely from car to car. Unlike the Powerglide, this TorqueFlite is nearly buried and much more difficult to remove. Begin by pulling pan and draining the fluid.

4. Disconnect the U-joint and pull the driveshaft while the fluid drains. This is a good time to check the U-joint for possible wear; if it needs rebuilding, this is the time to do it.

5. Pan goes back on after the fluid has drained. All you need is a single bolt on each side to hold pan in place. This precaution prevents damage to valve body when trans comes out.

6. Remove converter cover plate and let drain while you unhook linkage, cables, etc. Two bolts hold the linkage torque plate in place.

7. Getting at the starter motor for removal from the trans case is a slow and difficult job. Be sure to disconnect the battery ground cable first, then unhook the neutral start switch from the transmission.

8. The speedometer pinion adapter comes out when bolt and retainer are pulled. If you find trans fluid in the cable housing, the adapter seal should be replaced.

9. Disconnect the oil cooler lines at the trans case using two wrenches. Plug the lines to prevent a drippy mess and remove the oil filler tube.

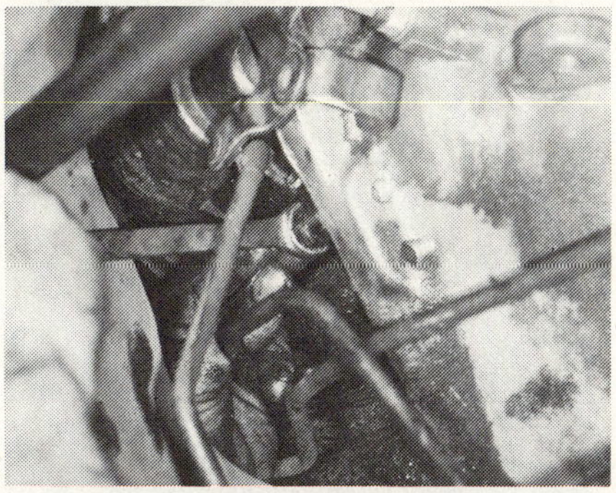

7

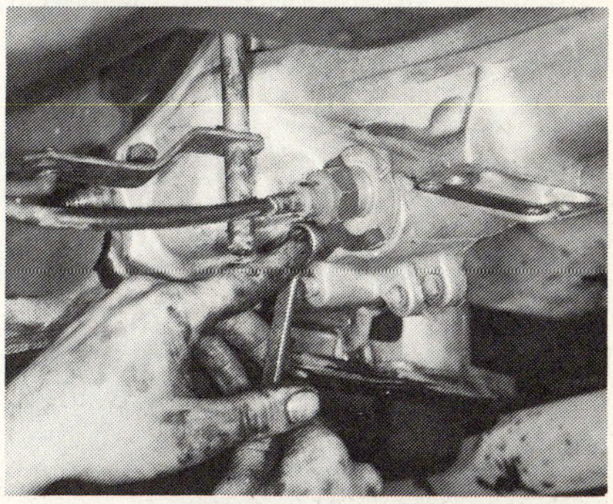

8

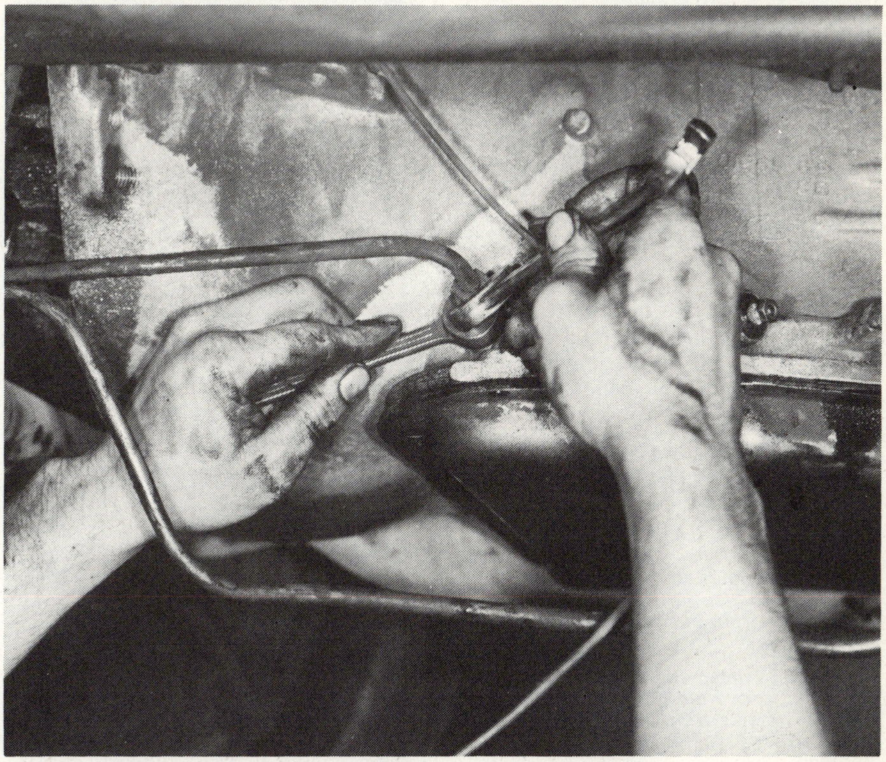

9

transmissions, the converter is removed after the transmission is removed, similar to the way a manual clutch is taken out. Check in your shop manual for the right procedure.

With the converter unhooked and all the other paraphernalia disconnected from the transmission, you are ready to start taking out the bellhousing bolts. The weight of the engine is still on the support that you built, and the transmission is hanging on the back of the engine. In order to get the bellhousing bolts out and free the transmission without bending everything, you must bring the jack and cradle up under the transmission so that they just barely take the weight of the transmission without raising the engine off its support. Then, go around the bellhousing and take all the bolts.

If you feel pressure on the bolts as you are removing them, then something is wrong, and when you take the last one out, everything is liable to give up with a crash. If there is any pressure on those bellhousing bolts at all, recheck your supports so that the transmission is supported independently of the engine.

Some manufacturers recommend that when you remove the bellhousing bolts, you insert guide pins, which are nothing but a bolt with the head cut off. The guide pins will force you to remove the transmission straight back from the engine and thus keep you from bending pump drives or any connection between the transmission and the engine. Once the bellhousing bolts are free, then all that is necessary to do is to roll the jack toward the rear of the car and the transmission along with it.

As soon as the transmission starts moving backwards, you must put your hand up through the opening in the bottom of the bellhousing and hold the converter against the transmission. If you don't, the converter may fall out of the transmission and either be damaged when it hits the ground or ruin the seal as it falls out. As we said before, you don't have to have the cradle and the jack—you can use some husky friends. The trouble is they not only have to be husky, but brave. If a couple of them decide that things aren't going well and they relax their hold, the transmission can go splatting down onto the ground.

You should definitely have a troubleshooting chart for your transmission. It not only tells you what to do, but it can save you lots of work by telling you what not to do. And, a troubleshooting chart might even tell you enough that you could fix the transmission without removing it or tearing it down.

However you do it, you will probably find that automatic transmissions are not nearly as complicated or as hard to work on as you thought.

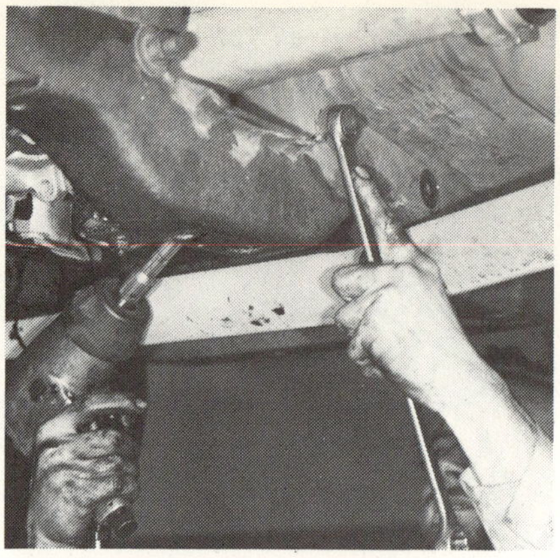

1

2

3

4

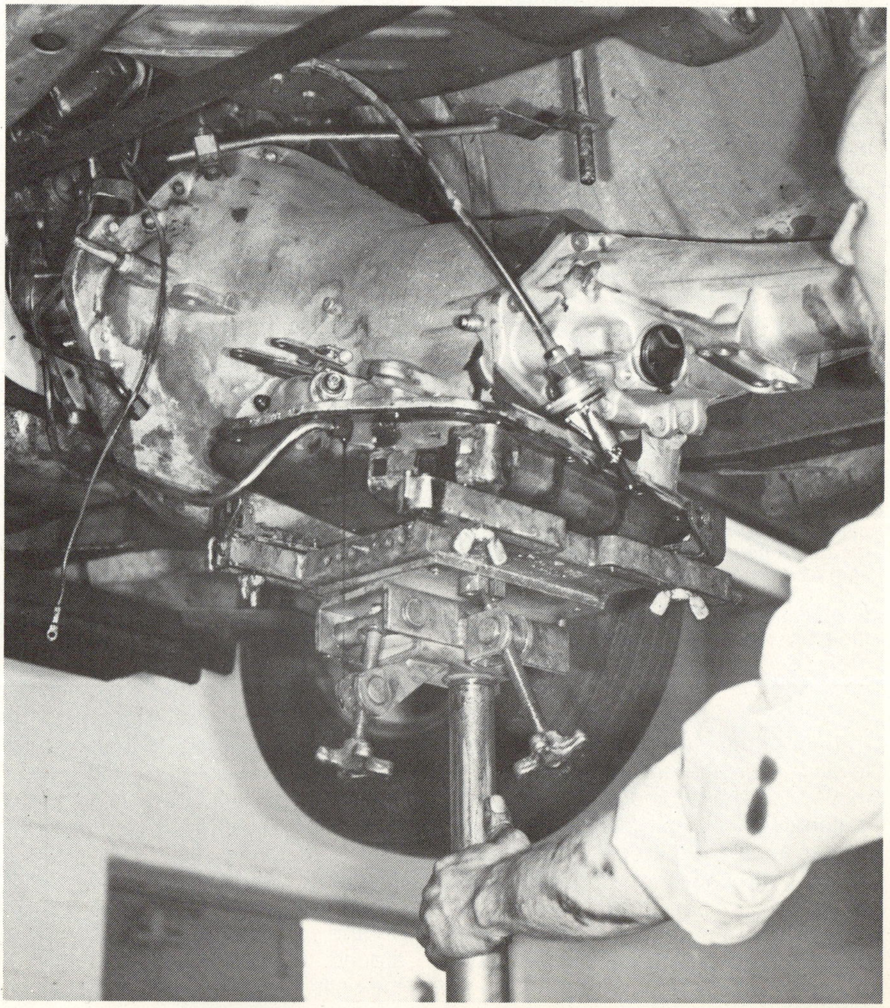

5

1. Power tools make removal of the crossmember attaching bolts easy. The last bolt should be loosened but left in place for now.

2. The rear mount-to-extension housing bolts are also hard to get at. Here a 4-in. extension socket is required.

3. The last crossmember bolt can come out now. Hold the transmission in place while you are pushing the bolt and housing together.

4. You're lost with simple hand wrenches here. The upper bellhousing retaining bolts can be reached only with a 2-ft. extension. Leave one bolt in place.

5. A trans jack will support trans weight while the last bolt is removed. Work the trans rearward off engine block dowels and pull hub from crank end. The real difficulty with an installation such as this is the inability to see what you're doing much of the time, and the requirement for long extensions.

1. Remove converter assembly by pulling toward you—firmly. Place converter face on clean surface. Protect it from dirt with clean cloth.

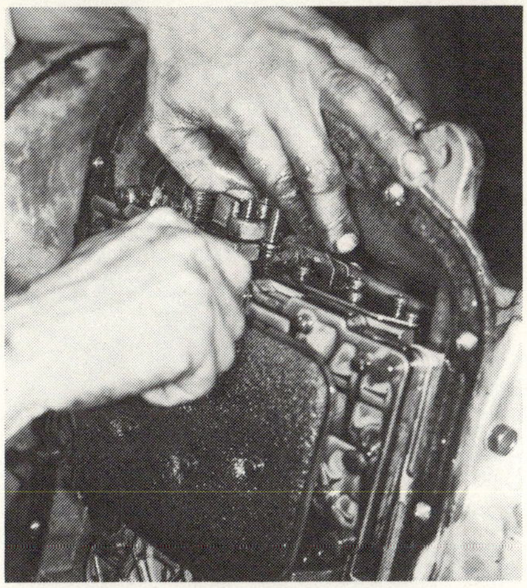

2. To disconnect throttle lever shaft from manual lever assembly, snap off E-clip; remove washer. Slide off by pushing throttle valve down.

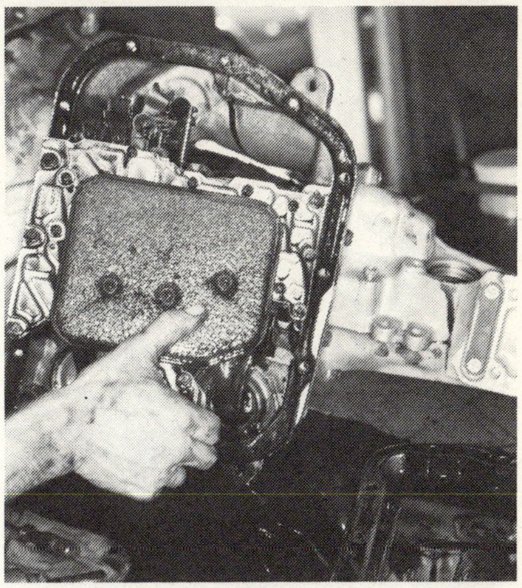

3. Lift entire valve body from trans housing, move to clean area for disassembly. Note dirty fluid filter, preventable by maintenance.

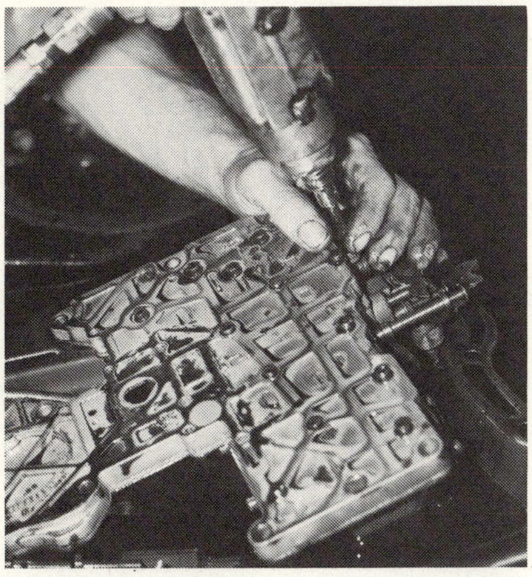

4. Remove three screws to free spring retainer, torque converter control valve spring, regulator valve spring with line pressure adjust screw.

5. Remove kickdown valve and detent, throttle valve and torque converter control valve. Watch order of parts removal—it's important.

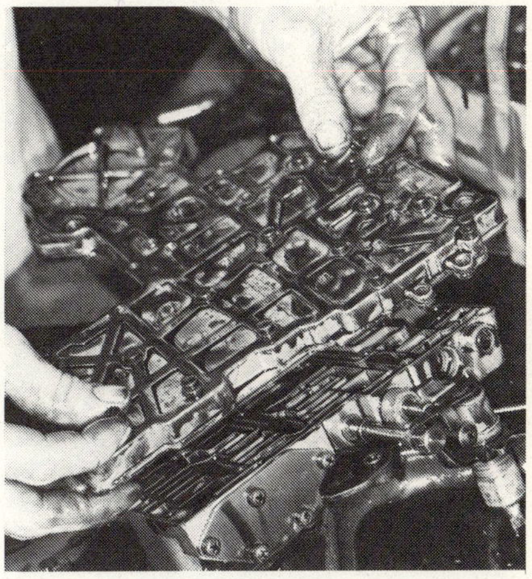

6. Rotate manual valve out of valve body. Transfer and separator plates are held to body by 17 screws; loosen, lift transfer plate, remaining screws.

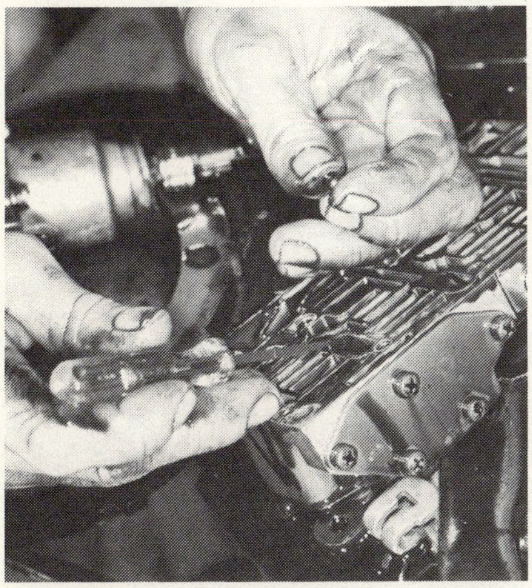

7. One of six steel balls in valve body is considerably larger than other five, and is located in the larger chamber. Take note of this.

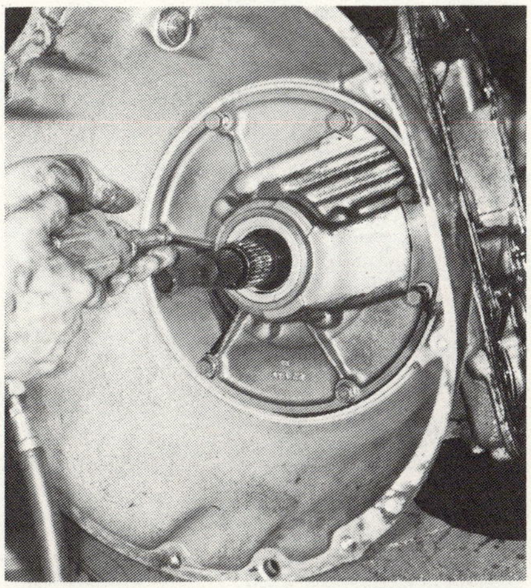

8. Tighten front band on front clutch retainer to prevent it coming out with oil pump and damaging clutches. Remove oil pump retaining bolts.

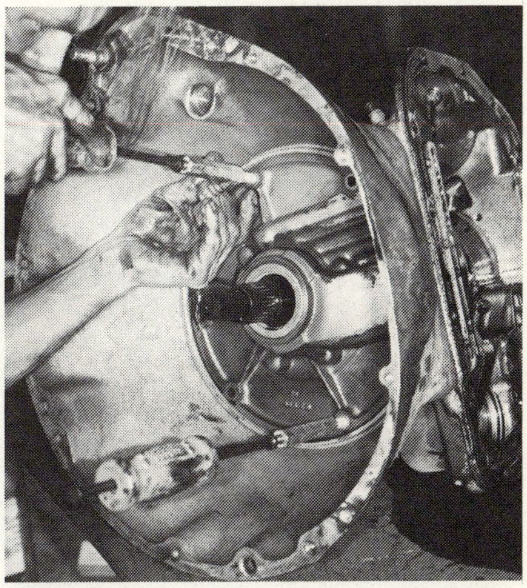

9. Thread two knocker weights in pump housing flange to remove oil pump, reaction shaft support assembly. Thread opposite sides for even pull.

HOW-TO: TorqueFlite Overhaul

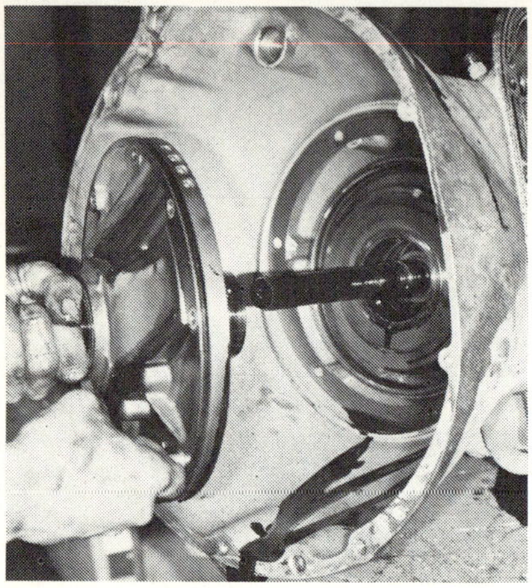

10. Bump outward on the knocker weights to withdraw assembly from trans case. It'll take more effort to free a previously untouched assembly.

11. Inspect pump body and shaft support for wear, blow dry and reinstall new seal as shown. Careful—tap in gently or seal may leak.

12. Turn oil pump body over and insert outer and inner rotors. A light coating of lubricant before fitting the outer rotor certainly won't hurt.

13. Now install reaction shaft support assembly, position vent baffle over vent opening. Thread the bolts and torque to 150 in.-lbs. exactly.

14. The interlocking seal rings on reaction shaft support must turn freely in grooves. Don't remove unless worn or broken. Replace O-ring.

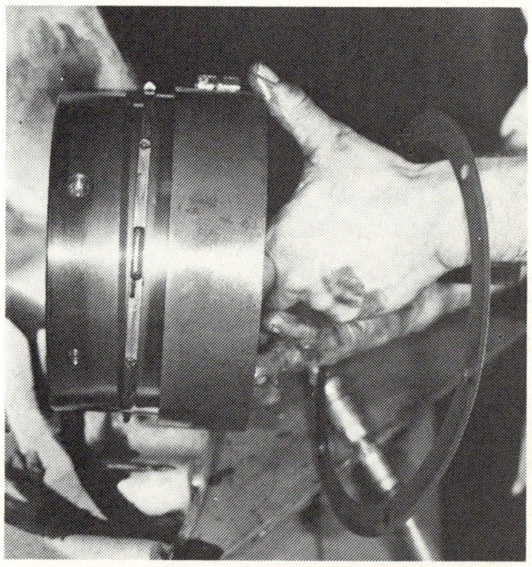

15. Now loosen front band adjuster, remove band strut and slide kickdown band, front clutch assembly out. May be trouble source if trans is dry.

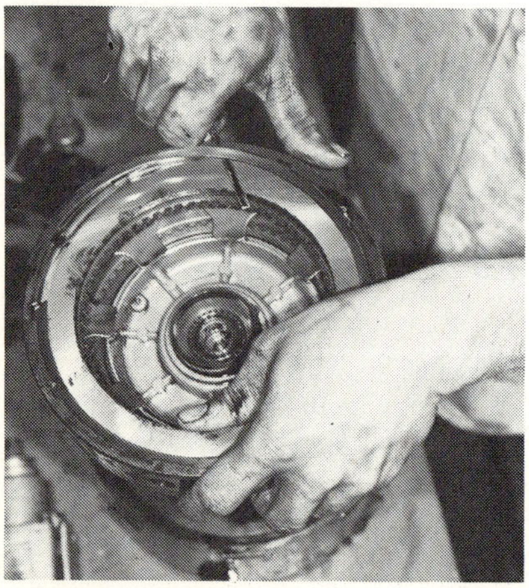

16. Remove large selective snap ring holding pressure plate in clutch piston retainer, lift both plates from retainer. Repeat up front; check.

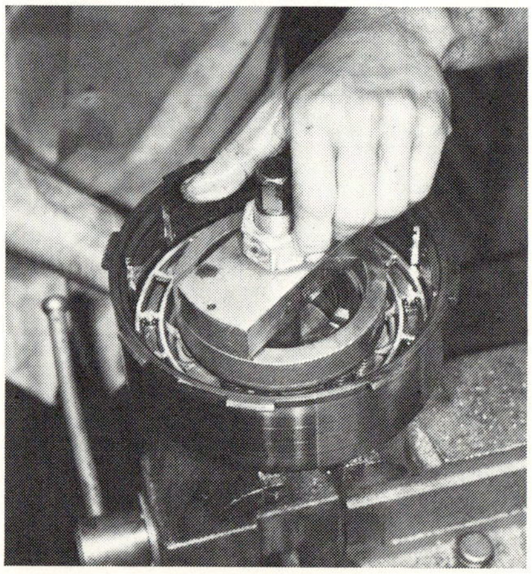

17. Fingernail-test driving disc surface for flaking. Use compression tool for front clutch dissassembly. Compress springs, remove snap ring.

18. After removing compression tool, lift out retainer, return spring. You'll see 8, 10 or 12 springs; 8 and 12 versions have 10 and 15 shafts.

19. Fit all seals carefully to avoid leaks. Now install inner, outer seal on front clutch piston. Place piston firmly, see that it moves freely.

20. Replace springs on piston shafts. If 8 springs, place 4 on each side. With 12 springs, make 3 equal groups separated by empty shaft.

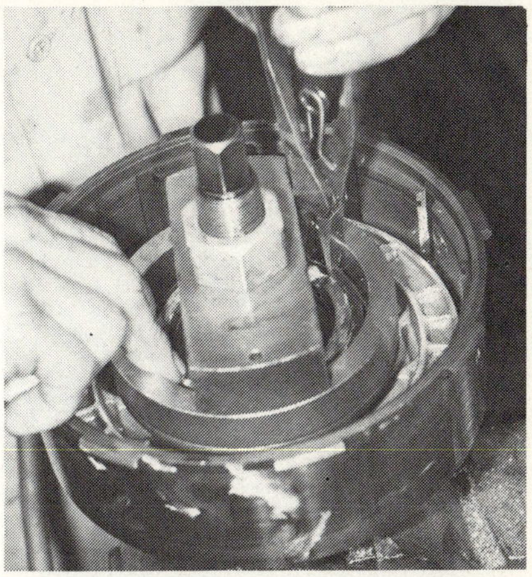

21. Insert spring retainer and snap ring, using compression tool to compress and seat. Snap ring pliers will help. Space here is limited.

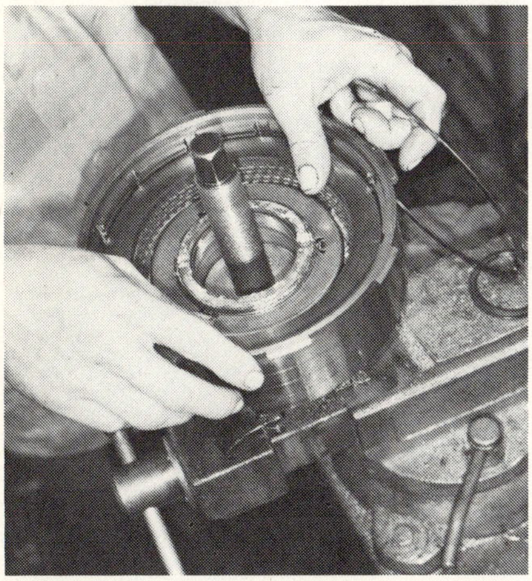

22. Lubricate the clutch plates, install one steel plate and then one lined disc until all are in place. Note: do not use snap ring for rear clutch.

23. After careful seating process, pry one end of wave spring from groove in rear clutch retainer, remove two springs, spacer, piston, seals.

24. Fit piston into piston retainer. Small screwdriver, twisting motion on piston will give big assist. Rotate while pushing down until snaps in.

25. Place clutch piston spring, spacer ring on top of piston in retainer. Both must be positioned in retainer groove, pushed progressively.

26. Replace inner pressure plate in retainer with raised portion resting on spring; install clutch plates, number depending on engine type.

27. Drop outer pressure plate in place, follow with selective snap ring. Work on each unit separately so front/rear snap rings won't mix.

HOW-TO: TorqueFlite Overhaul

28. Position the thrust washer, then replace two seal rings on input shaft prior to clutch mating. Set the unit aside until ready to assemble.

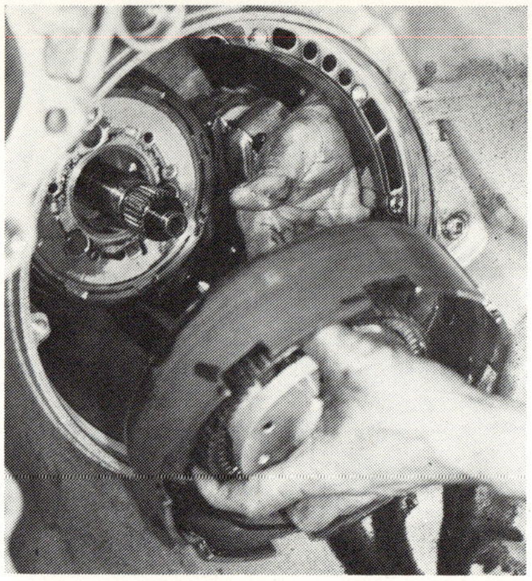

29. Remove small thrust washer and snap ring from output shaft end, slide off front planetary gear assembly and driving shell. Inspect.

30. Shaft and gear units should be OK unless trans is very sick. Now remove rear planetary gear assembly, rear annulus gear, low/reverse drum.

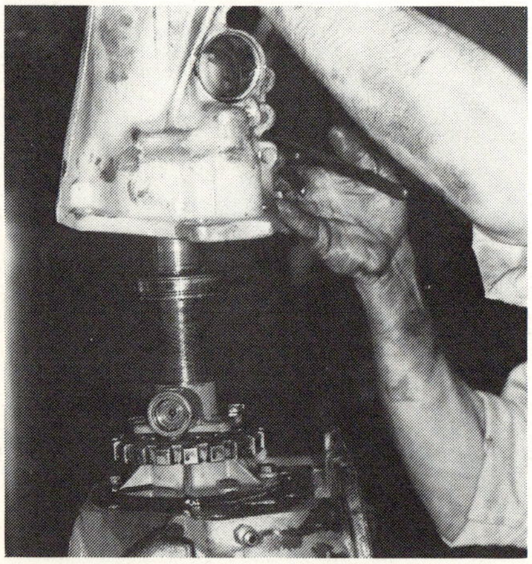

31. Inspect all removed parts while cleaning. Remove two screws, plate and gasket from housing mounting pad extension; spread snap ring.

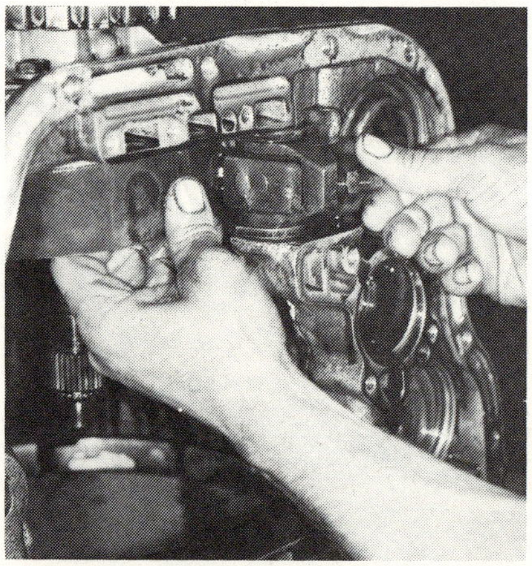

32. Remove low/reverse band by loosening adjustment screw to drop strut into hand. Long link, anchor assembly lifts out easily. Clean with solvent.

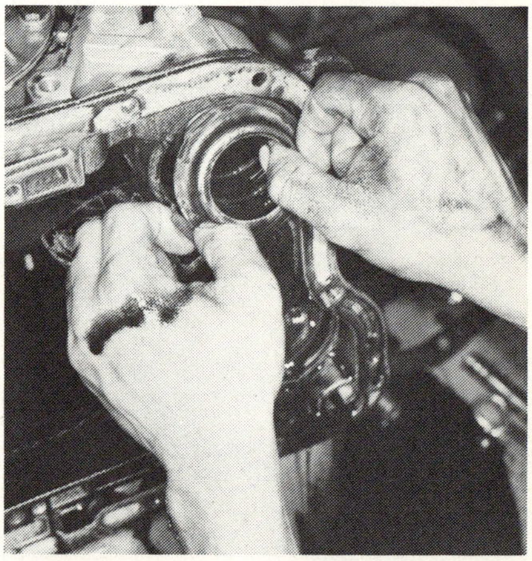

33. Inspect band, bond linings for wear, replace if grooves smooth. Then remove low/reverse servo, including springs, retainer, plug, piston.

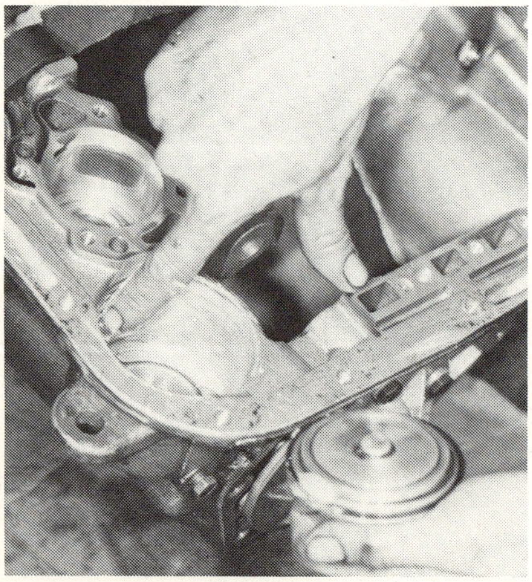

34. Accumulator piston must be removed; output shaft pulled, cleaned. Clean and dry trans case; begin reassembly, starting with seals. Lube.

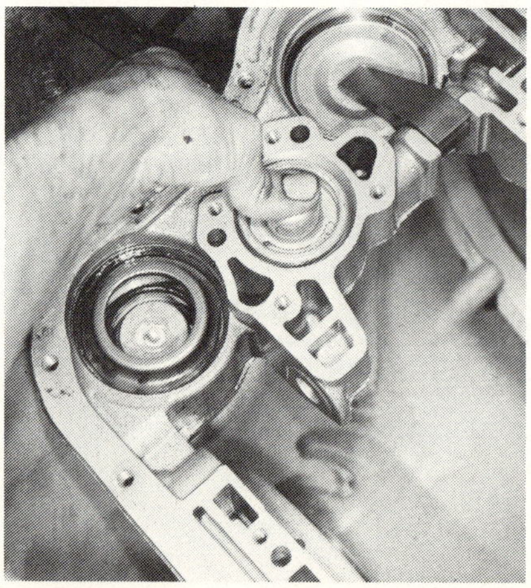

35. Replace accumulator piston and spring after fitting new seal rings. Snap ring holds piston in place, should engage inner groove.

36. Output shaft is reinstalled and low reverse band inserted—just handed onto shaft. Must fit linkage into place before securing band.

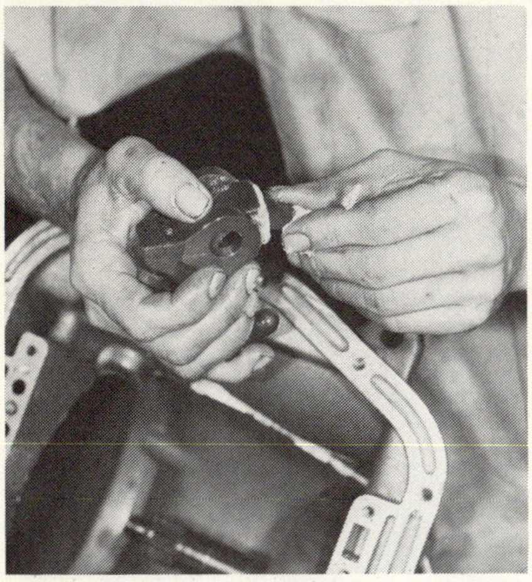

37. Joining low/reverse linkage with grease helps in getting parts in proper place. Place O-ring on band reaction pin before inserting.

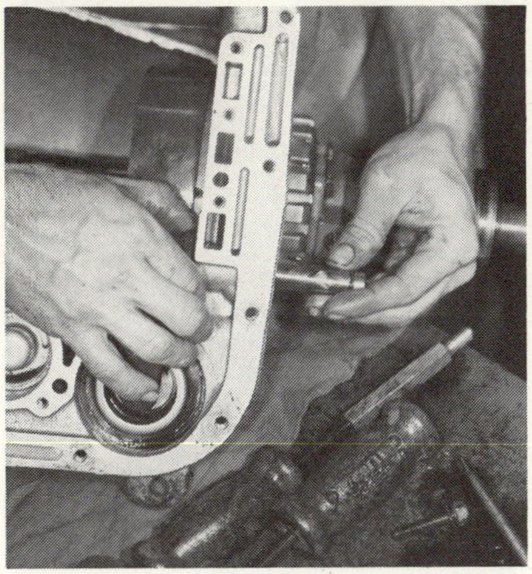

38. To secure the band in position, mate the linkage so that it fits around the band ends, using one hand to slide in the band reaction pin.

39. Fit rear planetary gear and rear annulus gear assembly around thrust plate. Washer goes atop assembly, is held in place with snap ring.

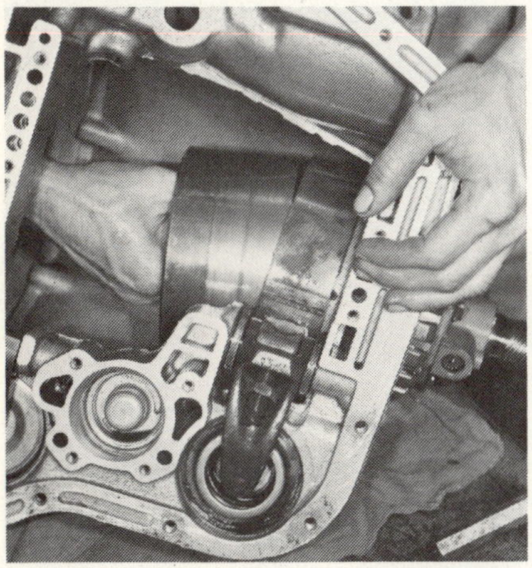

40. Fit low/reverse drum onto output shaft with rotating motion. You may have to loosen band adjustment for proper fit inside the drum.

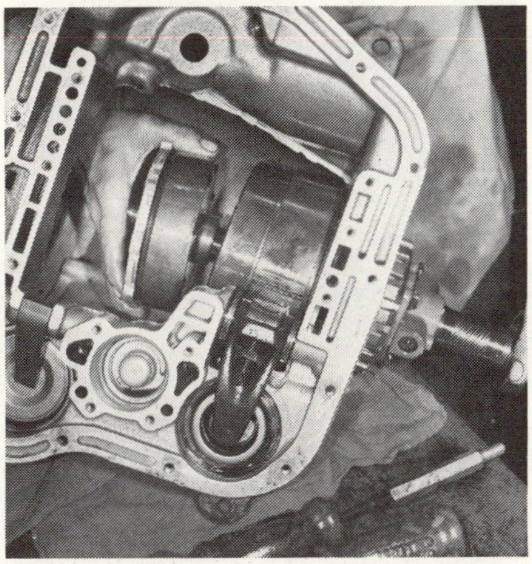

41. Rear planetary/annulus gear assembly is fitted over output shaft and inside low/reverse drum. Use a rotating motion to engage teeth.

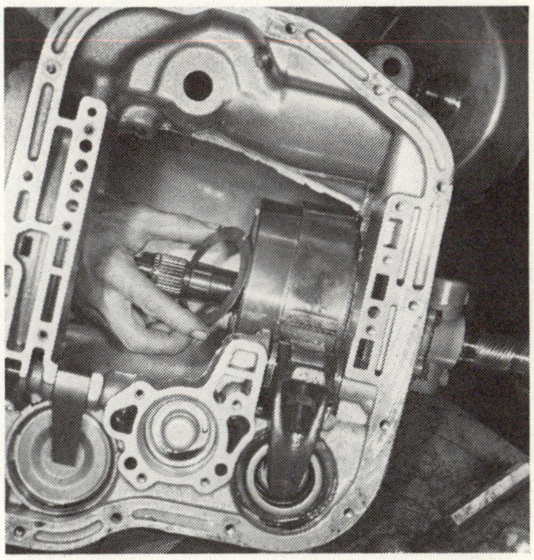

42. Fit another thrust washer at front of rear planetary gear assembly. Use light coating of lubricant—important for correct placement.

43. Slide driving shell and sun gear assembly onto shaft, rotating to match teeth. Insert thrust washer inside front of driving shell.

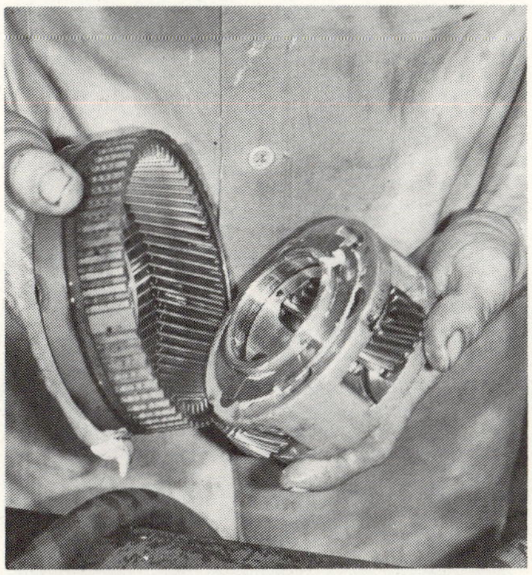

44. Thrust washer goes inside driving shell on rear hub of planetary gear set—with lube. Slide assembly into front annulus gear. Rotate to mate.

45. Work front planetary, annulus gear assemblies onto output shaft, meshing planetary and sun gear teeth. Gears may hesitate, but will mesh.

HOW-TO: TorqueFlite Overhaul

46. Now install selective snap ring on front end of output shaft. Measure end play here; replace snap ring if not within specs for thickness.

47. Thrust washer must be inserted on output shaft; position the front and rear clutch assemblies on splines. Front lugs must fully engage.

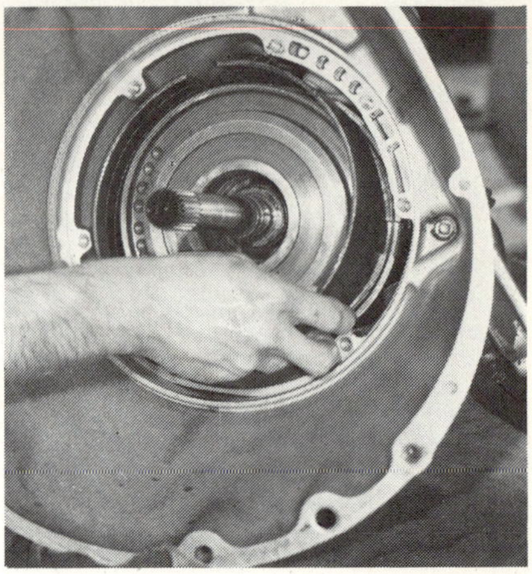

48. Slide kickdown band over front clutch assembly, rotate ends for placement of strut and anchor. Band will rest on front assembly.

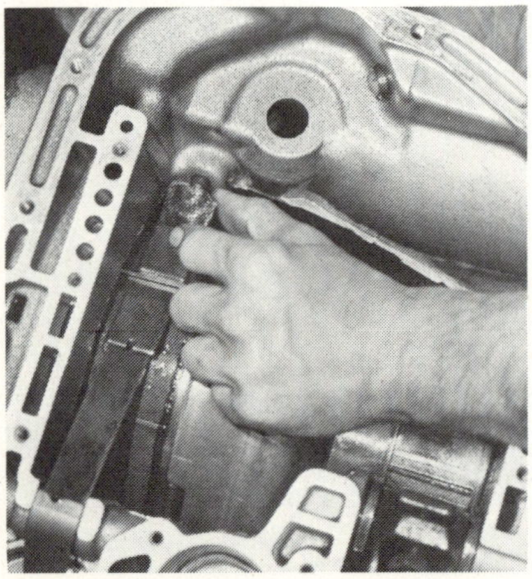

49. Set band strut in place; position anchor on opposite side, tighten adjuster. This locks front band until ready to adjust to specs.

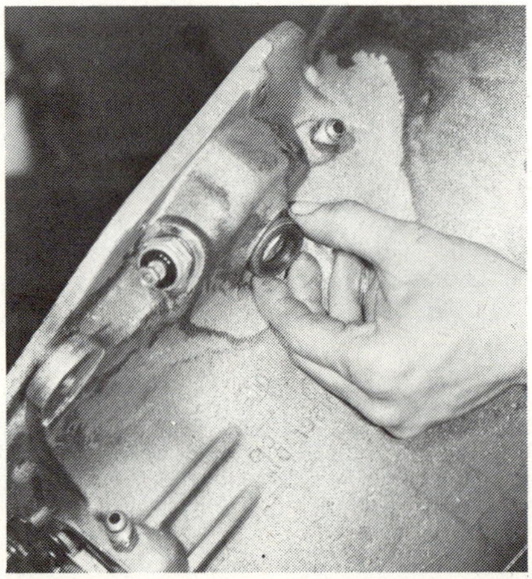

50. Pry out old seal from valve body manual lever shaft and replace. Use 15/16-in. socket and hammer to install seal, tapping gently, evenly.

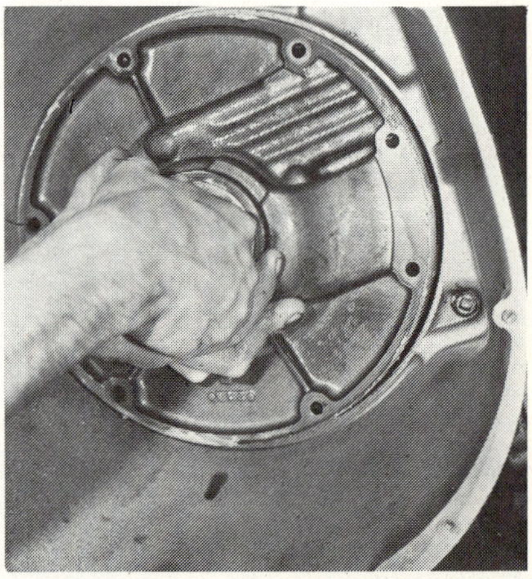

51. Next fit oil pump and reaction shaft assembly in place. Tap lightly with mallet, line up stud holes; replace studs. Torque to 150 in.-lbs.

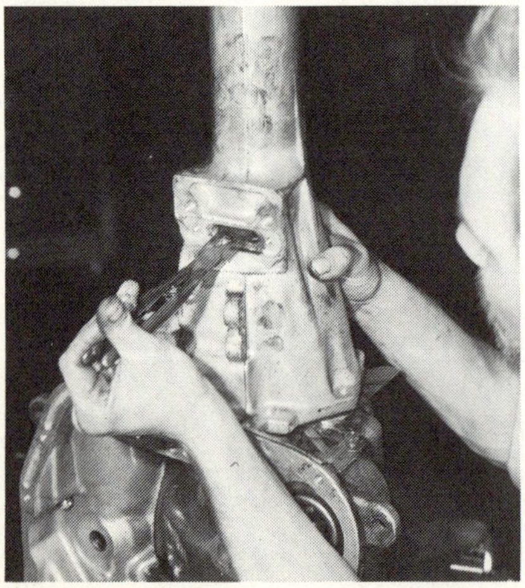

52. Fit new extension housing gasket on case, replace housing, spread output shaft bearing retaining snap ring; tap housing, ring into groove.

53. Torque housing bolts to 24 ft.-lbs., replace gasket. Adjust bands, back off low/reverse screw nut five turns; torque screw to 72 in.-lbs.

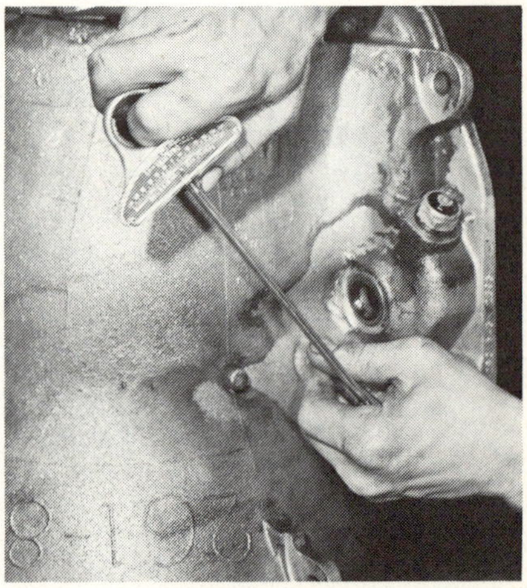

54. On above, torque lock nut to 35 ft.-lbs—exactly. Kickdown band adjust screw is backed off five turns, and then torqued to 72 in.-lbs.

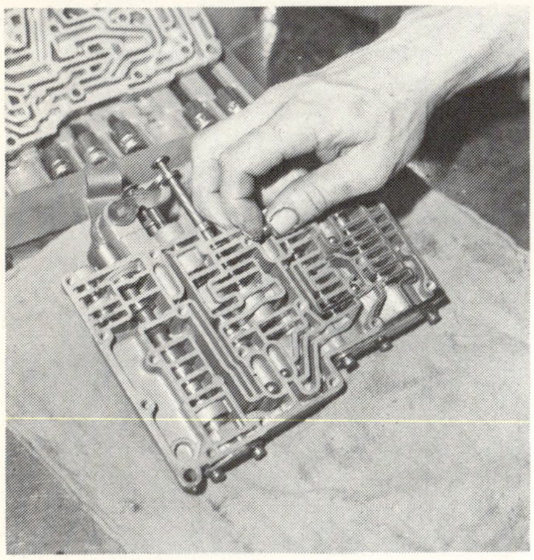

55. Torque lock nut to 29 ft.-lbs., completing factory specs. Now reassemble valve body (with cleaned passages). Reinsert six steel balls.

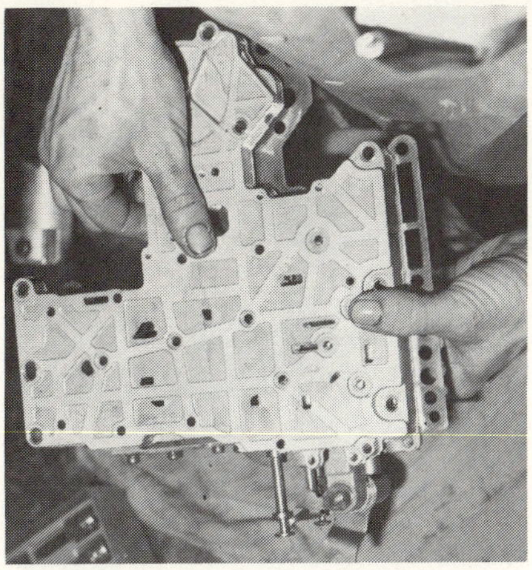

56. Position steel separator plate over valve body; set transfer plate in place, lining up screw holes. Pieces should mate with no difficulty.

57. Install 17 retaining screws, starting at center, working outward. Tighten to 35 in.-lbs., working same way. This assembles valve body.

58. Now replace torque converter control valve, spring, throttle, regulator, kickdown valves; fasten behind spring retainer. Replace screws.

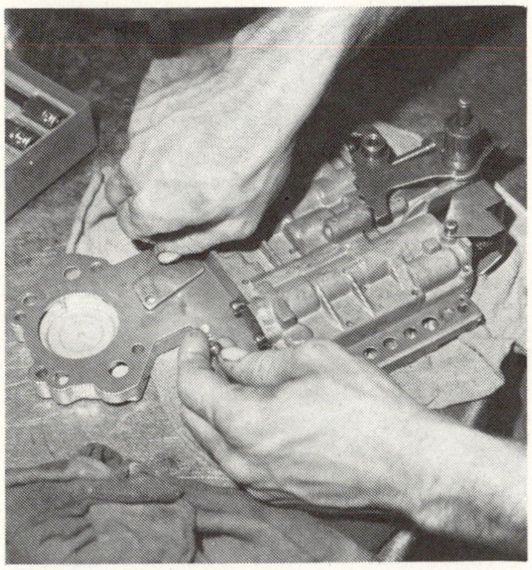

59. Replace shuttle valve cover plate; position stiffener plate on separator plate; torque to 28 in.-lbs. Use of torque wrench is critical.

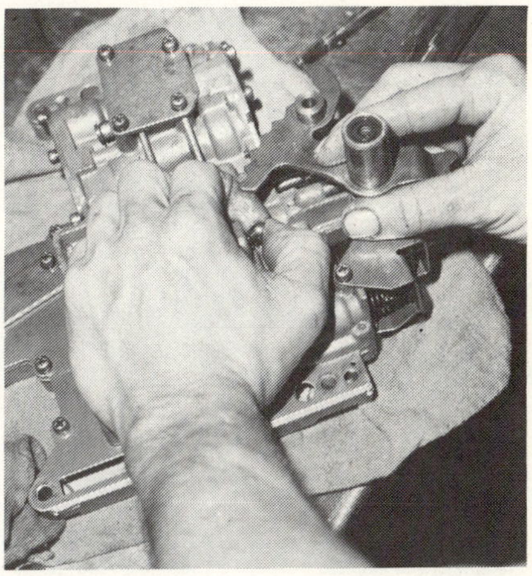

60. Install detent ball, spring, control levers, after replacing old throttle lever shaft seal. Insert detent ball, spring; engage valve and ball.

61. Fit retaining washer in place on throttle lever shaft, lock everything with E-clip. Work manual lever assembly, make sure operation is smooth.

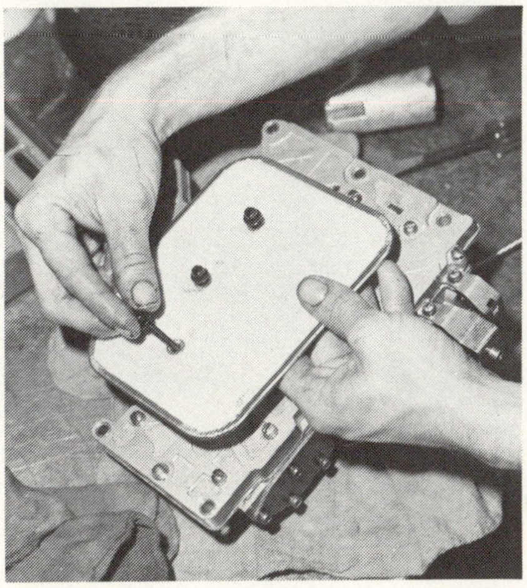

62. New trans fluid filter completes reassembly of valve body. Torque retaining screws to 35 in.-lbs. Notice the new/old filter difference.

63. Valve body is now ready to insert into trans case. Parking lock rod must connect to manual lever, E-clip slipped in. Bolt on pan—and done.

HOW-TO: Pinto C-4 Overhaul

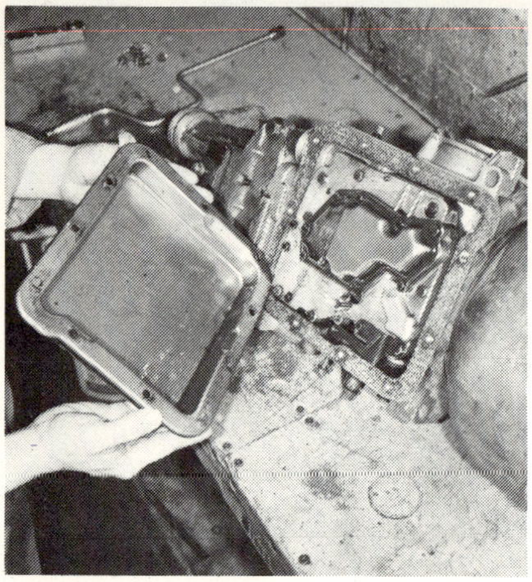

1. Begin disassembly of Pinto C-4 transmission by unbolting and removing the oil pan and old gasket. As you remove each part, clean with solvent.

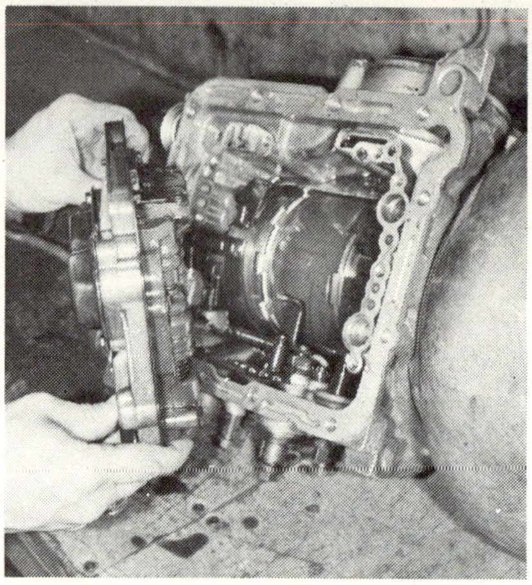

2. Remove the valve body screws holding the unit in place and lift it out. The valve body is a 3-piece unit and will be serviced later.

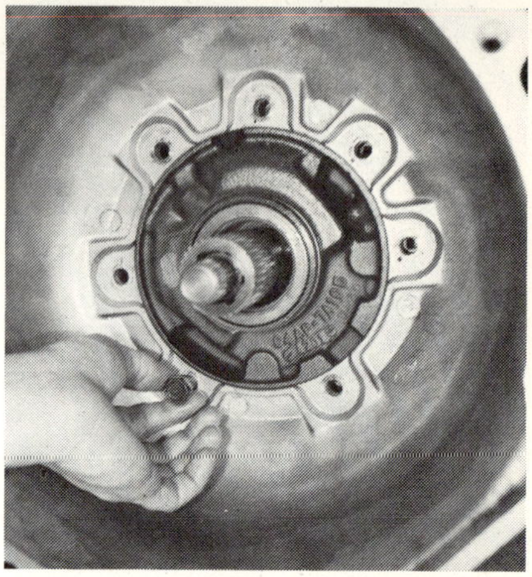

3. Loosen and remove the seven bolts that hold the converter housing to the transmission case, then lift housing off and set it to one side.

4. Back off intermediate band adjuster until band struts are loose. Remove them. This eliminates band tension for removal of other parts.

5. Gently pry between input shell and forward clutch cylinder. This will push the front pump assembly forward so that it can be removed easily.

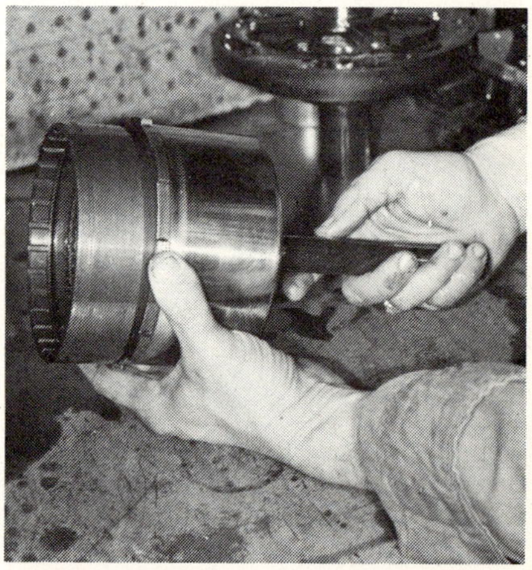

6. Now withdraw the input shaft. The reverse and high clutch drum and the forward clutch cylinder will come out as a unit. Set in solvent for cleaning.

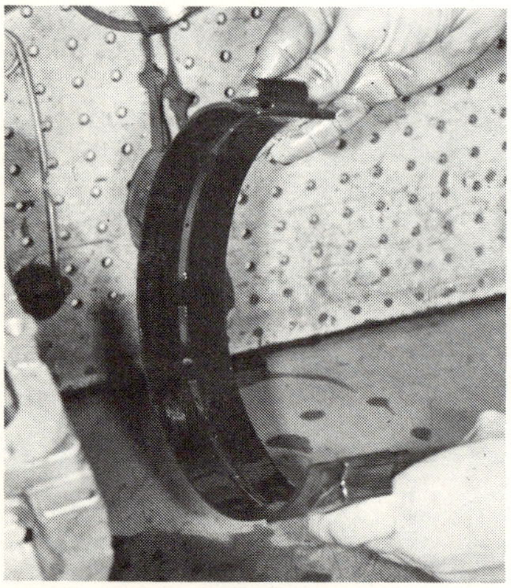

7. The intermediate band can be removed now. Inspect the band carefully for signs of cracking, burning, etc. This is a fairly common C-4 problem.

8. Lift out thrust washer and forward clutch hub. Remove each part carefully to prevent accidental nicking or burring of the bearing/mating surfaces.

9. Another thrust washer and the front planet carrier come out next. Planetary gears rarely malfunction, but inspect the carrier just the same.

10. Lift out the input shell with sun gear and 2 thrust washers. The C-4 uses 10 thrust washers between subassemblies; note their position carefully.

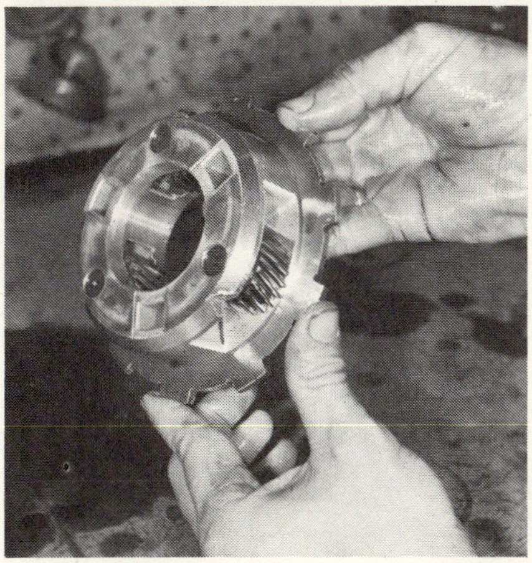

11. The reverse planet carrier and the No. 7 thrust washer are withdrawn from position inside the low/reverse drum. Clean and inspect carefully.

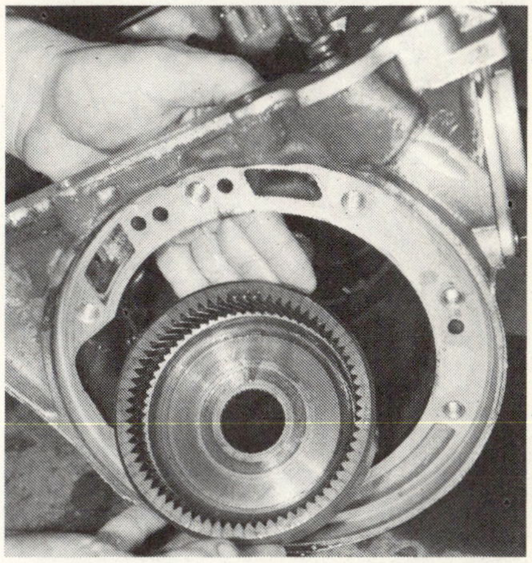

12. A snap ring must be removed from the output shaft before the reverse ring gear and hub can be taken out. Check gear and hub for defects.

13. After backing off the low/reverse band adjuster, the low/reverse drum can be removed from the input shaft. Place in solvent for cleaning.

14. Pull the band struts out of position and withdraw the low/reverse band from the transmission case. Inspect its surface for defects.

15. Taking the transmission vacuum unit off is made easy if you have the special tool shown; if not, you'll have to work at it a little more.

16. After removing the vacuum unit gasket and control rod, the primary throttle valve is removed. Use a pencil-type magnet to speed up the process.

17. Unbolt and remove the extension housing from the trans case; then pull off and throw the old gasket away. Always use new gaskets for reassembly.

18. Withdraw the output shaft with governor and distributor assembly attached by pulling straight back on the shaft. It should come out easily.

HOW-TO: Pinto C-4 Overhaul

19. After unbolting the governor distributor sleeve, carefully pry it away from the trans case and remove to expose the parking gear.

20. The parking gear and another thrust washer simply lift out of place. Note that the washer fits between the gear and the transmission case.

21. Lift the parking pawl and its return spring off the retaining pin. If you try to pry the spring off, it will pop into the air and may get lost.

22. Remove the low/reverse servo cover and lift out the piston and return spring. Piston seal is bonded; if defective, entire piston must be replaced.

23. The intermediate servo on the opposite side of the trans case is also removed. Piston separates as shown, and all seals should be replaced.

24. Only the one-way clutch assembly remains in trans case. Inner and outer races will separate when removed, with bearings and springs falling out.

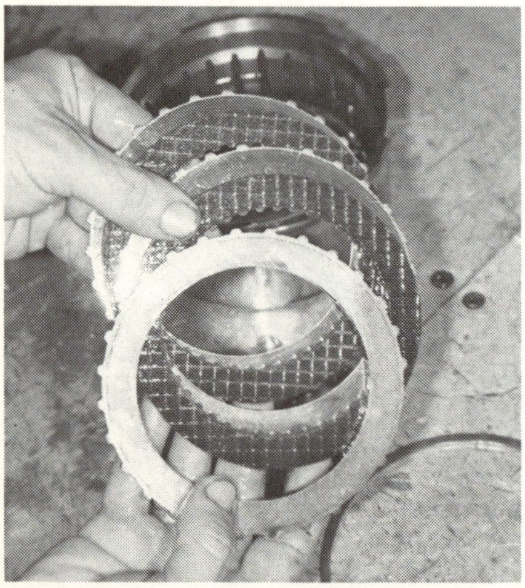

25. Remove retaining snap ring from reverse/high clutch and lift out the drive and driven clutch plates. Inspect all surfaces carefully.

26. You'll need a spring compressor to remove the piston spring snap ring from the clutch hub because the pressure exerted by return springs is great.

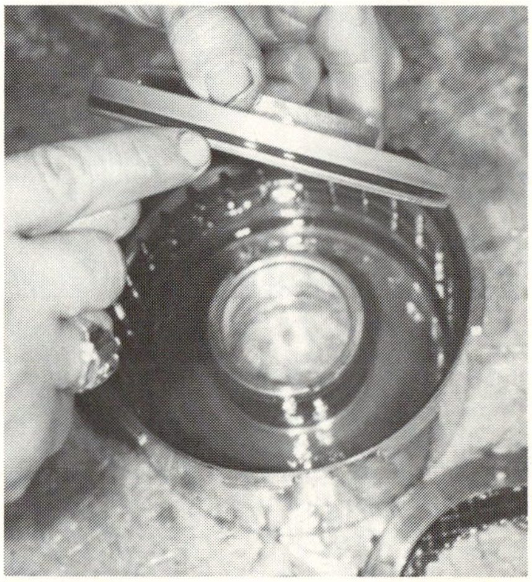

27. Lift piston out, remove and replace the outer seal from the piston and the inner seal from the clutch drum before reassembly.

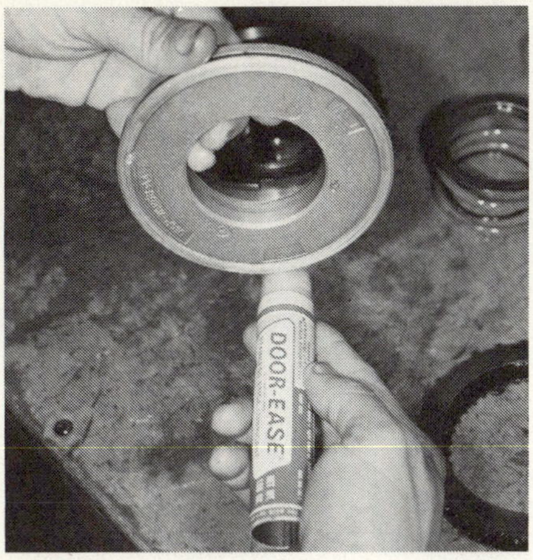

28. An application of dry stick lubricant, such as Door-Ease, to the new seal surfaces will make the replacement of the piston easier.

29. Compress the return springs, replace the snap ring and insert clutch plates alternately starting with a steel plate. Then install pressure plate.

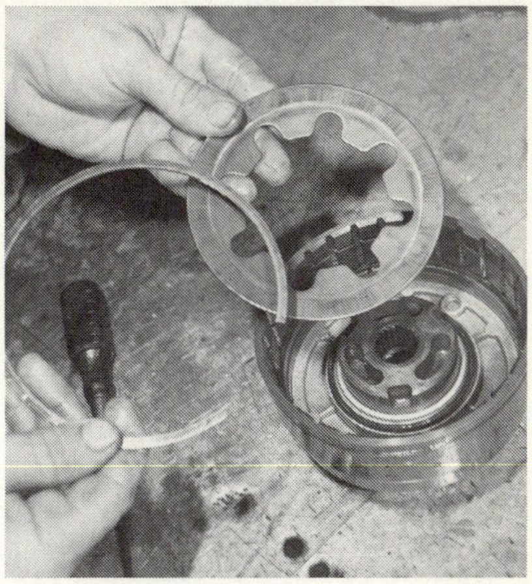

30. Remove forward clutch snap ring and plates, and then inspect. Depress disc spring to remove the second snap ring and lift disc spring out.

31. Next, you should lift out the clutch piston. Remove and replace the outer seal from piston and the inner seal from the clutch hub.

32. Install piston with dry stick lubricant; replace disc spring and snap ring. Insert thin pressure plate, clutches, thick pressure plate and snap ring.

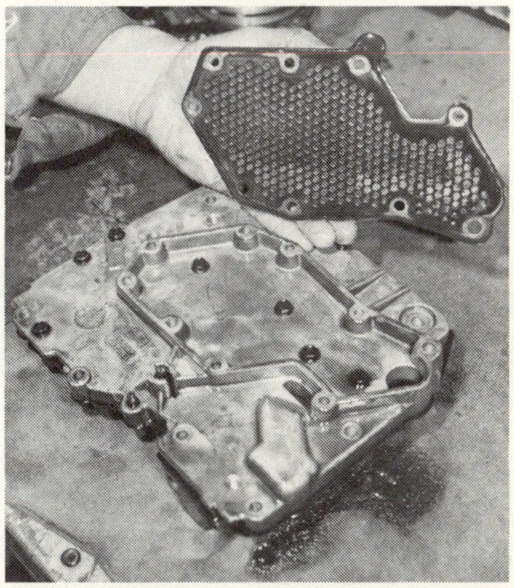

33. Unbolt the oil screen from the valve body and lift off both screen and gasket. Don't lose the throttle pressure limit valve and spring.

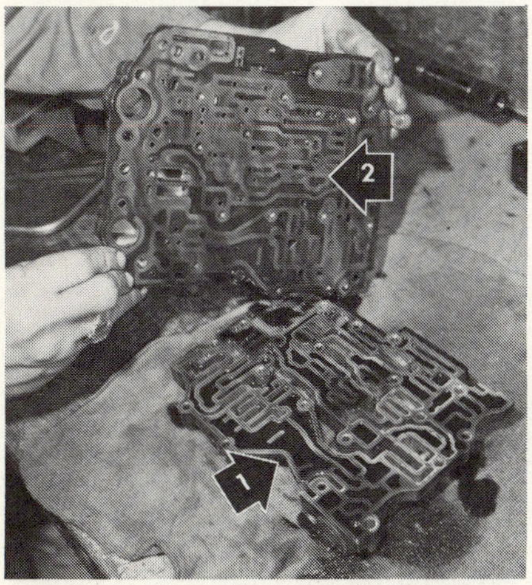

34. Two attaching screws must be taken from the upper valve body (1) and nine screws from the lower section (2) before separating the two.

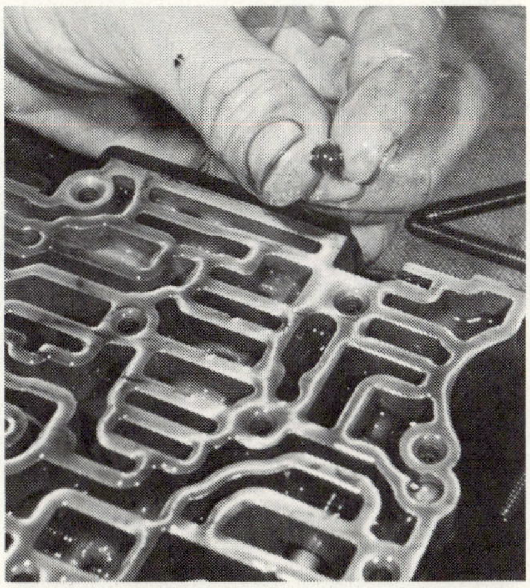

35. The valve body contains three shuttle valves and a servo check valve. Because all of these are rubber, they cannot be removed with a magnet.

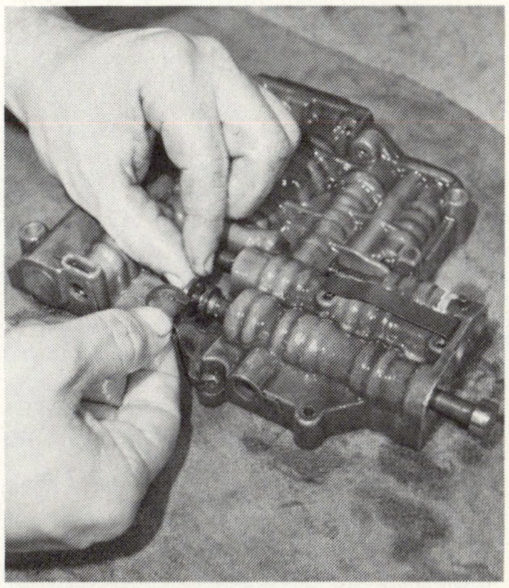

36. Unclip manual valve retaining ring and slide manual valve from the body. Ring holds valve in bore during shipment and should be discarded.

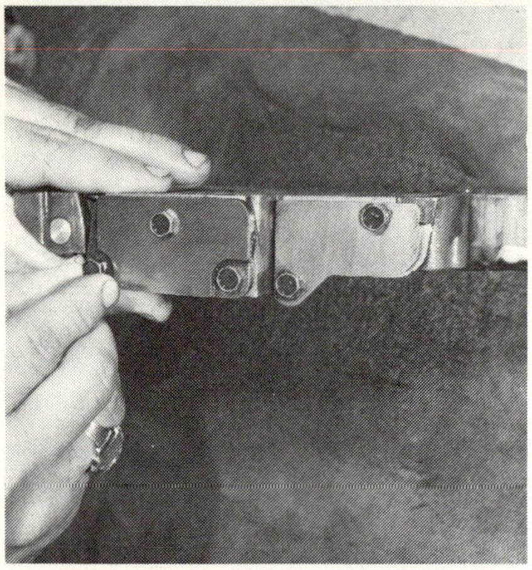

37. These two cover plates must be taken off for access to the shift, transition and cut-back valves. Once off, all five come out easily.

38. Start removal with the cut-back valve. As each valve is removed, line them up beside the valve body in their correct order of removal.

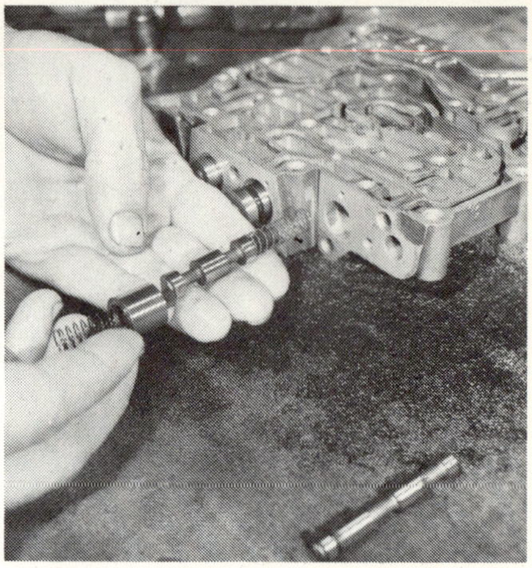

39. A spring, transition valve, the 2-3 back-out valve and spring is the correct order here. Keeping the sequence is absolutely necessary.

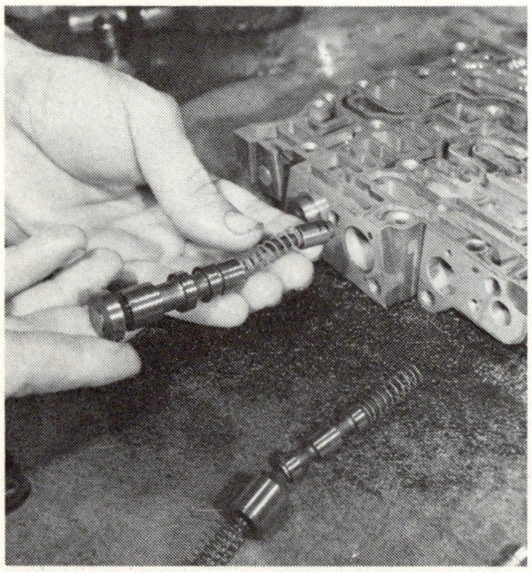

40. The 2-3 shift valve and throttle modulator valve are separated by a spring. When replacing, make sure the long neck fits into the spring.

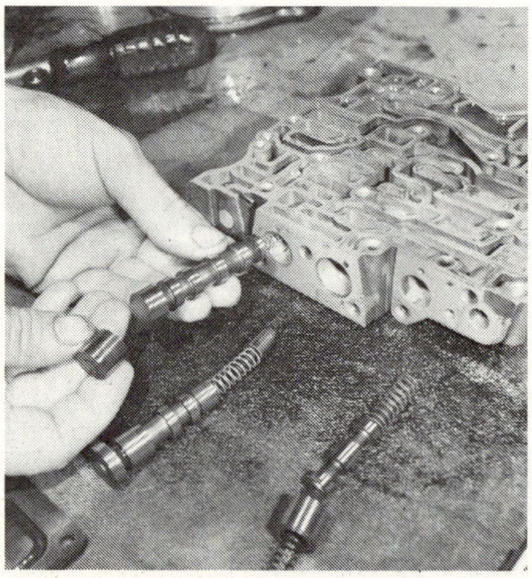

41. The 1-2 shift valve and D-2 shift valve mate end-to-end and are followed by a spring. Don't confuse springs, as their tension values are different.

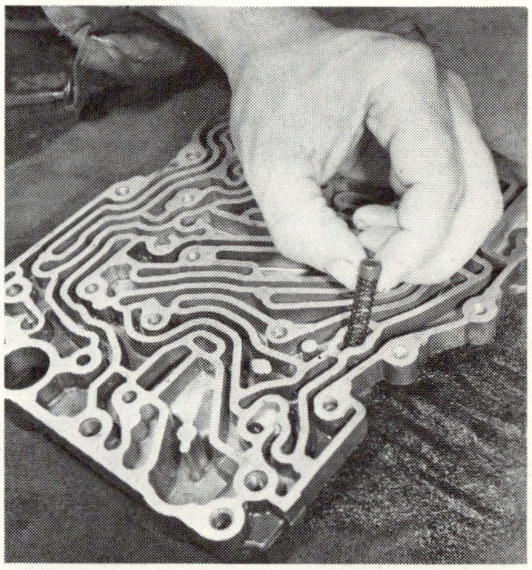

42. Don't overlook the throttle pressure limit valve and spring when reassembling the valve body. It fits into place right here.

43. Unbolt the stator support from the front pump housing and remove it. Check the support bushings in each end of the stator for wear.

44. Remove the drive and driven gears from the pump housing and check the teeth and surfaces. These seldom give trouble and should be in good shape.

45. The old gasket must be cleaned off completely. It is a good idea to use a scraper or putty knife and work carefully and thoroughly.

46. Remove and replace the four seal rings from the stator support. You'll need a snap ring tool to do the job right—these are sharp and brittle.

47. Use a bit of grease around the pump housing as it will let you position the new gasket correctly and hold it in place during reassembly.

48. Start trans reassembly by installing the one-way clutch. Insert both races, then alternate bearings with springs until all 12 are in place.

49. Position the parking pawl over its retaining pin and engage the short end of the return spring in cut-out. Hold and secure the other end of spring.

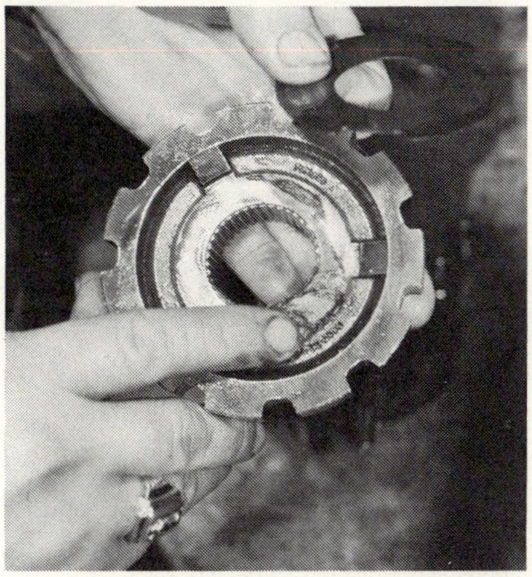

50. Grease the inside surface of the parking gear and fit the proper thrust washer in position; then flip over and insert gear in the case.

51. Fit the governor distributor sleeve in place and tap into position with a hammer. Be sure that the two tubes have sufficient clearance in the case.

52. After governor distributor sleeve is bolted down, remove and replace seal rings on the governor distributor assembly before replacing unit.

53. Insert the output shaft into the transmission case. New seal rings may make the fit a tight one, so work it in place as carefully as possible.

54. Replace the extension housing and bolt into place. Torque limits specified by Ford are 28 ft.-lbs. and bolts should be torqued in that range.

HOW-TO: Pinto C-4 Overhaul

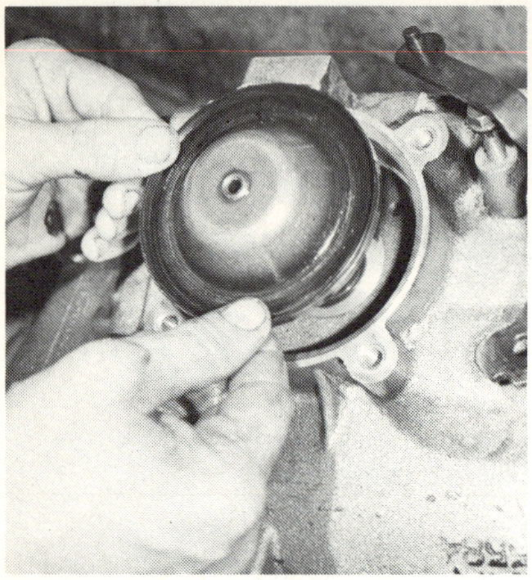

55. You can lubricate the low/reverse servo with clean transmission fluid and set it in place in the case bore right over the return spring.

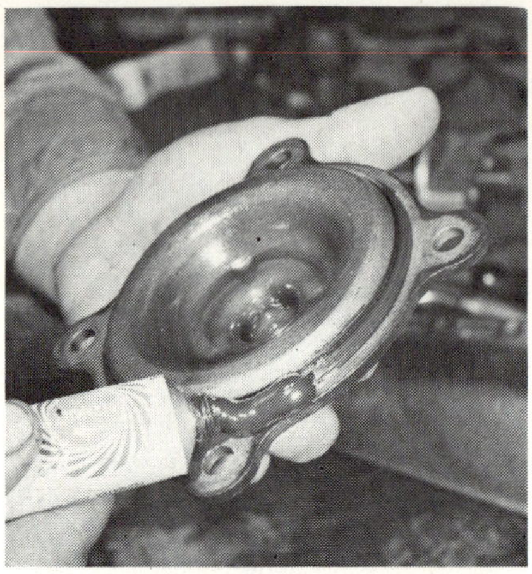

56. A cover seal is used, but adding a bit of Permatex around the cover flange is also good insurance against an unwanted leak developing later.

57. Install a new seal on intermediate servo cover and lubricate with clean trans fluid. Insert return spring and servo; torque bolts to 20 ft.-lbs.

58. Slip the low/reverse band in the case and fit low/reverse drum over output shaft. Fit band struts in place, and tighten the adjustment to hold.

59. After placing a thrust washer into the low/reverse drum, insert the reverse ring gear and hub; then clip the snap ring into place.

60. A thrust washer is used at each end of the reverse planet carrier. The washer with tangs goes at rear of carrier and fits into reverse gear.

61. After installing the reverse planet carrier into the low/reverse drum, insert the input shaft with the sun gear and work it into place.

62. The front planet carrier also uses a thrust washer at each end and fits into the input shell. The carrier must be seated properly.

63. The next step is to position a thrust washer in forward clutch hub and ring gear in place, then set assembly into reverse/high drum.

64. Holding the gear train package together with the input shaft inserted, install entire assembly into trans case and engage the output shaft.

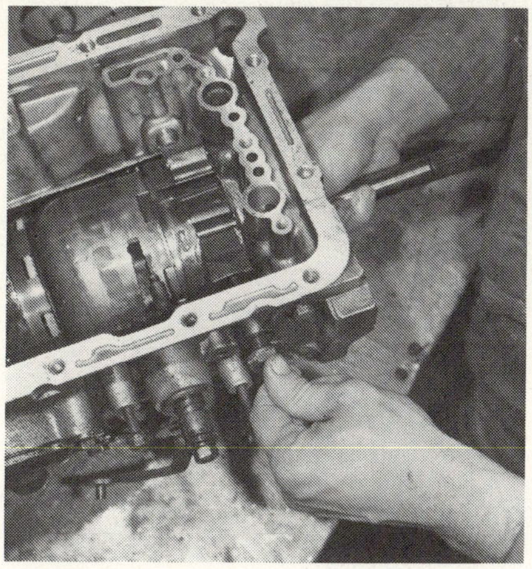

65. Work the intermediate band in position and set band struts in place; then you should tighten the adjuster to hold the band correctly.

66. The remaining two thrust washers fit at the rear of the front pump assembly, which is fitted in place over the input shaft against the trans case.

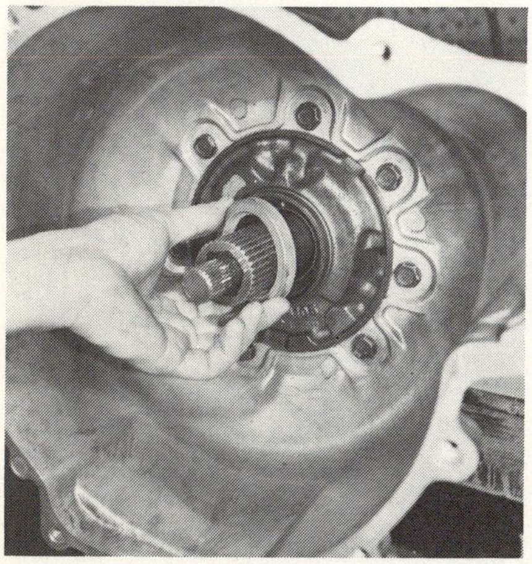

67. After bolting the converter housing to the transmission case, torque to 28-40 ft.-lbs.; then remove and replace the front pump seal.

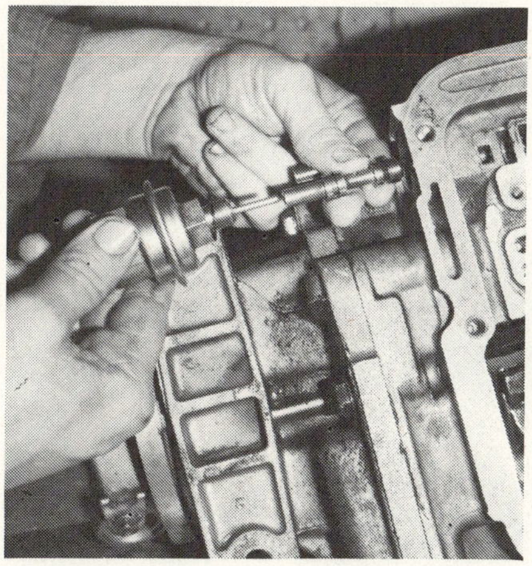

68. Reinsert the primary throttle valve, control rod, gasket and trans vacuum unit at the back of the case, tightening with the special tool.

69. Position the inner downshift lever between the downshift lever stop and valve when reinstalling the control valve body. Torque bolts to 80-120 in.-lbs.

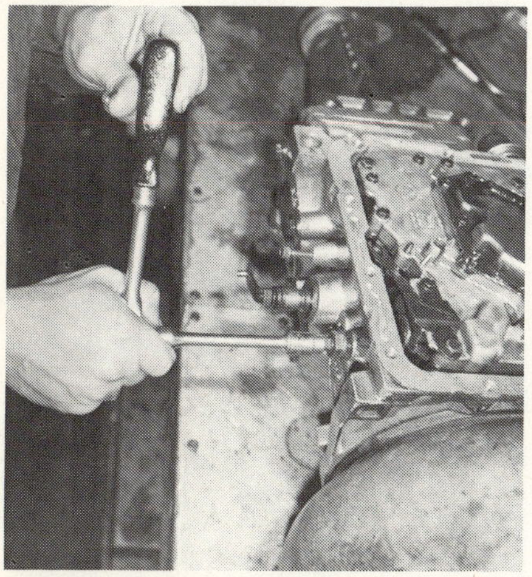

70. The intermediate band adjustment is made by carefully tightening the adjuster all the way in and then backing it off exactly 1¾ turns.

71. The adjustment of the low/reverse band is made in exactly the same way. Adjustment screw lock nuts should be torqued to 35-45 ft.-lbs.

72. Next, install the new oil pan gasket, and fit pan in position, torquing pan bolts to 12-16 ft.-lbs. The C-4 overhaul is now completed.

HOW-TO: Turbo Hydra-Matic Overhaul

1. Begin Turbo Hydra-Matic 350 overhaul by unbolting and removing the bottom pan. Pan gasket should be taken off and discarded.

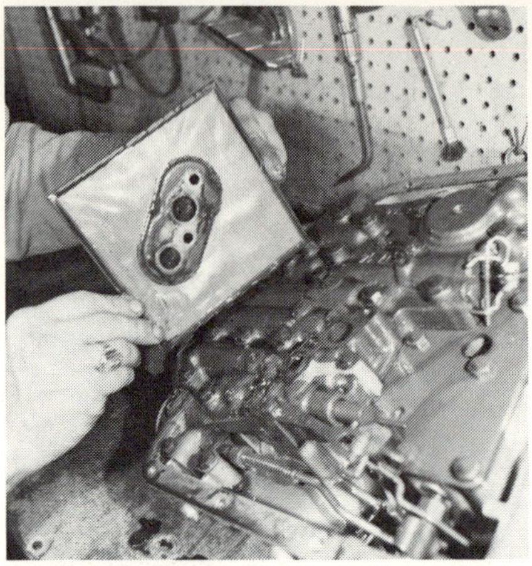

2. Two retainer bolts hold oil pump strainer assembly. Size and shape of strainer differs, depending upon model and year of transmission.

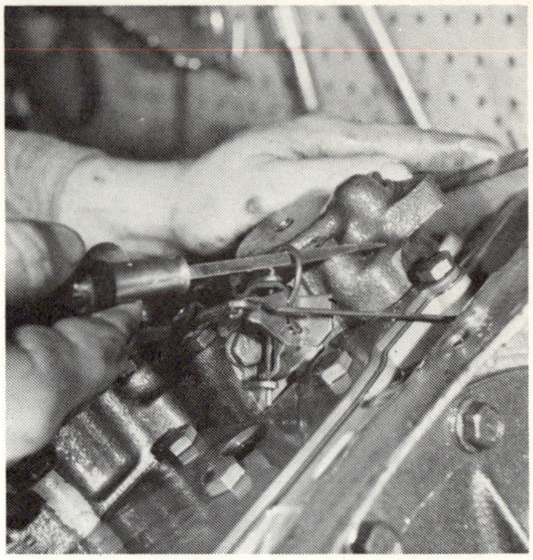

3. Unclip and remove detent control valve wire. Wire extends through side of trans body and a seal fitting must be unscrewed to pull wire out.

4. Unbolt and remove the control valve assembly. The exact configuration of this unit also differs according to year and model of transmission.

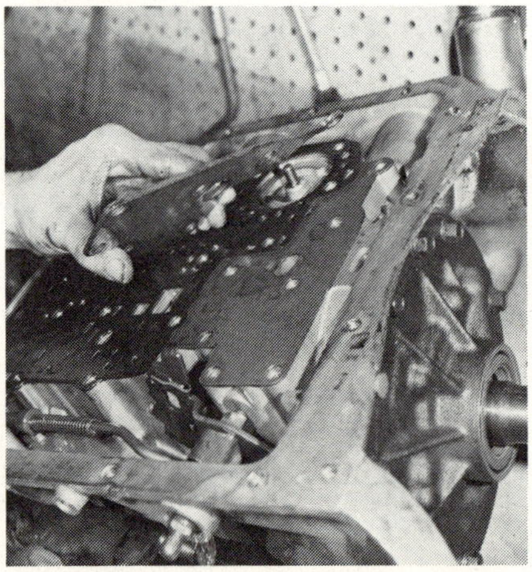

5. Transfer plate is a 2-piece affair on Hydra-Matic 350. The support plate is removed first and then the larger spacer plate comes off.

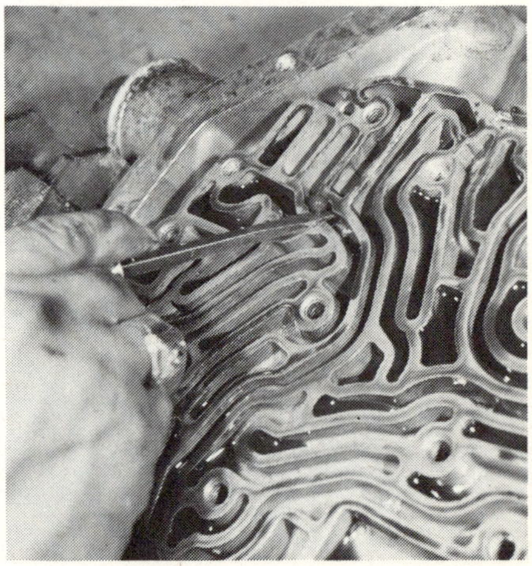

6. They're using rubber instead of steel check balls these days and a magnet proves useless here. Locate and pick all four from cored passages.

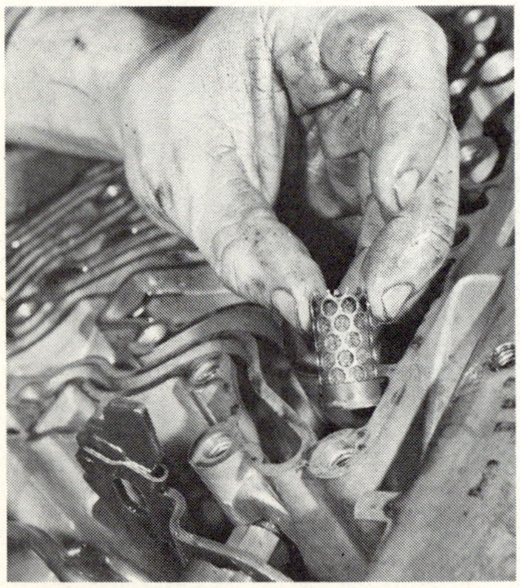

7. Oil pump pressure screen lifts out from the oil pump pressure hole in trans case. Clean all parts with solvent and air dry—do not wipe with a cloth.

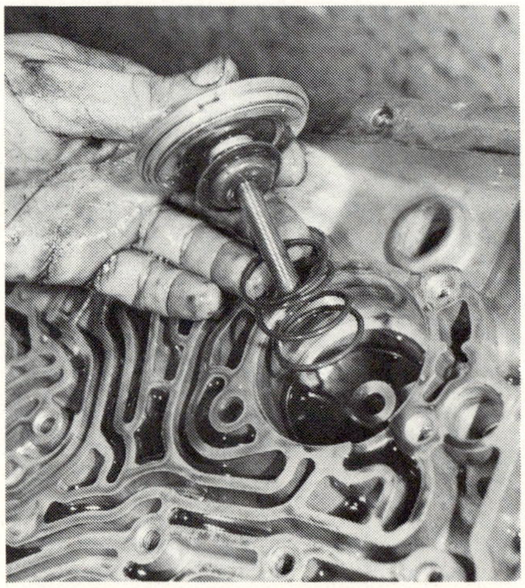

8. Remove front servo piston, pin and spring. Check piston for cracks, porosity and damaged oil ring groove; make sure servo pin is not damaged.

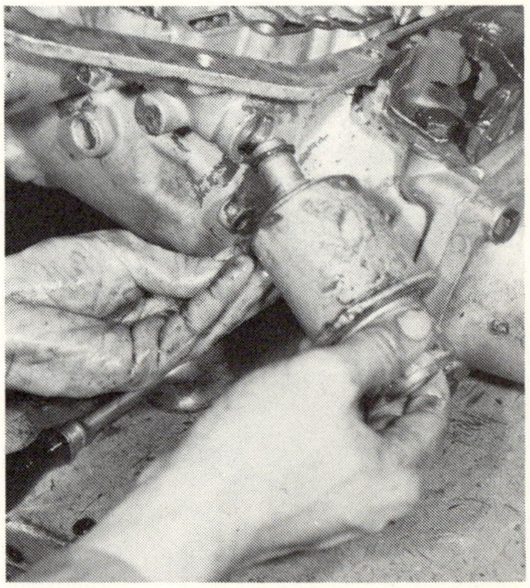

9. The vacuum modulator assembly pulls out when the retainer is removed. The "O"-ring seal should be replaced with a new one.

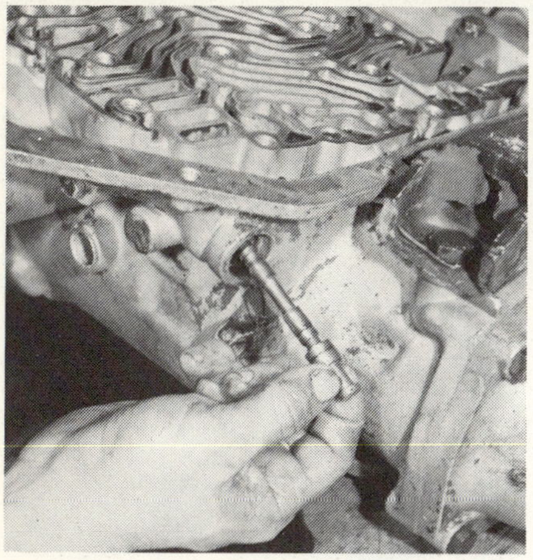

10. The slip-fit vacuum modulator valve can now be withdrawn from the trans case. Clean and inspect it carefully for burrs and other defects.

11. Next you can unbolt and remove the extension housing. Peel the old gasket off and discard it. A new one should be used for reassembly.

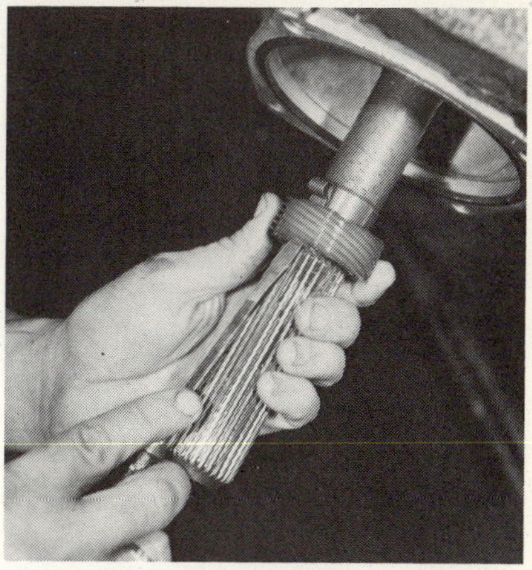

12. Speedometer drive gear is held in place by a spring clip. Depress rear of clip with screwdriver and gear can be pulled off easily.

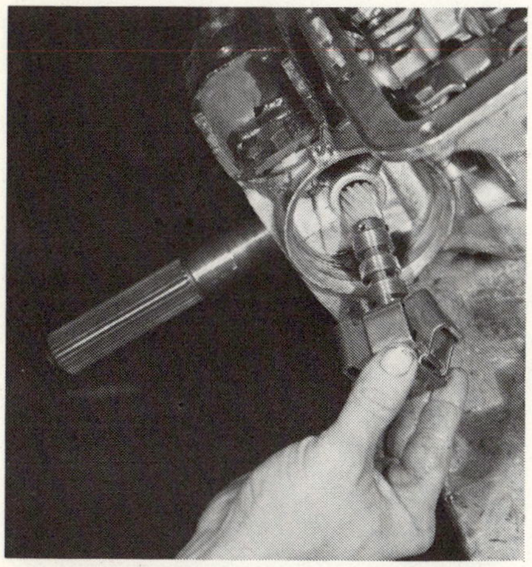

13. Pry off large spring clip holding governor cover in place. Remove cover and pull governor out of case. Larger Hydra-Matics use a bolted-on cover.

14. To remove the pump assembly, you'll need a slide hammer of some sort. This one threads directly into pump; others use separate bolts.

15. Draw pump off the main shaft; then remove and discard the pump-to-case seal ring and gasket. Place to one side for further diassembly later.

16. Remove intermediate clutch cushion spring, clutch faced and steel reaction plates and clutch pressure plate. Check clutch plate condition.

17. The intermediate overrun brake band is pulled out next. Check band surface carefully for signs of burning, scoring or cracked areas.

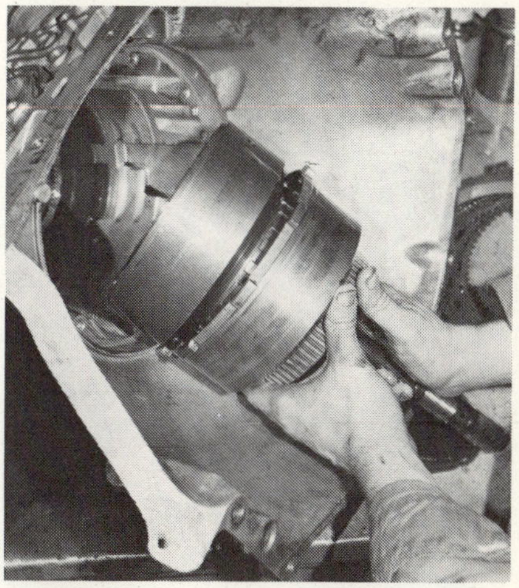

18. Direct and forward clutch assemblies are removed by withdrawing the input shaft. Set to one side for further disassembly later.

HOW-TO: Turbo Hydra-Matic Overhaul

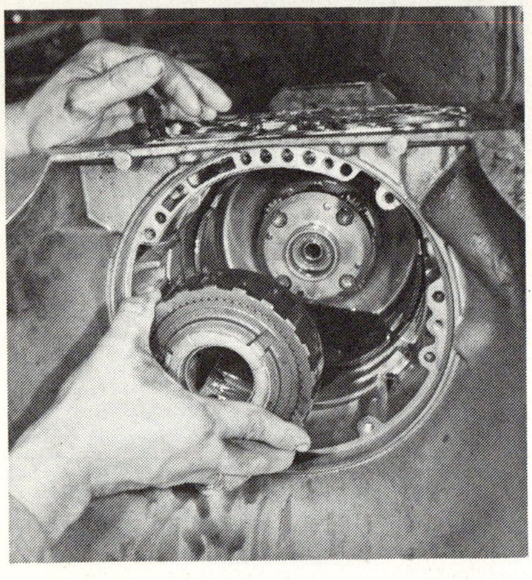

19. Planetary gear train disassembly begins with the input ring gear. There's a 3-tang thrust washer at the front of the ring gear.

20. The next step in disassembly requires the removal of a snap ring holding the output carrier in position on the output shaft.

21. With the snap ring removed, the output carrier lifts out fairly easily. Be sure to inspect all the gear teeth for worn or damaged areas.

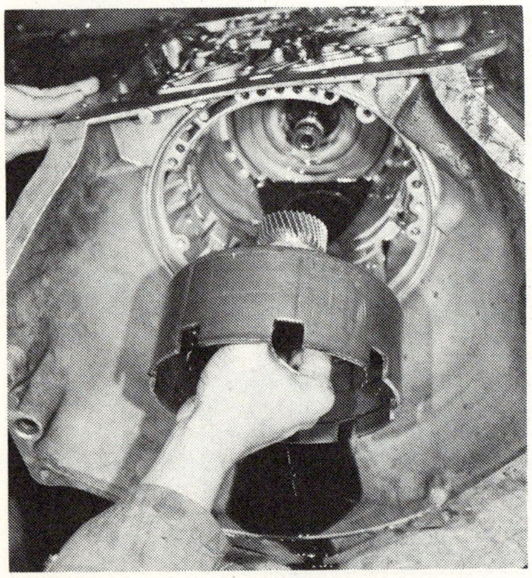

22. The sun gear drive shell with sun gear assembly comes off output shaft. If gear is damaged, remove from shell by unclipping snap ring.

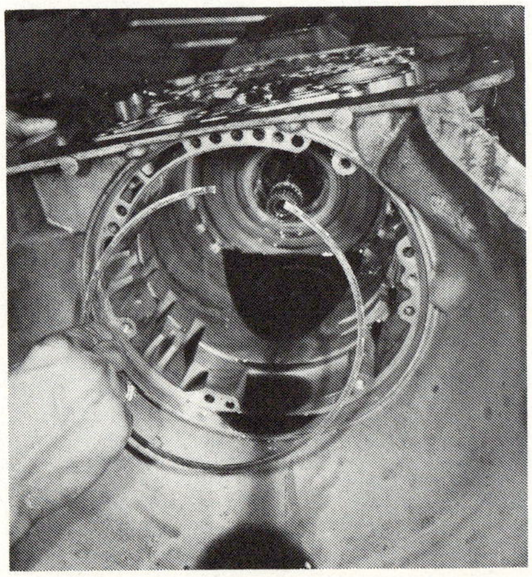

23. A large selective snap ring holds the low/reverse clutch assembly in place and must be removed before the clutch can be taken out.

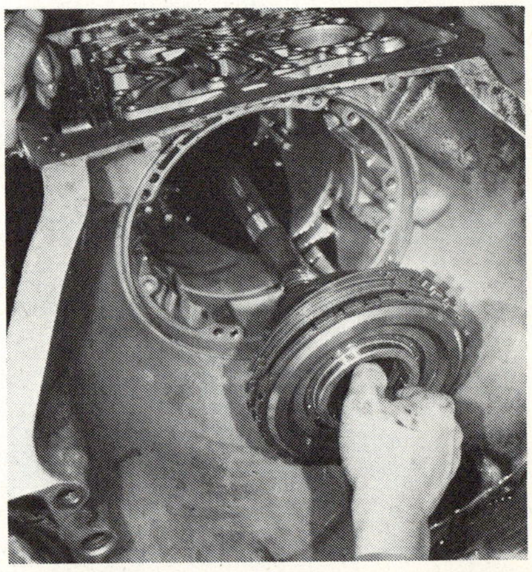

24. Low/reverse clutch assembly and support are final units to be removed from trans case. Place to one side and begin work on direct clutch.

25. Retaining ring must be removed from direct clutch in order to get to clutch plates. Pry ring up and out with a screwdriver.

26. Remove the pressure plate, faced and steel separator plates and cushion spring. Inspect all plates carefully as this is a common trouble area.

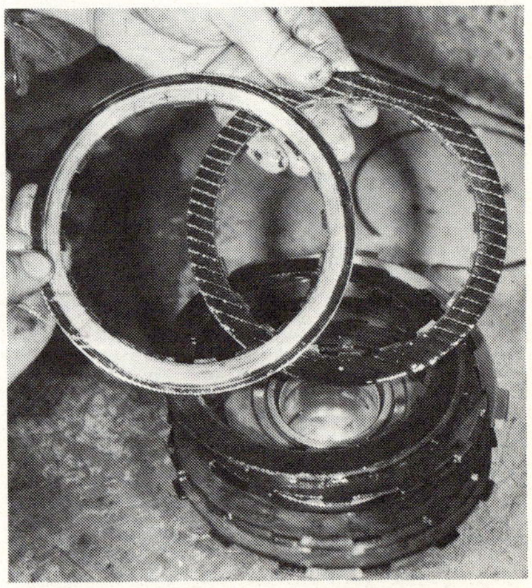

27. While both of the plates shown are worn, one side is really gone, pointing up the fact that a thorough examination should be made.

28. To remove the direct clutch piston assembly, you'll need a spring compressor that will let you remove the snap ring without difficulty.

29. Once the snap ring is off and spring compressor removed, the piston return spring seat lifts off, exposing the 17 clutch return coil springs.

30. Remove and replace the outer and inner seal on the piston, as well as the center seal on the direct clutch drum. A little trans fluid helps here.

31. The next step is to reseat the clutch piston in the drum by revolving a feeler gauge between the piston inner seal and the drum center seal.

32. Replace return springs and seat, then use compressor to reinstall snap ring. Install cushion spring and one separator plate, then alternate others.

33. The same procedure applies to forward clutch assembly. Inspect surfaces carefully for burning, scoring or wear. If one is bad, all must be replaced.

34. A hydraulic press can take the place of a compressor to help in snap ring removal, as well as pressing out the output shaft.

35. The piston return seat is lifted out after snap ring removal, exposing 21 springs. Be sure and check to see if any are distorted or collapsed.

36. The forward clutch piston can be removed once springs are taken off its top. As with direct clutch piston, seal rings are important here.

HOW-TO: Turbo Hydra-Matic Overhaul

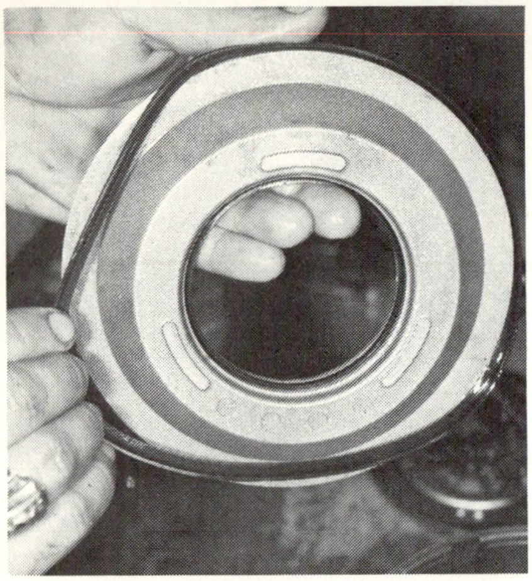

37. Remove and replace forward clutch inner and outer piston seal rings. Inspect piston for cracks and check housing for wear or scoring.

38. When reassembling the direct and forward clutch units, this cushion spring is fitted in place before inserting the faced and steel plates.

39. Loosen oil pump cover bolts and lower the pump body to the bench. Remove spring retainer and the 30 intermediate clutch return springs.

40. Intermediate clutch piston uses an inner and outer seal ring. Remove and replace both with new seal rings dipped in transmission fluid.

41. Lift the pump cover/stator shaft assembly off. Check the drive and driven gear teeth for nicks or other damage and the body for nicks or scoring.

42. Oil pump priming valve and spring are important, as are cooler by-pass valve seat, spring and check ball (1). Remember where they fit.

43. Replace gears, pump cover, piston and springs. Lift up body and insert attaching bolts through spring retainer. Torque to 17 ft.-lbs.

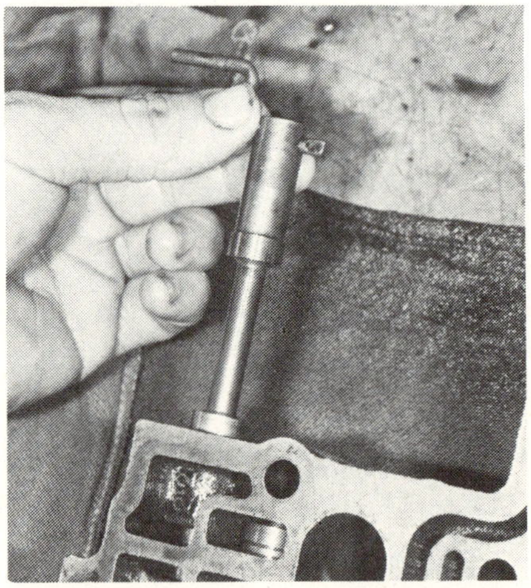

44. The manual valve is a slip-fit and can be lifted out of valve body as shown. Inspect this and other valves for cracks and free movement.

45. Valve body uses retaining pins to hold valve parts in bore. Press in slightly on valve and lift each pin out with magnetic pencil.

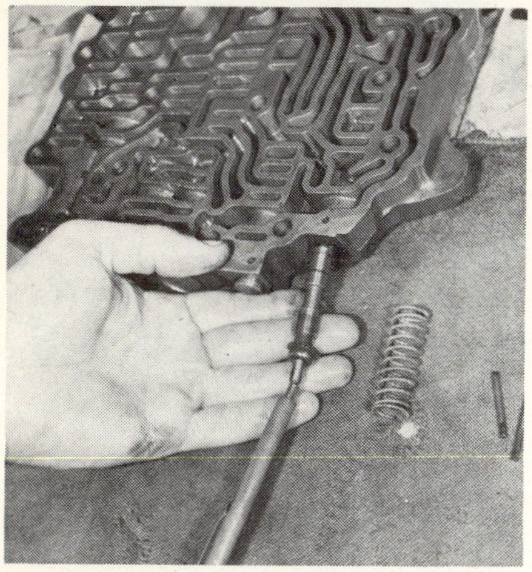

46. Detent regulator valve, spring seat and spring are located here. Line up each set as removed to avoid confusion during reassembly.

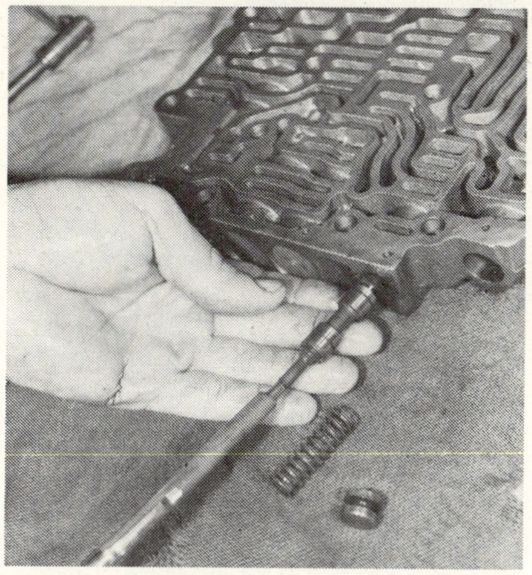

47. The manual low control valve, spring and plug come out next. Check carefully for scoring after cleaning valves thoroughly in solvent.

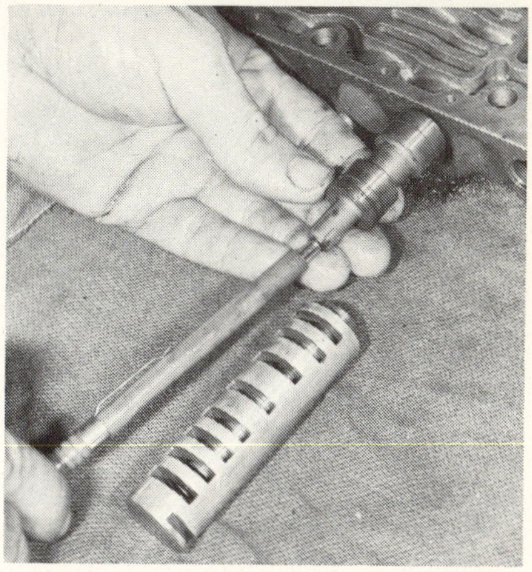

48. The 1-2 shift valve, shift control valve, spring and sleeve are the third set to be removed. Magnet shown is handy to draw valves out of body.

49. If you have difficulty in removing the sleeves, carefully insert a screwdriver as shown and pry from one notch to the next until sleeve is free.

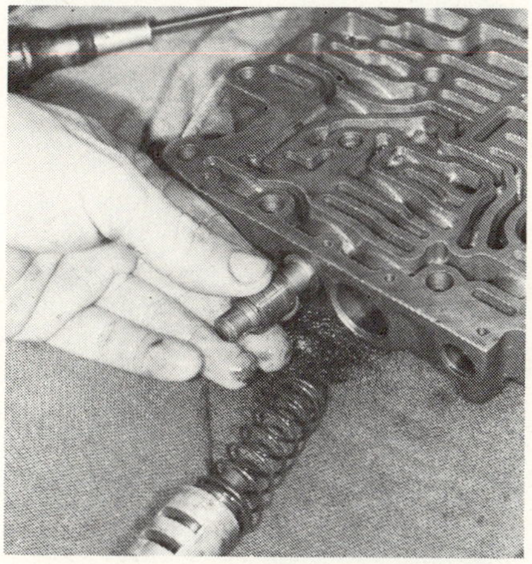

50. The 2-3 shift valve, spring, control valve, spring and sleeve come out here. Sequence can get a little confusing with this and next valve.

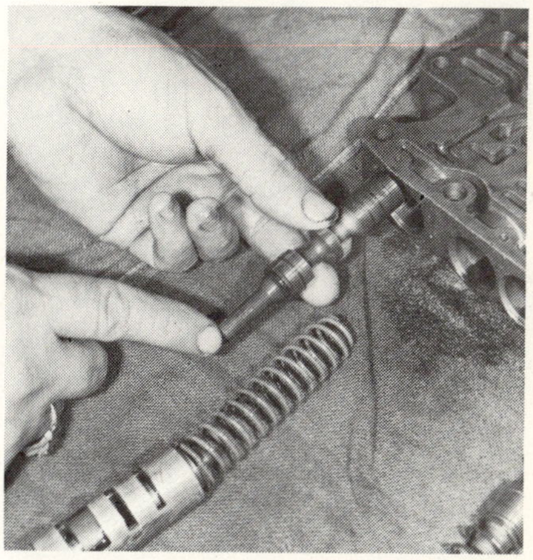

51. Remove the pressure regulator valve, spring, reverse and modulator boost valve, intermediate boost valve and sleeve. Clean and reinstall all valves.

52. The output ring gear and output shaft are inserted in trans case to begin reassembly sequence. Be sure tanged thrust washer is in place at front.

53. Notched washer will only fit into case one way. Rotate until notches line up with grooves inside case and fit against ring gear.

54. Now install the reaction carrier and set up against the large notched washer until the rear planet assembly fits into ring gear.

HOW-TO: Turbo Hydra-Matic Overhaul

55. Low/reverse roller clutch assembly is seated into position next. Rotate until notches fit into case grooves, and it'll go in place easily.

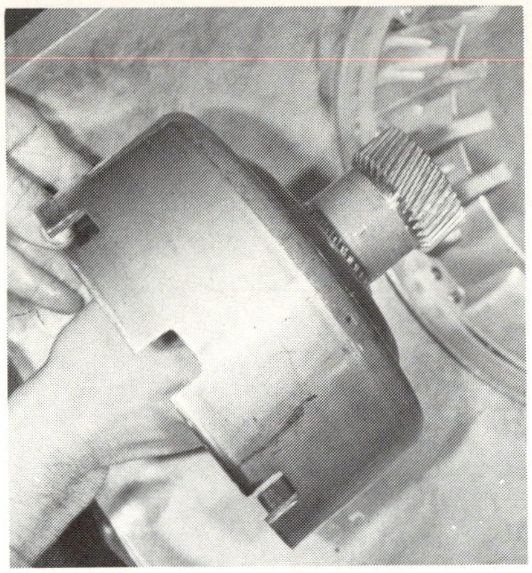

56. Install the sun gear and drive shell on output shaft. Rotate if necessary to mesh the sun gear properly and push back into place.

57. Replace the output carrier-to-output shaft snap ring. Make sure that snap ring is seated completely before proceeding with the next step.

58. Now you can fit the output carrier into the input ring gear and position in the trans case until it meshes with the sun gear assembly.

59. Position tanged thrust washer on the input ring gear; then couple the forward and direct clutch assemblies together and insert.

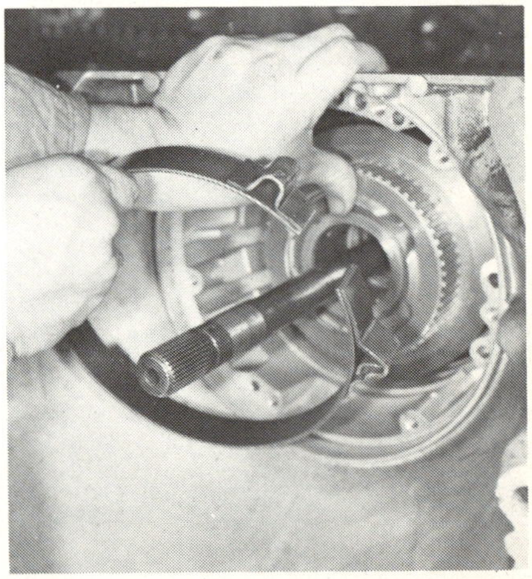

60. Slip the intermediate overrun brake band over the clutch assemblies and slide it into position. No band struts are necessary here.

61. Intermediate clutch is reassembled by installing cushion spring and steel plate, then alternate faced and steel plates with pressure plate last.

62. Replace oil seal rings on stator shaft if necessary and install oil pump/stator shaft assembly over input shaft. Torque to 18½ ft.-lbs.

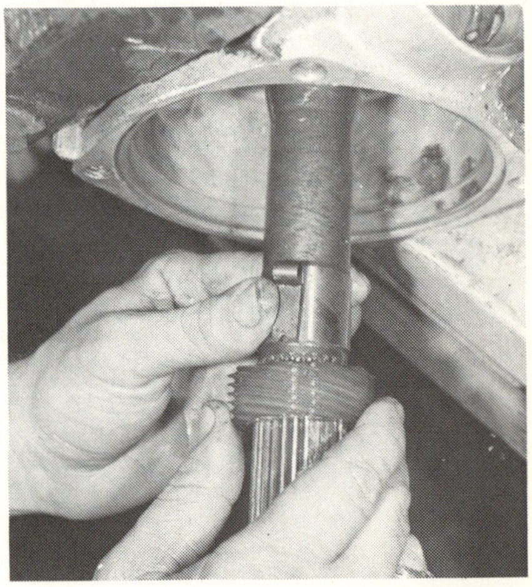

63. Fit pegged end of spring clip into hole in output shaft and slide speedometer drive gear on and over clip for accurate positioning.

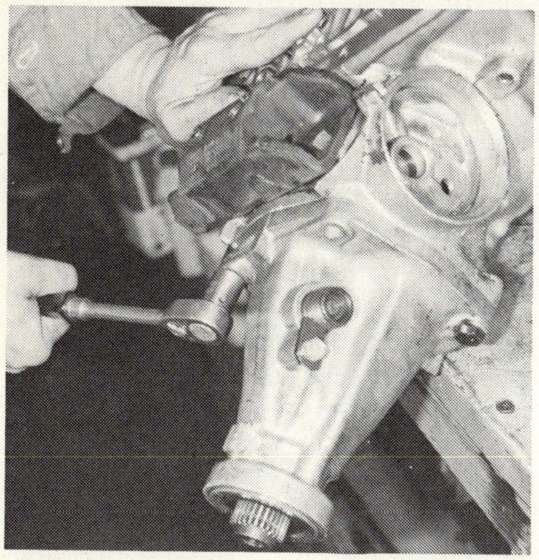

64. The next step is to replace extension housing on output shaft and bolt to transmission case. Bolts should be torqued to 25 ft.-lbs.

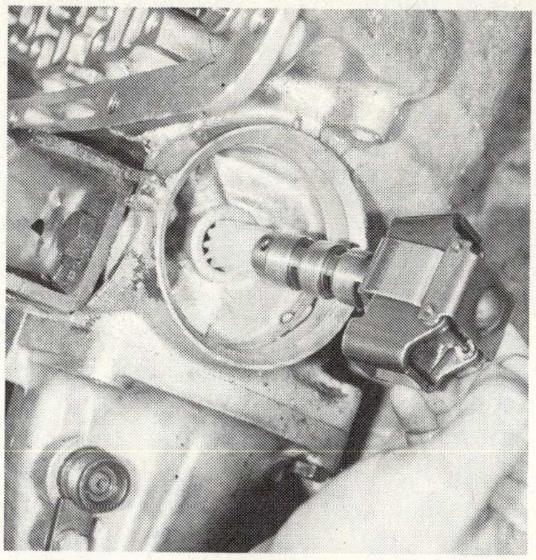

65. Check governor bore and sleeve for scoring and reinstall governor. Fit cover in place (use new "O"-ring seal) and attach spring clip.

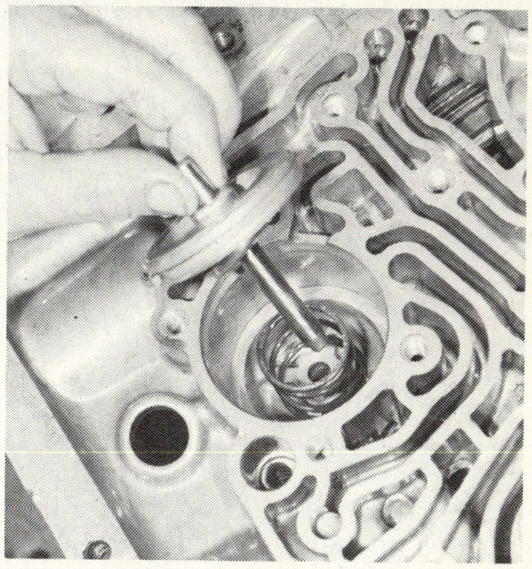

66. Intermediate servo piston, apply pin and spring seat are reinstalled over spring. It's a good idea to replace the piston seal first.

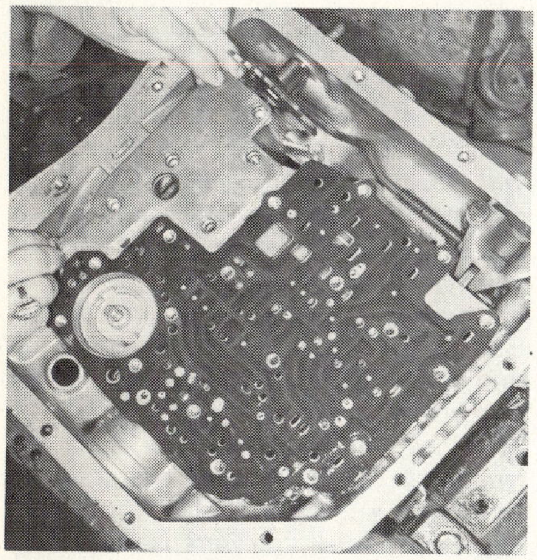

67. Replace the valve body spacer plate using a new gasket; then you can add spacer support plate and torque everything to 130 in.-lbs.

68. Reinstall valve body so that manual control valve link engages range selector inner lever. This should be torqued to 130 in.-lbs.

69. Replace oil pump suction screen or strainer to the valve body and torque to 40 in.-lbs. Screen is marked "Front" to prevent improper installation.

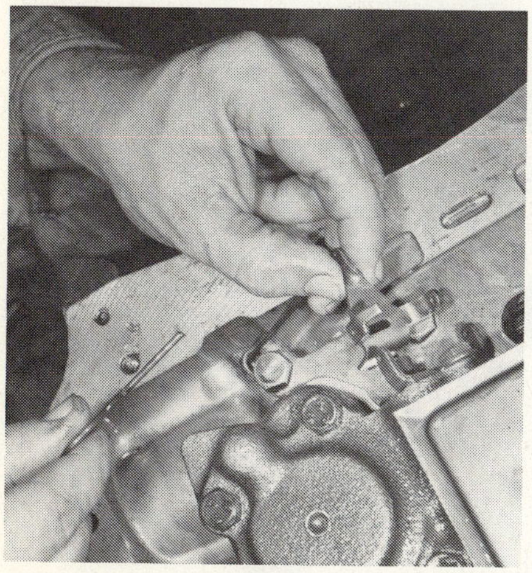

70. The detent control valve wire is inserted in trans case and reconnected to actuating lever. Retaining pin holds lever to bracket on valve body.

71. Vacuum modulator valve is inserted at rear of case, with modulator assembly and "O"-ring seal following. Tighten holding bracket.

72. Install new pan gasket and rebolt the oil pan to trans case. Torque to 130 in.-lbs. and Turbo Hydra-Matic 350 overhaul is complete.

Bringing Overdrive Back

The once-outdated OD's are a "new" road to fuel economy. Here is how one was put in a new Gremlin V-8.

The late model fuel consumption blues that people are singing these days are going to cause a lot of switching from the big, slinky boulevard plush-mobiles to something closer to motorized skateboards. Whether the actual shortage of fuel remains a reality or not, it's a safe bet that gas prices are going to balloon sky-high. A hasty switch from a big car to a smaller, more economical one will save fuel dollars, obviously, but is it worth the reduction of comfort? This is something to muse while thinking about an interesting fact; until as late as 1969 there actually were big American-built, V-8-engined cars that delivered as high as 30 miles per gallon, day after day after day. Of course, this was before power- and gas-robbing emission devices came on the scene, but it was also due to an interesting little device which, unfortunately, can no longer be opted for; an overdrive transmission.

The overdrive unit has kind of a funny history in that it's been on again, off again since it was first previewed by the Chrysler Corporation on those bulbous Airflow models of 1934. The Airflow met an almost instant death because of its radical styling but the overdrive lived, although dwindling, until it was finally discontinued by Chrysler in 1942.

Because of the nature of the thing, overdrive was a system better suited to an accessory gear manufacturer than a primary automobile builder, and one of the best-remembered OD makers was the Columbia Gear Company. They offered units adaptable to many makes of cars, but their biggest users were Ford owners. Although the Columbia overdrive was never actually offered factory-installed by the Ford Motor Co., (any Ford dealer could install it for a nominal charge, of course) in a Ford (or Mercury) beginning about 1935. Despite their idiocyncrasies (the vacuum-operated controls left much to be desired, especially in the shift time lag and severe gear clash if it went "in" at the wrong time), they were loved by hundreds of thousands of Ford-ists.

After World War II, when Ford Motor Co. left the horse and buggy age for good with the elimination of its cross-spring automobile, a "good" overdrive appeared on a Ford. That was in 1949. It was not until 1952 that Chrysler Corp. got back on the overdrive band wagon. But that didn't last very long, and in 1959 overdrive appeared on Chrysler Corp. cars for the last time. Ford stuck it out until 1967 and then finally gave up. Chevrolet, which was without overdrive for many years and still sold more cars than anybody else, eventually got with it in 1955 but dropped it with their 1969 line.

American Motors was another big user of overdrives, their units supplied by the Borg-Warner people. But alas, 1969 also marked the last year that AM could sell you its gas saver.

Several things lead to the OD's demise. One was the popularity of the mighty musclecars and the 4-speed gearbox where mph meant nothing and super acceleration and high engine revs were definitely the "in" things.

The main disadvantage of an overdrive is its free-wheeling feature. In a manual transmission car the driver expects the engine to perform a certain amount of braking when he takes his foot off the gas pedal. Below approximately 25 mph, a car with an overdrive free wheels. This means that engine compression has no effect on slowing the car. It's just like putting the transmission in neutral and coasting. That alone meant its phasing out, at least in high-performance applications.

But because the use of OD means improved gas mileage, since when it is engaged the engine turns as much as 30% slower at a given rate of road speed, overdrive is once again the way to go. Not that any manufacturer is (yet!) again offering OD as an option. But listen: since optional OD's were dropped in '69, many of these now-old cars are languishing in wrecking yards. Why not, then, nab a unit and bolt it in a present-day car? Why not indeed, and that's the real point of this tale. As an example of what can be done, given the impetus, we'll describe the relatively easy chore of installing an older Borg-Warner unit (from a 1969 AM Classic sedan) into a V-8-powered Gremlin. If you're not interested, you should be. Before the swap the owner was getting 16.4 mpg average during normal city Los Angeles) driving. After the changeover to OD, the otherwise-identical car in the same state of tune, same rear-end ratio, same driving route, etc.), delivered 21.7 mpg. *Now* do you want to read on?

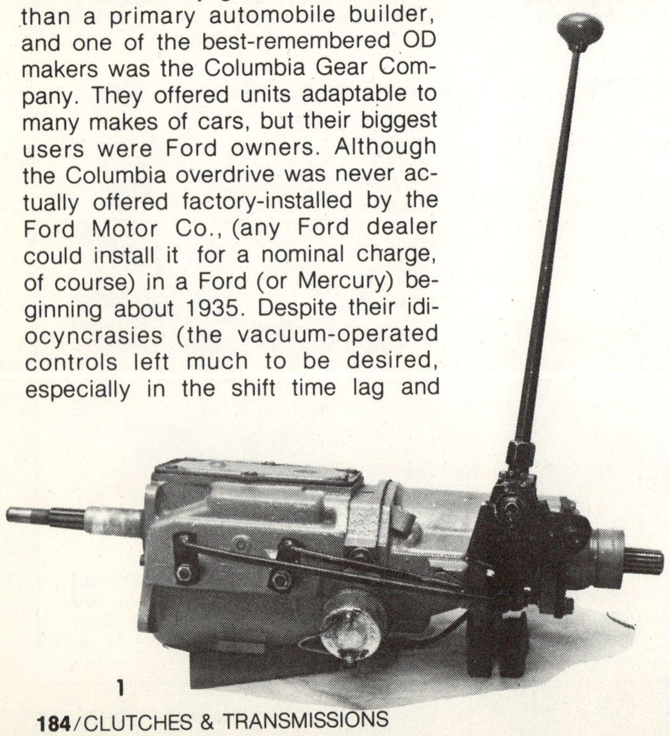

1

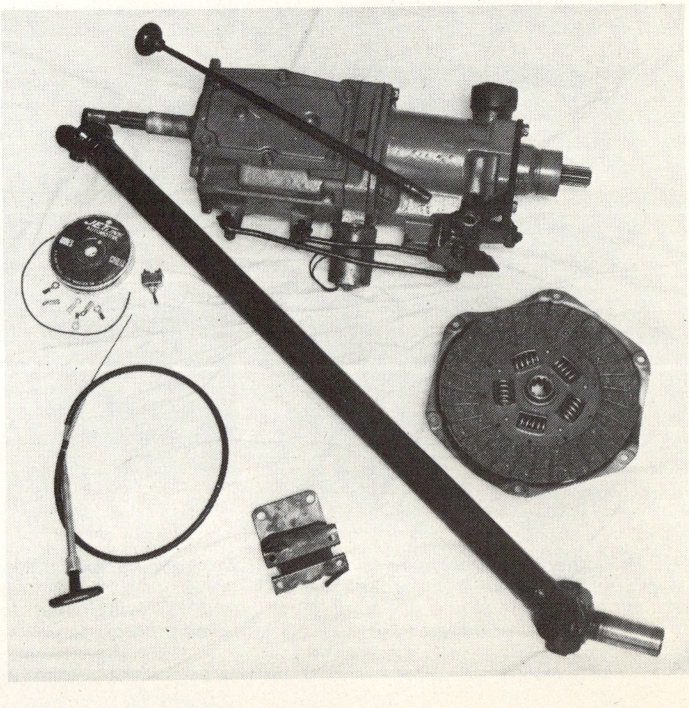

2

GETTING IT TOGETHER

Before making the purchase of a used overdrive transmission that has the correct bolt pattern to fit your engine, take the measurements of the overall transmission size to see if the drivetrain tunnel will accommodate the needed clearance around the overdrive, the governor and the solenoid. A transmission shifter will probably have to be built because overdrive floor conversion shifters are almost nonexistent. You must also modify the stock transmission crossmember and mount, or be able to make them. Find the proper yoke, and, if the driveshaft must be modified, it is better to shorten than to add on to one, because a one piece is sturdier. Unless it is a low mileage, heavy-duty unit, it is a good idea to replace your clutch and pressure plate to insure a positive hookup. Either a cable or linkage shifter may be used on the overdrive lever, depending upon the type of placement that is desired.

At least four spools of color-coded wire, connectors, some type of a four-attachment disconnect switch, electrical tape and a wiring schematic of the car that the OD came from, are needed in order to connect the electrical system. This job is tricky and may be frustrating if you don't know how to do it.

Finding the items needed for the installation may be a hassle. A person like Bill Hohman (Schugg's Auto Salvage, 1310 E. Lomita Blvd., Wilmington, Calif., (213) 830-6880) in this case was a real help. Because of his knowledge of AMC/Rambler parts, he was able to find which components were going to work for installing the overdrive in a 1972 Gremlin, equipped with a 304-cu.-in. V-8 and manual 3-speed transmission.

INSTALLATION

Before unbolting the transmission, but after the driveshaft is pulled out, be sure the output shaft is plugged so that no gear oil leaks out onto the floor or on your face when tilting the transmission for removal. The other choice is to drain the transmission before bringing it out. A hydraulic jack and wooden block under the engine keeps the weight off the motor mounts. The unbolted crossmember may be slid back on the exhaust pipes, out of the way, thus leaving the exhaust system intact. In the Gremlin the aluminum bellhousing and transmission may come out as one piece from the back of the engine. At this point the clutch and

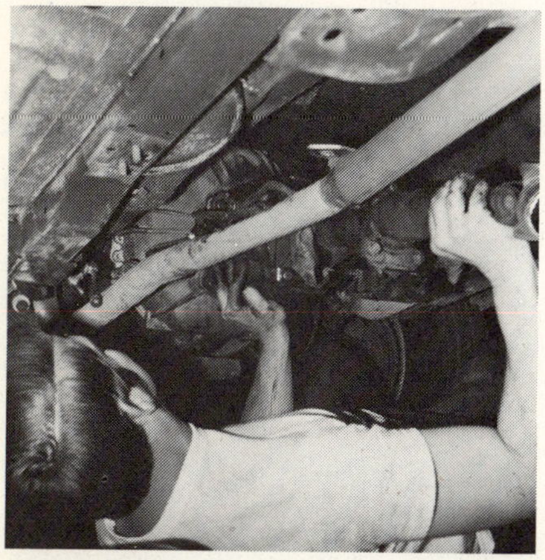

3

4

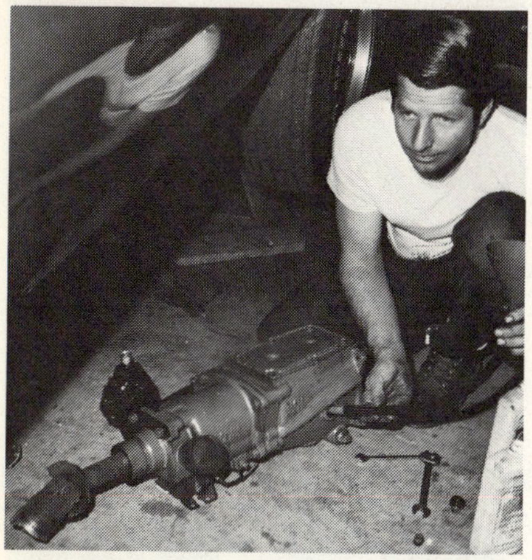

5

6

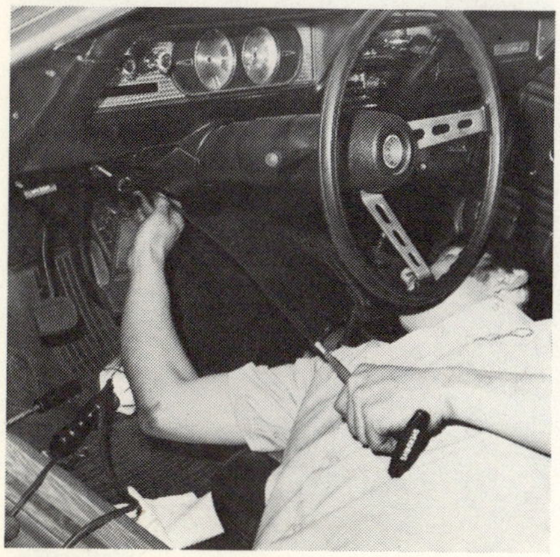

7

8

1. The Borg-Warner T-14 overdrive and transmission is a single-unit, all-synchro 3-speed with 30% overdrive ratio. Last produced in 1969 and available as an option on some AMC models, they are a little hard to find, but worth it for performance economy.

2. We're illustrating the installation of the older '69 OD/trans unit into a late Gremlin (which already had a 3-speed trans), and these were the parts used. Clutch and pressure plate were replaced as a matter of course.

3. Our parts and info source was Bill Hohman, here removing the Gremlin 3-speed manual box that was bolted to the 304 V-8. Trans and bellhousing came out as a unit. Engine is steadied with a jack.

4. Moon Mullins (8217 E. Compton Blvd., Paramount, CA) specializes in unusual shifters, and he's the one that rigged up our overdrive.

5. Bill demonstrates how to top-up the trans if you want to make the job easier than doing it on the car. A discarded yoke and a rag (which doesn't show here) keeps lube from spilling out around output shaft. Other openings (the speedo drive and assorted bolt holes) should also be stopped up.

6. We were able to keep Gremlin's exhaust system intact, though on many cars it has to come off for rear crossmember removal. Wood block and floor jack keep the engine where it belongs.

7. A '67-'69 Rambler American OD cable is run down beside steering column and back under floor to trans lever.

8. Pull-handle was bolted to right of the Gremlin steering column for convenience.

pressure plate are removed and checked. It is a good idea to have an old spline to realign the clutch when putting the unit back together.

Unless there is a way to pump gear oil into the overdrive and transmission, both housings must be filled on the floor before installation. Don't forget to plug any openings to prevent oil loss.

With the transmission bolted up, the shifter can be put on, after a hole is opened up in the floor. Bill used an overdrive mount from a 1969 Ambassador, "because it bolts right up to the Gremlin crossmember and does not require modifying." When it was time to find a driveshaft, Bill's know how really came through. All he had to do was to go out into the yard and get a 1970-71 Gremlin *automatic* transmission driveshaft. It is exactly the right length with the yoke and spline teeth matching perfectly. To allow the stock crossmember to slide forward the necessary ¼- to ½-in., in order to bolt to the mount, slots must be made from the holes where it bolts up to the frame. Since the transmission-to-tunnel clearance is critically close, two aluminum pads are made to bring down the tail of the transmission. They are .300-in. ins. thick and measure 3 ins. long by 3 ins. wide. Bill suggests that you use different weight pieces of cardboard as templates to find the right thickness plate to be needed, after marking the bolt distances. The template should be offset ¼-in. toward one end of the plates before drilling 7/16-in. holes. This is to compensate for the amount the crossmember will be moved. A piece of the crossmember must be cut out at the point where the shifter levers interfere. Now that the angle of the drivetrain is changed slightly, the fan shroud must be moved down on the radiator to prevent interference. This is done by holding the unbolted shroud in position and drilling new holes.

CABLES AND WIRING

The Gremlin's speedometer cable reaches and reattaches with no problem. Bill decided to use an overdrive cable and bracket from a 1967-69 Rambler American and needed no modifying whatsoever It attaches to the right of the steering column, at the bottom of the dash, and runs through the column's dust cover into the engine compartment and under the floorboard, where it connects onto the overdrive lever.

Wiring, a word that usually brings the jitters, is not too difficult to make up on the overdrive as long as a schematic is followed and the wires

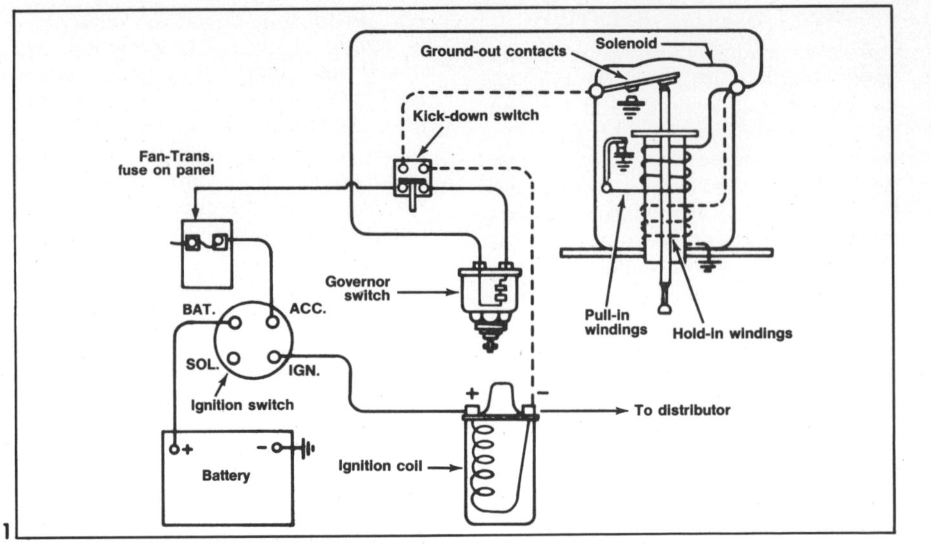

1

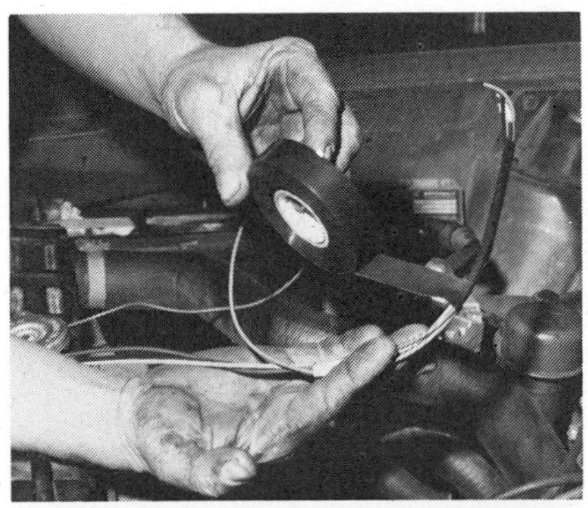

2

3

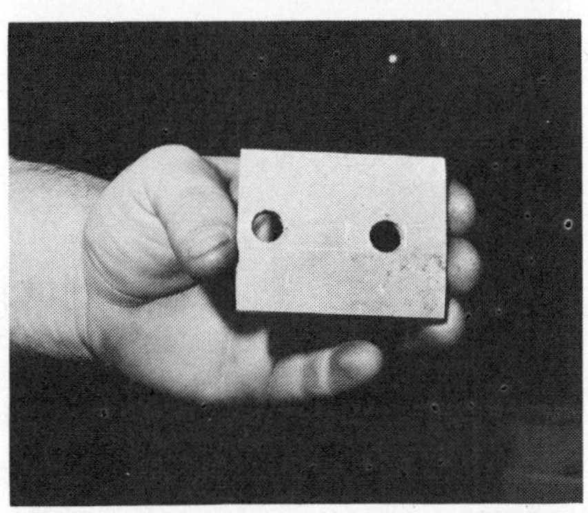

4

5

6

are color coded for their purpose. The only item which must be added to the system is a 4-point hookup control switch so that the overdrive may be disengaged when passing gear is needed. A '67-69 American throttle "kickdown" switch is used on the Gremlin, using a special bracket mounted on the left rear of the carburetor base plate. Leading down from the distributor to one of the contact points on the de-energizing side of the "kickdown," and then from the other contact to the ground out contact, is one color of wire. The second wire goes from the distributor to one contact on the energizing side of the "kickdown" and then to the governor. The third wire is from the heater fuse to the other energizing contact of the "kickdown." The fourth wire runs from the governor to the "hold-in" wiring contact on the solenoid. A wiring harness may be made with electrical tape, making a quality job of bunching the wires together. Tabs should be used to hold the harness securely in place under the floorboard and on the firewall.

ALL SYSTEMS GO

The Moon Mullins shifting assembly comes with a shifter lever which can be easily removed, cut and bent to the needs of each driver's position. A sheet of pliable aluminum is pop riveted over the old floorshift hole.

The best way to test the overdrive system is out on the road where the power is to the pavement. Using the aforementioned operating procedure, the Gremlin was taken up to about 38 mph in second gear, where Bill backed off the throttle. After a hesitation and a click, the overdrive caught immediately. The sensation is like losing 15 mph of speed, but it was just rpm dropping because the speedometer still read nearly 40 mph. Next, the transmission was shifted into third gear direct drive) at 55 mph, where the overdrive remained engaged as it should. The miles per hour remained the same, but the rpm's were lower again. This results from the 3.54:1 ring and pinion ratio dropping to a theoretical 2.48:1 due to the effect of 30% overdrive.

Around-town driving with the overdrive engaged isn't recommended unless there are long stretches between stops. Every time the car is slowed below 32 mph the overdrive disengages and freewheels, providing there is no torque applied to the drivetrain. It is also hard on the balk ring and sun gear to constantly engage and disengage them. The whole operating procedure must be done over when power is again used until the speed reaches 38 mph.

To prove the benefits of overdrive, the Gremlin V-8 was put on a real-life mileage test. After 350 miles of around-town and freeway driving in Southern California, the figure was 21.7 miles per gallon. Using the same driving pattern, the best mileage previous to the overdrive swap was 16.4 miles per gallon. This is a gain of 5.3 miles per gallon *without* a featherfoot driver. Think what the overdrive could do for a street machine running a 4.56:1 rear-end. By engaging the overdrive, the equivalent of a 3.19:1 ratio could be used for obtaining highway economy.

At the Edelbrock Equipment Company, El Segundo, Calif., a speedometer and rpm check showed that with the overdrive disengaged, 55 mph equalled 2650 rpm's. Engaged at 55 mph, it read 1850 rpm. With an 800 rpm drop, the engine is obviously working easier on less fuel. It is also saving wear and tear on the entire drivetrain by reducing heat and friction. With the decreases in wear and fuel consumption, the overdrive is the way to beat the gas crunch.

7

1. It may be necessary to scrounge up an OD wiring schematic from the shop manual for the car from which the OD unit came. This one was used for our '72 Gremlin, is for the '69 Rambler series.

2. Once wiring is properly routed, it is neatly bundled into a loom using plastic tape. It not only looks better than a lot of loose wires, it adds insulation protection.

3. A bracket secures the "kickdown switch" to the carburetor. When trottle is depressed, switch is closed and the OD is deactivated. In a sense, it's like a passing gear kickdown.

4. Spacer was made to fit between trans mount and crossmember to bring driveline into proper alignment.

5. Area of Gremlin's floorpan had to be cut out to provide for trans shifter. It's often necessary to butcher things like this as a starter, but later a more closely-fitting, sheetmetal patch can be pop-riveted over the hole.

6. It's a tight squeeze without a hoist, but it's the way a lot of us have to work. Bill's winding the job up, which involves repositioning the radiator shroud; engine angle was changed somewhat.

7. It's close in the Gremlin, but the older AMC overdrive just clears underside of floorpan and bracing. Driveshaft and yoke from a '70-'71 Gremlin with automatic trans are a straight bolt-in.

8. It's in and it works, to amazement of AMC dealership mechanics who "didn't know Gremlins came out of Kenosha with overdrive as an option".

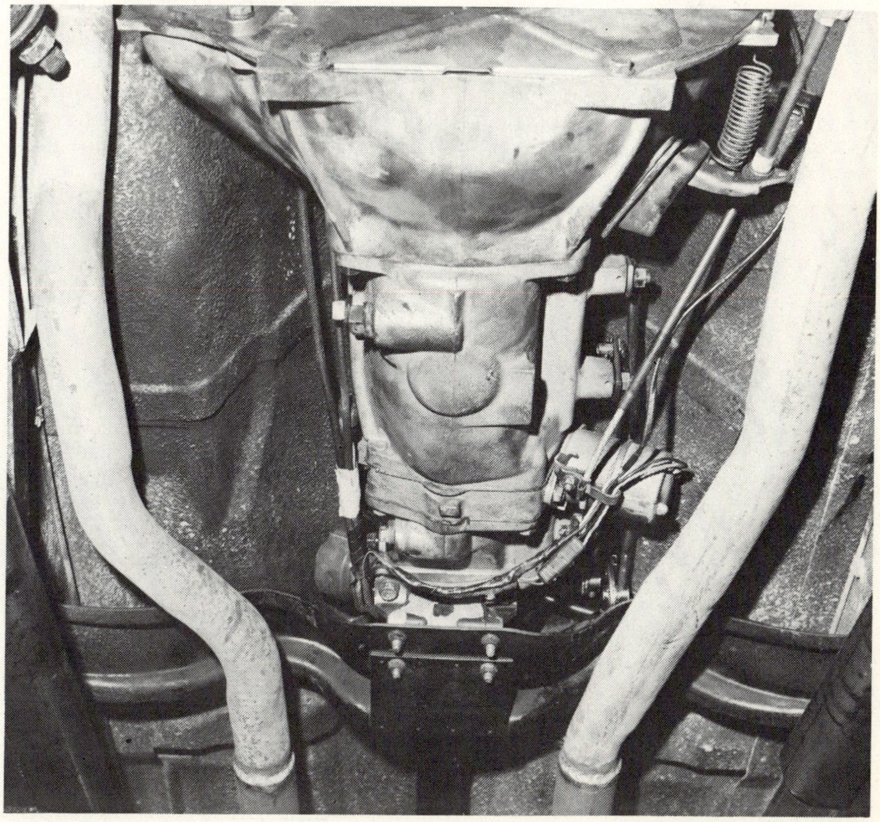
8

Pinto 4-Speed - Plus One!

Five-On-The-Floor for Ford's Ponycar.

One of America's premiere engine builders has decided to take the bull by the horns in the American economy car market.

By his own admission, Ak Miller in Pico Rivera, Calif., has always been "very interested in economy." He went to a 2000cc Pinto 4, put on some engine goodies for more power, but really laid it all out with the installation of a Toyota 5-speed. Ak thinks his 25 mpg will be an absolute minimum, which is pretty good, considering that he's running a blower and 2200cc's. But with a 3.18:1 Pinto automatic trans rear end, hooked up to the Toyota 5-speed trans, Ak estimates that his Pinto will be turning only some 2400 rpm at 70 mph. You might bend an ear, therefore, if power *and* mileage is your thing.

The Miller Pinto originally employed standard 4-speed gear ratios of 3.65:1 1st; 1.968:1 2nd; 1.368:1 3rd, and 1:1 4th. Into the car went the 3.18:1 rear; and the Toyota box. The particular trans is available either in the Corona and Celica, or the smaller Corolla—with an eyelash lower gearing, broken down as follows: 3.587 1st; 2.022 2nd; 1.384 3rd and a 1:1 4th. The big news, of course, is the .0861 5th, roughly a 27% reduction!

There are several advantages with the Toyota layout. It has an aluminum case, split vertically, and easy to open. The direct linkage 5th gear is something that Detroit would do well to study. Formerly, cost has precluded mass production, until the Japanese showed the way. To gain power and mileage Detroit found the accessory overdrive setup cheaper. The solenoid and rather undesirable free-wheeling effect carried the same 20% to 27% reduction, but needed a 4.11:1 or 4.27:1 rear end. All ratios were unbearably low. It was, in retrospect, the best of both possible worlds when working right; but easily degenerated into the worst of same when the solenoid quit—a not uncommon occurrence. Had Ak gone to such a layout, he would have had to make a new hind shaft and other modifications, a proposition so outweighed by the new Toyota 5-speed that there was only one way to go, and go he did.

MORE ADVANTAGES

The Toyota gears remain in the lower half of the case—another plus. The design uses servo-type synchros, rather than the more common cone-type, which will mean faster shifts. Normally, the Toyota 5-speed is hooked up to a 4:1 rear axle, so with the 3.18:1, there will be outstanding mileage. There is a slight whine in the Toyota 5th with the only-slightly-helical cut, but definitely of a minor nature. Some Toyota 5-speeds use iron cases when installed in trucks and for some types of racing, but the aluminum case is perfect in the Pinto.

Like all standard transmissions, there is a separate bellhousing, although removeable, with a throwout bearing and fork. Since Ak knew he was going to need an adapter plate and the Toyota shaft was a little long, he compared the bolt patterns of the Toyota bellhousing with the Pinto and came up with an adapter pattern. This was flamecut out of steel plate and Blanchard-ground for precise fit. Since the Pinto starter and flywheel were to be used, the ½-in. plate adapter soaks up the shock loads of the starter.

After the plate was cut, it was drilled and attached to the Pinto block with countersunk Allen screws. Dowel pins were pressed in to hold the bellhousing in alignment and bolts went into the bellhousing from the engine side of the adapter plate.

For a clutch, Ak had to go to a hybrid, taking the splined center from a Toyota clutch and machining it to fit into the center of a Pinto clutch disc, where it was heliarced into place. A stock Pinto pressure plate was utilized also. In order to use the Pinto throwout fork and bearing, Ak had to heliarc an aluminum boss inside the Toyota bellhousing for the pivot point. The only other modification necessary was to cut the starter-mounting area from an aluminum Pinto bellhousing and weld it to the Toyota housing.

This is a precisely built, well-engineered transmission, and adapts nicely to its new home in Ak's turbocharged, propane-burning Pinto. Most gas-conscious people have to go to an old 3-speed from a junkyard to get overdrive, but with the Toyota gearbox you still have the driveability of a 4-speed. What more could one want?

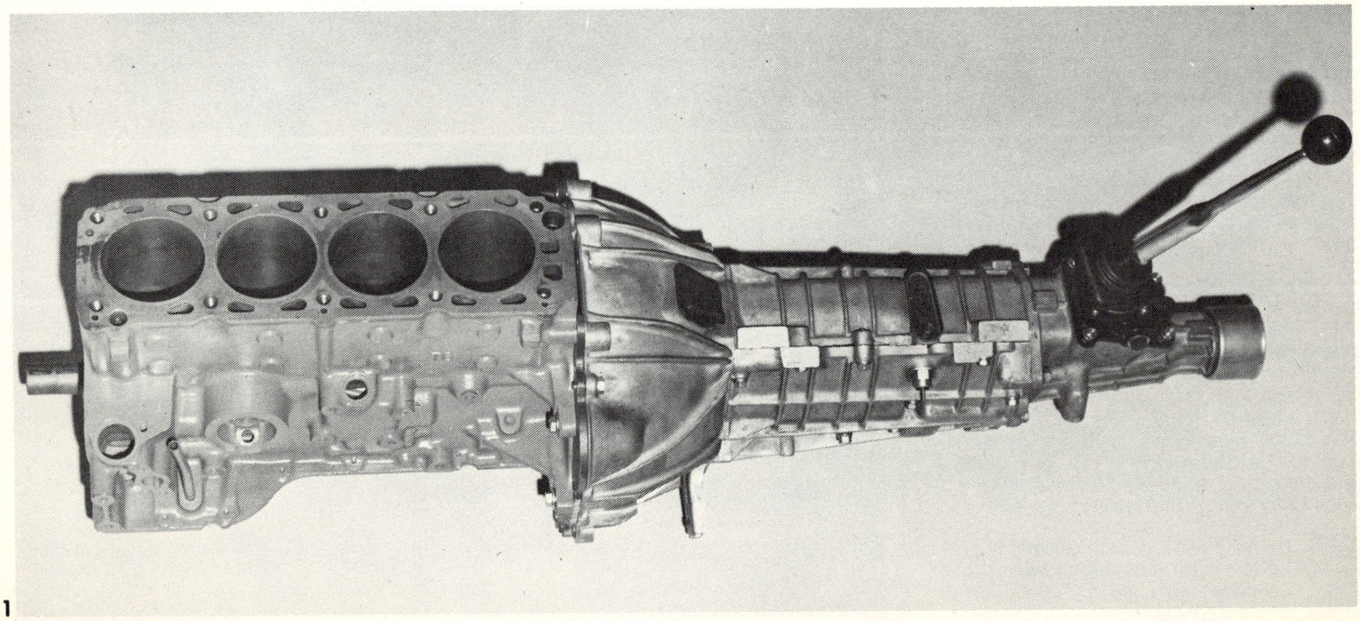

1

1. Top view of Ak Miller's Toyota 5-speed transmission, shows bell-housing mate to 2200cc Pinto. Fit is a thing of beauty.

2. Toyota 5-speed is from a Corolla and is split in two halves. That's a big plus for the guy who likes to know what's going on with a minimum of fuss. Note the improved clearance hole for the throwout fork.

3. The adapter plate was cut, and a superimposed Toyota housing is on top. Bolt patterns were countersunk.

4. Steel plate is ½-in. thick. It's secured with countersunk Allen screws. Note the new starter motor cut-out on lower right of photo.

5. The Pinto starter was used, and the stock Pinto cover is retained and welded to the bellhousing. Toyota position is directly opposite.

6. Good close-up of the pivot for the throwout fork. Notice the lineup of the bolt holes.

7. Top view of Ak Miller's hybrid Toyota trans with Pinto block. He was careful not to weaken the separate aluminum bellhousing. End result is a mileage-making runner with a blower that puts out 25 mpg-plus!

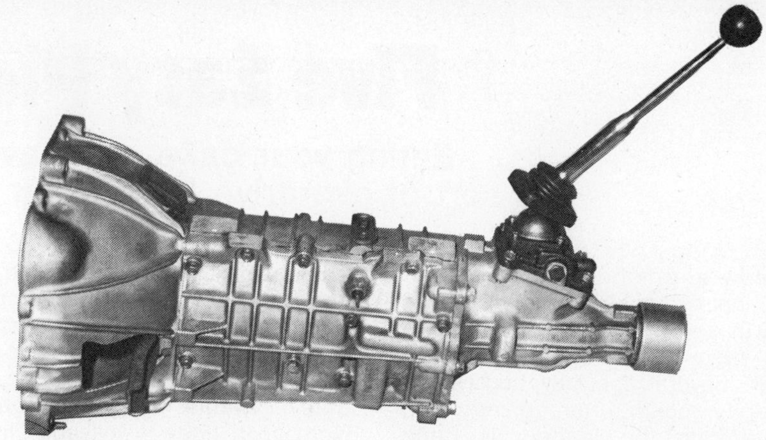

2

3

4

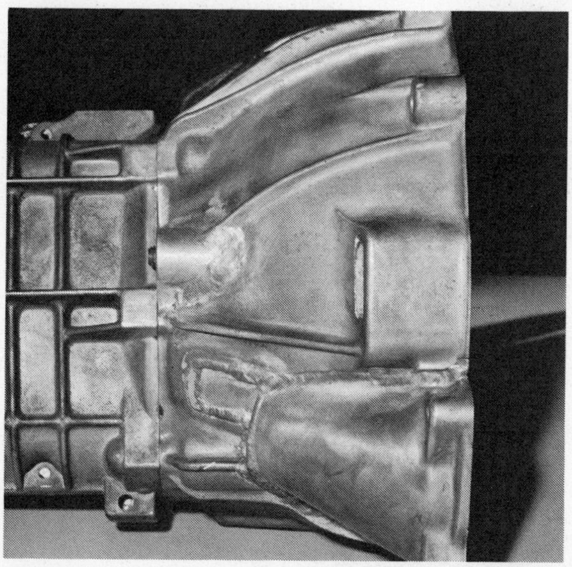

5

6

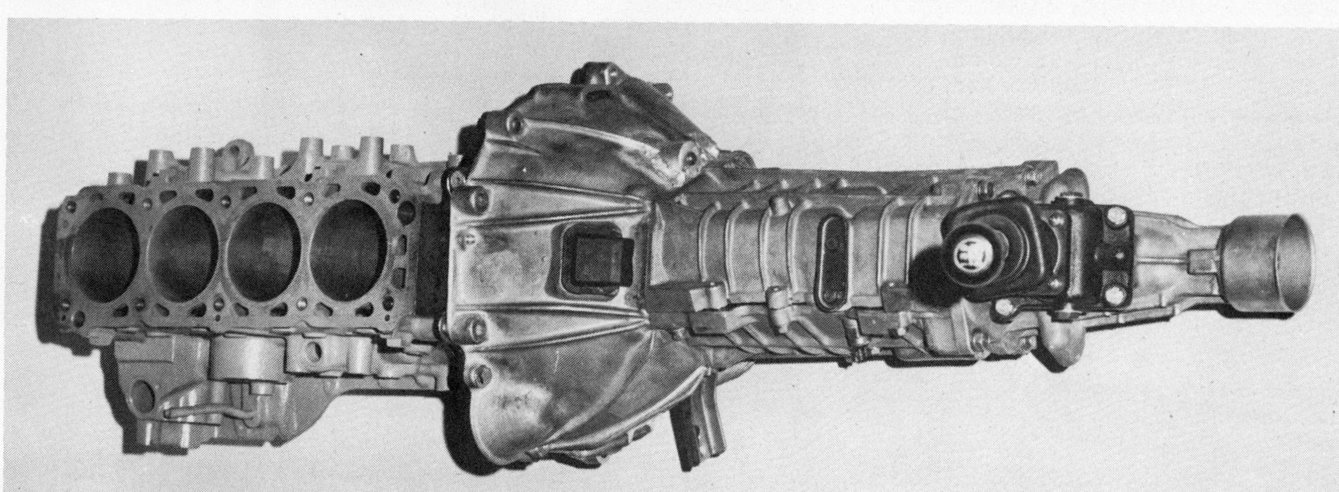

7

Five Easy Pieces

Eating your cake and having it too. A sort of overdrive for Warner T-10 4-Speed.

A new and improved variation of the well-known Warner T-10 is in the works for those who must have more low gear pulling power. Several specialty houses are currently making the conversions, and among the leaders is Pepe Estrada in Paramount, Calif.

The Borg-Warner T-10 is a preferred trans for the guy who must have a good 4-speed for hauling his boat or trailer, and a must in the heavy going for the big camper set. It depends where you're sitting, of course, as to whether you go for beefed hydros or 4-speeds, but Pepe got into the Warner modification following a not-too-friendly situation on the way to Lake Tahoe recently.

"I was pulling well up a pretty long grade, with my big V-8 truck, upon which was a very large and heavy camper."

It didn't make much difference that Pepe said he was toting a lot of gear and what he called "six, very fat people." The point is that his automatic began to overheat. It's not unusual for torque converter fluid to boil under brutal use.

"It's not very comforting to think that my automatic was cooled by water from the radiator, which itself was running about 200° F, and the trans was between 375 and 475°."

Following a nice 30-minute wait in the hot sun, waiting for angry stator blades and cooling reservoirs to cool, Pepe was convinced that the time was ready to take a giant design step forward. The trouble with hauling heavy boats and campers with automatics, is that most factory combinations have 2.78:1 and similar ratios in the back, which invariably challenges an automatic's ability to multiply enough torque in the rough going. Most big campers try to strike a happy medium with factory installation of 3.00 or 3.30:1 rear ends, but the lack of hauling poop still exists.

Invariably, many owners in the know opt for the Warner T-10 4-speed, available with GM, Ford and Chrysler-built rigs. But still, the ratios are too high. The improvement in trans reliability is largely negated by the conservative Detroit ratios. So, how in the heck can the owner get gutsy, reliable low gear power and high-speed freeway mileage? It takes time to get to most backwoods spas, and that consumes gasoline. Most big campers are lucky to get an overall 9 mpg, with averages around 7 mpg not uncommon when you're talking heavy gross weights.

THE BEST ANSWER

The solution, Pepe found, was to rebuild a Warner T-10 in the 1st, 2nd and 3rd gear departments, keeping the same reverse ratio, and, of course, 4th with its 1:1. This combination is a pleasure to play with when you're stuck somewhere, miles from a tow line.

A typical Warner T-10 comes with a 2.54:1 1st; a 1.88 2nd; a 1.47 3rd and a 1:1 4th. Reverse, not a ballyhooed ratio, is usually around 2.50 or 2.70:1. This setup, in most stock forms, is hooked to a 2.78 or 3.00:1 differential. The trouble is that the 2.54 1st doesn't have the total pull with either rear end in the heavy load categories. The only other alternative means that fuel economy suffers. Pepe has made the change by installing 3.44:1 gears in 1st; 2.28's in 2nd; and 1.46 in 3rd. Fourth has the same 1:1 ratio as before. The change results in a slightly longer spread between 2nd and 3rd, but enough people are switching, so evidently it can be lived with—and you're not looking for neck snapping shifts with a camper or truck, anyway. Actually, Pepe's modification is a beautiful setup.

A 4-speed is going to hike the 9 mpg figure another 6 mpg, and will provide all the low-end poop needed. It's compatible with any rear-end ratio up to 3.55, before the mileage falls off again. But without doubt, the best thing is to leave the factory rear end at 2.78:1 or related, and go to the trans modification.

The Warner T-10 is intended for small and large block V-8's. The bellhousing bolt patterns aren't compatible with small engines, but of course, a small engine with the beefy T-10 is like burning down your house to get rid of the rats. So, the end result of Pepe's modification is the Borg-Warner Ultra Low Ratio 4-speed. You can't

1

2

3

4

5

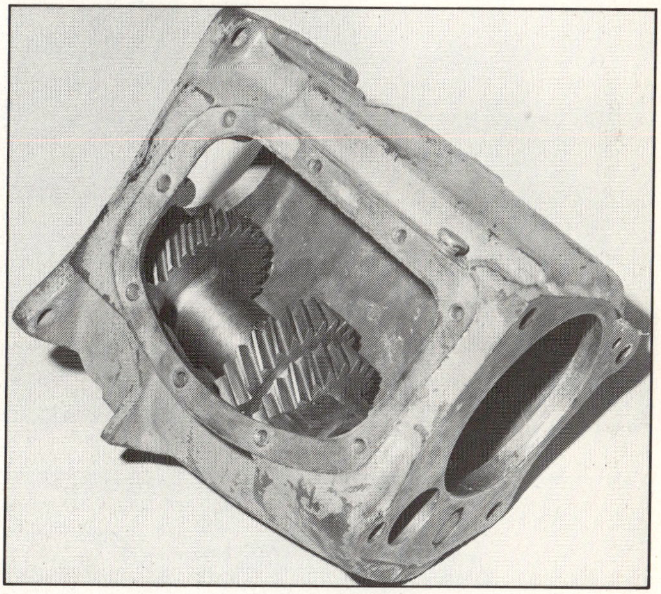

1. The main components of the Borg-Warner super-low-gear "kit" include new input shaft, countergear group, and new first, second and third gears. Combine these with your T-10 or Super T-10 and you've just about got what amounts to a 3-speed with overdrive; for pulling power when you need it, and mileage on the highway.

2. For longer gear life, Pepe Estrada uses a high-speed hand grinder to cut four oil grooves in the gear cones.

3. The inside diameter of the gears also receives the grooving treatment. Here they are cut from the oil slits, a factory addition with later T-10's, to the edges. All the surfaces are then polished to remove any grinding burrs left near the new oil grooves.

4. A 3-sec. cut with aluminum oxide stone produces this ¼-in. wide oil groove, which eliminates welding of the gears under the hardest use.

5. One half of the team of new parts that makes the T-10 a super-low-gear is the countergear setup. Compare the size of these gears with stockers.

get it as a complete trans; only the gears and new main shaft. Even though a lot of people still don't get into the real heavy hauling loads, there are enough who do, and it's great to have the Low Ratio trans.

SECRET WEAPON

Pepe has added the "ultra touch" in the Warner Ultra box, with the installation of oiling grooves on the cone faces. He's applied for a patent on it, claiming that continuous lubrication on the cone faces greatly increases life of the synchros. Under super-high loads out in the boonies, main transmission shafts have been known to suffer welding due to lack of oiling when temperatures have reached 2000° F and faces tend to fuse together. Reportedly, the early T-10's and Super T-10's had a bad habit of doing this—again in brutal conditions—even though a factory attempt at oiling slits helped somewhat. On the other hand, oiling grooves are always rotating with gears, and provide continuous lubrication. Because so much of a transmission depends on operation of friction surfaces, it's a concept hard to fault.

In the finished Warner Ultra trans, Pepe uses Torco 140 or Valvoline 90-140 wt., following a schedule of dumping the lube the next day—from a freshly built box—and then every 10,000 miles.

For the AMC owner (Cherokees, Wagoneers, Jeeps) it is necessary to use GM clutch discs—no problem, since all U.S. manufacturers use 10-, 10½- or 11-in. discs.

The complete package uses the same T-10 box, with the addition of 4 gears and main shaft; hence the term "five easy pieces." The input shaft is longer and has to have the famous flange; and the cluster gear is bigger, as well. Both the T-10 and up-dated

Five Easy Pieces

Super T-10 boxes showed a thinner flange on the main shaft, which had a bad habit of breaking off under heavy pulling loads. But that has apparently been solved with a longer shaft which Pepe uses. Early clutch problems are reportedly no longer a threat.

The Warner Ultra Low Ratio 4-speed uses the same 8620 steel and there's nothing trick about it; only strengthened guts. For those who want part numbers and prices, the new 2nd gear has compatibility with four main shafts. The new 2nd gear is B-W Part No. T10S-31A, and is compatible with main shaft numbers: T10Q-2; T10S-2; T10P-2 and T10P-2C. Later model 1972 main shaft numbers are: T10S-2A; T10S-2B; T10S-2C and T10S-2D. For the early Super T-10, the main drive gear is No. T10U-16; the cluster gear, T10U-8; 3rd, T10U-11; with 1st having a Part No. T10P-12, and 2nd, T10S-31. All numbers here are compatible with the new Ultra Low Ratio setup.

In order to get the 3.44:1 1st gear, the teeth were increased from 26 to 31. The gears are harder to put into the case, but if done right, it's no problem. The only thing is that the main drive has to go in first; and the Christmas Tree (upper assembly; input shaft; complete synchronizers) is a little tight—but once inside the case, very much at home. It's also important to use the proper synchros. For 1st and 2nd synchro assembly: Part No. AT10-80HP, and for 3rd and 4th: Part No. AT10-2½HP. This is mandatory with hollow synchro dogs—or strut keys (which separate the two blocker rings).

The same case is used, along with bearings, sleeves, retainers, clips, seals, gaskets, etc.—if usable. If not, they're replaced. The new Borg-Warner parts—complete—run a little more than $250 and Pepe charges $60 for labor plus parts. The complete job of modifying a Borg-Warner T-10 or Super T-10 4-speed is a worthwhile enterprise. If Pepe does it, he'll need 10 days, but he has a long line of credits on street and track. Predictably, the Ultra Low Ratio 4-speed will do all that is asked of it. And that, good friends, will be a lot. So keep your 2.78:1 or 3.00:1, or even 3.30:1 rear end. You need it for gas mileage. But you'll also find that the Ultra Low Ratio 4-speed is a much welcomed addition. In fact, you might even say that it's a necessity.

1

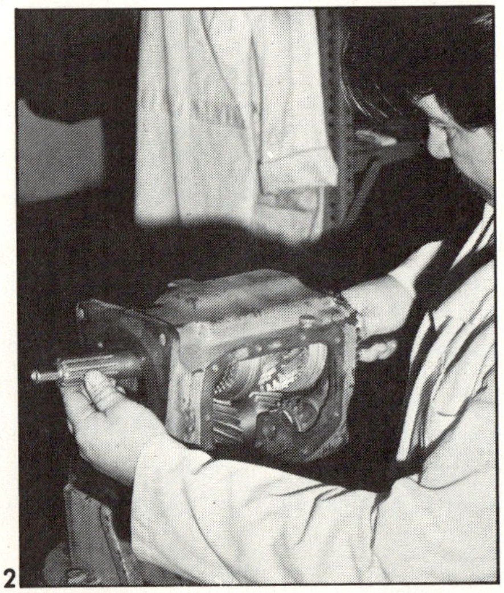

2

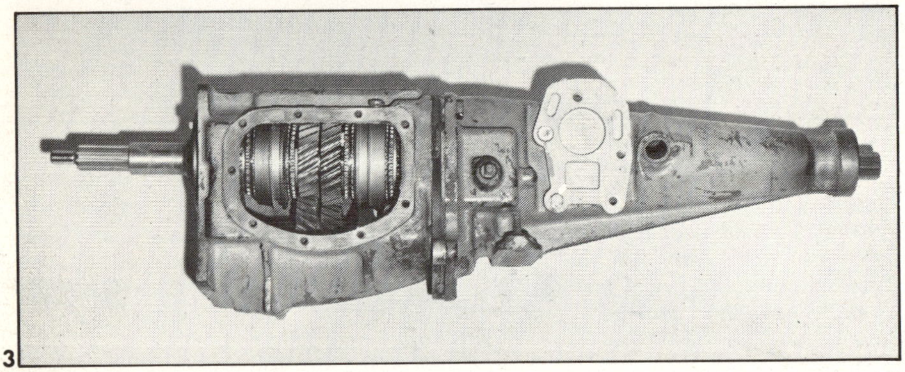

3

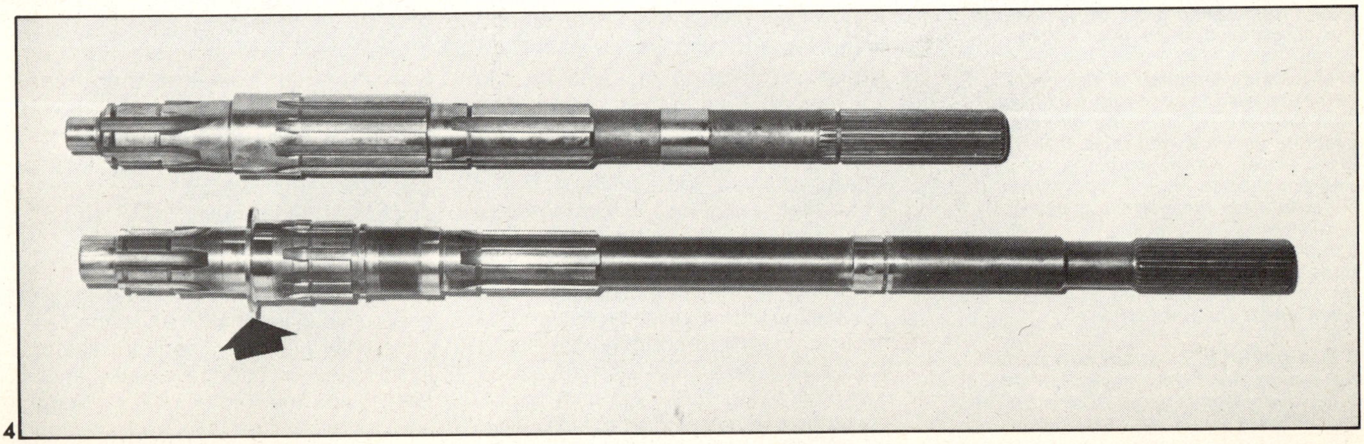

4

1. The new input shaft (left) looks smaller than the original T-10, and it is, since the new countergear is so much larger than the old one.

2. Pepe is assembling a super-low T-10 for a boat-hauling customer. The new Christmas Tree is bigger and harder to put into the case than the stock one, so put the input shaft in first, then the Christmas Tree.

3. Completed minus the side cover, our super-low-gear T-10 looks just like a stocker, except for the big gears. At around $300 for the whole package, it's the best way out for late-model pickups with "smog" rear-end ratios like 3.23's and 2.78's. This way you get the pulling power of a stiff rear gear in your first three speeds, yet economy in fourth.

4. If yours is a Super T-10 already, then you're in luck; you don't need a new main shaft. Regular T-10's need a new Super T-10 main shaft (below) which has this flange (arrow).